THE TIDE-DOMINATED HAN RIVER DELTA, KOREA

THE TIDE-DOMINATED HAN RIVER DELTA, KOREA

Geomorphology, Sedimentology, and Stratigraphic Architecture

DON I. CUMMINGS
DCGeo Consulting, Aylmer, Quebec, Canada

ROBERT W. DALRYMPLE
Queen's University, Kingston, Ontario, Canada

KYUNGSIK CHOI
Seoul National University, Seoul, Korea

JAE HWA JIN
Korea Institute of Geoscience and Mineral Resources, Daejeon, Korea

ELSEVIER

AMSTERDAM • BOSTON • HEIDELBERG • LONDON • NEW YORK • OXFORD
PARIS • SAN DIEGO • SAN FRANCISCO • SINGAPORE • SYDNEY • TOKYO

Elsevier
Radarweg 29, PO Box 211, 1000 AE Amsterdam, Netherlands
The Boulevard, Langford Lane, Kidlington, Oxford OX5 1GB, UK
225 Wyman Street, Waltham, MA 02451, USA

Notices
Knowledge and best practice in this field are constantly changing. As new research and experience broaden our understanding, changes in research methods, professional practices, or medical treatment may become necessary.

Practitioners and researchers must always rely on their own experience and knowledge in evaluating and using any information, methods, compounds, or experiments described herein. In using such information or methods they should be mindful of their own safety and the safety of others, including parties for whom they have a professional responsibility.

To the fullest extent of the law, neither the Publisher nor the authors, contributors, or editors, assume any liability for any injury and/or damage to persons or property as a matter of products liability, negligence or otherwise, or from any use or operation of any methods, products, instructions, or ideas contained in the material herein.

ISBN: 978-0-12-800768-6

British Library Cataloguing in Publication Data
A catalogue record for this book is available from the British Library

Library of Congress Cataloging-in-Publication Data
A catalog record for this book is available from the Library of Congress

For Information on all Elsevier publications
visit our website at http://store.elsevier.com/

Working together
to grow libraries in
developing countries

www.elsevier.com • www.bookaid.org

CONTENTS

Please find the companion website at http://booksite.elsevier.com/9780128007686/.

CHAPTER 1

Introduction

Chapter Points

- Tide-dominated deltas are the most common type of large delta in the world today
- Few have been studied
- Within them, proximal–distal trends in hydraulic energy, facies, and architecture are complex
- A well-accepted facies model does not exist
- The Han River delta was studied to help fill this knowledge gap
- It has the largest tidal range of any delta studied to date

River deltas are some of the most beautiful, complex features on Earth. Viewed from space, they almost look alive, their channels branching across the land surface like air passages in a lung or the limbs of a tree. In addition to looking alive, deltas help sustain life. Hundreds of millions of people live on them, commonly within meters of sea level (Goodbred and Saito, 2012). To the farmer, the delta is the field; to the engineer, the substrate of the city; to the geologist, the reservoir for drinking water and petroleum. To the ocean, the delta is like a security gate, a macroscopic physical, biological, and chemical filter through which most sediment, nutrients, and water-borne pollutants from the continent must pass. Deltas were one of the first sedimentary environments to be studied in detail (Gilbert, 1885; Fisk, 1944) and they remain the focus of intense study today (Bhattacharya, 2010).

All deltas share two fundamental traits. First, they consist primarily of fluvial sediment: deltas owe their existence to deposition from decelerating, expanding river flow issuing into a body of water. The sediment may be reworked by non-fluvial processes following initial deposition (Wright and Nittrouer, 1995), but ultimately the delta exists because a river supplied it with sediment. Second, deltas are progradational: they build outward into the body of water. The outbuilding typically occurs by accretion of sediment onto the distal (i.e., seaward) face of the delta, which in mature systems takes the form of a sloping depositional surface known as a clinoform. The term progradational is used somewhat loosely here. Deltaic outbuilding typically involves shoreline progradation, but for many of the larger and more strongly tidally influenced deltas, most of the outbuilding takes place far from shore, beneath the water surface (Figure 1.1). In these systems, the top ("topset") of the delta, across which most sediment is bypassed, is largely subaqueous, the prograding *clinoform** displaced 40–120 km offshore and submerged in 10–50 m of water (Friedrichs and Wright, 2004; Walsh et al., 2004).

---

* Words that are italicized are defined more fully in the Glossary at the end of the book.

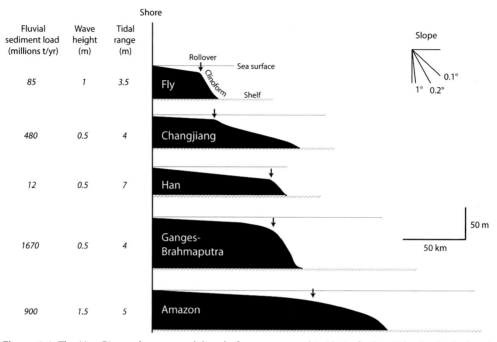

Figure 1.1 The Han River subaqueous delta platform compared to that of other tide-dominated and strongly tide-influenced deltas. Basinward is to the right. The sediment load, tidal range, and wave height values quoted should be thought of as average and approximate because they are inherently spatially and temporally variable. *Data from Milliman and Meade (1983), Hong et al. (2002), and Hori et al. (2002).*

Beyond these two common traits, variability is the norm. Grain size, basin physiography, tectonic setting, climate, and physical oceanography all play a role in determining delta physiography (Coleman and Wright, 1975), leading to a conceivably infinite number of possible outcomes. However, for large marine deltas, which tend to be fine grained because of inland trapping of coarse sediment, variability is often primarily a function of the physical oceanography of the receiving basin, and in particular the degree to which waves versus tidal currents rework the fluvial sediment in the coastal zone. Recognizing this, Galloway (1975) proposed a geomorphological classification of deltas with three end-member types: (1) river-dominated deltas, whose *mouth bars*—the fundamental deposit of river outflow (Bates, 1953)—experience little reworking by waves or tides; (2) wave-dominated deltas, within which waves and wave-generated currents perform most geomorphic work, reworking the mouth-bar sediment alongshore into beaches; and (3) tide-dominated deltas, within which tidal currents perform most geomorphic work, reworking mouth bars into shore-normal tidal bars. Significant advances in understanding have been made since Galloway's publication, thanks in part to numerical modeling (Pirmez et al., 1998; Swenson et al., 2005), flume experiments (Southard, 1991; Dumas et al., 2005), integration of ichnology (MacEachern et al., 2005) and hydrodynamic data (Kineke and Steinberg, 1995;

Geyer et al., 2004), and, perhaps most importantly, the study of deltas themselves (Wright and Nittrouer, 1995; Goodbred and Saito, 2012). However, Galloway's simple yet elegant scheme remains the *de facto* starting point for discussion.

Of Galloway's three end-member delta types, tide-dominated deltas remain the least well understood, despite the fact that five of the 10 largest rivers in the world today have tide-dominated or strongly tide-influenced deltas (Middleton, 1991; Goodbred and Saito, 2012). There are several reasons for this. At the time of Galloway's publication, several of the tide-dominated deltas he listed had not been studied in detail. Subsequent work has shown that some of these may actually be drowned river mouths still undergoing transgression (i.e., they are "*estuaries*" *sensu* Dalrymple et al. (1992)), not progradational sediment bodies that are actively outbuilding onto the shelf (i.e., *deltas*). The presence of very broad shallow water areas on the subaqueous delta plain (Figure 1.1), coupled with the presence of strong tidal currents, makes access difficult and dangerous. Thus, to this day, few modern examples of tide-dominated deltas have been studied in detail (for a summary, see Goodbred and Saito (2012)). Furthermore, within tide-dominated deltaic environments, proximal–distal variations in hydraulic energy, and thus facies, tend to be more complex and less well understood than in their river- and wave-dominated counterparts (Dalrymple and Choi, 2007). The morphologic complexity of tide-dominated deltas with their multitude of channels and bars further compounds the problem of characterizing such deltas. Consequently, a comprehensive tide-dominated delta facies model has not taken root in the geological community. This may explain why so few ancient examples have been identified despite their abundance in modern settings (Willis, 2005; Bhattacharya, 2010).

To help fill this knowledge gap, the Han River delta, a tide-dominated delta along Korea's west coast, was studied using a large integrated dataset of short cores, long drill cores, and seismic data (Figure 1.2). The study represents the most comprehensive investigation of a tide-dominated delta to date. Straddling the border between North and South Korea, the Han is the Korean peninsula's largest river. It debouches into Gyeonggi Bay (alternative spelling: *Kyunggi Bay*), a shallow (~40 m depth), rocky, wide-mouthed embayment fringed by tidal flats, dotted by bedrock islands, and characterized by an extreme tidal range (Figure 1.3), which exceeds 9 m at the very head of the bay during spring tides. The river mouth lacks a distinct shoreline protruberance, which has led some to refer to it as an estuary (Lee et al., 2013). However, as is common to all tide-dominated deltas studied to date (Figure 1.1), most of the action has taken place out of sight, beneath the water surface: previous data show that an aerially extensive, heterolithic package of sediment has aggraded and then prograded offshore of the river mouth, with upward of 60 m of sediment having accumulated since the last glacial maximum lowstand (Jin, 2001). The Han's subaqueous delta platform is not flat-topped like those of other deltas. Rather, it consists of several broad, shore-attached, finger-like protrusions. The protrusions are larger—and especially wider—than "typical" tidal bars observed in previously studied deltas. We therefore refer to them as large tidal bars. In the

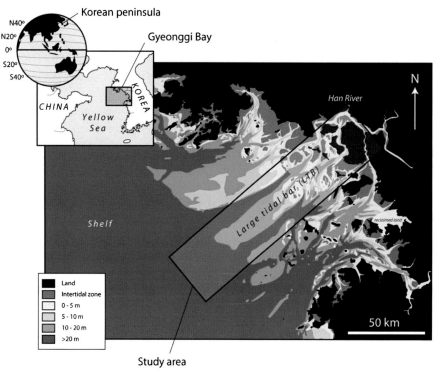

Figure 1.2 Location of the study area, Han River delta, Gyeonggi Bay, Korea. The Han River at this location demarcates the boundary between North and South Korea.

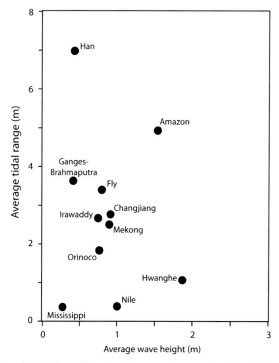

Figure 1.3 The Han River delta's depositional setting, as characterized by tidal range and wave height, relative to that of other major deltas *(in part from Hori et al. (2002))*. The Han River delta has by far the largest tidal range of any delta studied to date.

mid-2000s, an extensive dataset was collected from the centermost large tidal bar by a government–industry–academic consortium. These data, along with older, unpublished data from the inner large tidal bar (Jin, 2001), form the focus of this book. In later chapters, data from other subenvironments, such as *distributary channels* and *tidal flats* will be folded into the analysis, the overarching objective being to provide the reader with a holistic view of the geomorphology, sedimentology, and stratigraphic architecture of the delta. Specifically, readers will be provided with an outline of the "rules" governing the geomorphology of such systems; a comprehensive description of sediment facies, from the landward limit of the tidal flats along the shoreline seaward to the *prodelta* and shelf; information on the architecture of the deltaic deposits; and the sequence stratigraphic organization of the thick wedge of sediment that underlies the seafloor.

CHAPTER 2

Depositional Setting

Contents

Chapter Points

- Korean climate is monsoonal: summers are warm and wet; winters are cold, dry, and windy
- Precipitation events are intense and are associated with summer storms
- The Han River catchment is small and steep, with limited floodplains
- Fluvial discharge is seasonal and flashy; sediment yield is moderately high
- Extreme tides characterize the receiving basin
- Wave energy is low except during winter storms; the dominant wave energy flux is southward

The depositional setting imposes a set of conditions on a sedimentary system that, assuming similar internal (autocyclic) behavior, will lead to a repeatable (predictable) stratigraphic response in other locations under similar conditions. For the Han River delta, the key conditions are as follows (Figure 2.1).

- Han River water discharge is $23\,km^3$/year, 70% of which is delivered in summer. Maximum discharge is $30,000\,m^3$/s.
- Han River sediment discharge is 12.4 million tons/year, 10% of which is sand, and 90% of which is silt and clay. Sediment yield is $475\,tons/km^2$/year.
- The catchment is small, steep, commonly bedrock-confined, with limited floodplains.
- The receiving basin is a shallow epicontinental sea with a low-gradient seafloor, flanked by steep, bedrock-cored hills on the Korean peninsula.
- The tidal range is extreme (maximum 9 m). Tidal currents are almost rectilinear in Gyeonggi Bay. Current speeds exceed 2 m/s locally.
- Winds are weak and northward in summer, except during typhoons, and they are strong and southeastward in winter.
- Waves are small in summer, and bigger in winter but not huge (i.e., the bay is fetch limited). Waves decrease in size landward.

- The water column is well mixed due to strong tidal currents, except possibly during river floods.
- The *Coriolis effect** is substantial, potentially leading to mud advection to the north.
- The precipitation is high in summer (intense rainfalls) and very low in winter.
- Climate is temperate and monsoonal. Climate may have become increasingly arid since the start of the Holocene.
- Sediment provenance is mixed (fluvial and shelf), but is fluvially dominant now. Comparatively more sediment may have been derived from the shelf during the early–mid-Holocene transgression.
- Bedrock consists of granite and metamorphic rocks that form small, steep mountains and coastal islands.
- The area is tectonically stable; a few meters of subsidence may possibly have occurred over the last 100 ka.
- In terms of sea-level history, the floor of the Yellow Sea was exposed during the last glacial maximum (LGM) lowstand, followed by a rapid rise (14 mm/year) from 17 to 8 ka BP, then a slow rise (1 mm/year) from 8 ka BP to present.

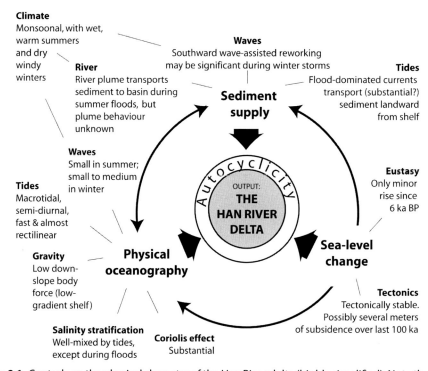

Figure 2.1 Controls on the physical character of the Han River delta (highly simplified). Note that three major controls exist: sea-level change (independent variable), physical oceanography (dependent variable), and sediment supply (dependent variable).

* Expanded definitions of italicized words appear in the Glossary.

- Sequence stratigraphically, the system is in the early highstand systems tract.
- Anthropogenic influences before Japanese occupation (i.e., pre-1910) include deforestation, dyking, and tidal-flat reclamation. Post-1910 influences include extensive dyking and an increased rate of tidal flat reclamation, construction of dams and concrete levees, deforestation, urbanization of floodplains, and extensive mining of sand bars in the lower Han River for concrete production.

2.1 PLATE TECTONIC SETTING

The tectonic framework of the Korean peninsula, and of East Asia in general, is characterized by two major factors: northwestward-directed subduction of oceanic crust beneath Japan (Schellart and Lister, 2005) and eastward expulsion of continental crust by the collision of India and Eurasia (Molnar and Trapponier, 1975) (Figure 2.2). Although the Korean peninsula now sits in a tectonically quiescent zone in the middle of stable, eastward-moving continental crust (Chiu and Kim, 2004), it has a complex tectonic history. Korea was assembled in the Mesozoic when three Precambrian terrains collided and sutured (Chough et al.,

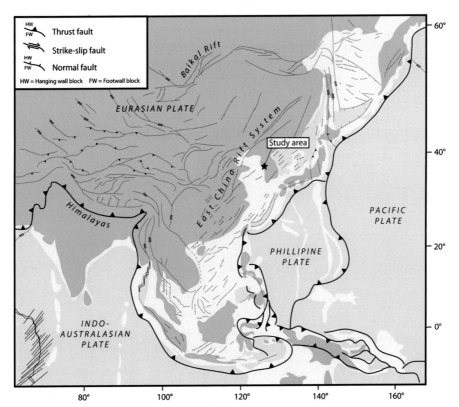

Figure 2.2 Tectonic map of East Asia. *Modified from Schellart and Lister (2005).*

2000). Since then, sedimentary, plutonic, and volcanic rocks have been deposited and emplaced as a result of westerly directed subduction (Figure 2.3), and the Japan/East Sea and Yellow Sea depressions have formed on either side of the Korean peninsula, likely as a result of backarc extension associated with slab-rollback (Schellart and Lister, 2005). Yellow Sea rifting and subsidence occurred first, between 65 and 30 Ma (Ren et al., 2002), whereas rifting in the Japan/East Sea started later, at 30 Ma, and proceeded to a seafloor-spreading stage before extension stopped at 10 Ma (Jolivet et al., 1994). The orientation of Gyeonggi Bay corresponds to the NE–SW structural grain of the Korean peninsula and Yellow Sea (e.g., Chough et al., 2000), suggesting its location may be structurally controlled.

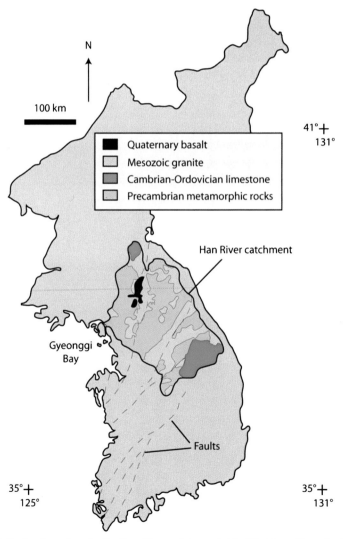

Figure 2.3 Bedrock geology in the Han River catchment. *Modified from Chough et al. (2000).*

2.2 CLIMATE

The climate in Korea is temperate and monsoonal (Nestor, 1977) (Figure 2.4). In summer, the Asian landmass heats faster than the ocean to the southeast, generating a pressure gradient that causes weak, warm, moisture-laden winds to blow northward from the ocean (high pressure) to the continent (low pressure). Conversely, in winter, the continent cools faster than the ocean, causing strong, cold, dry winds to blow southward from the cool continent (high pressure) to the warmer ocean (low pressure). Summers in Korea are therefore warm, wet, and generally calm, and winters are cool, dry, and windy. Mean annual precipitation in Korea is ~1300 mm (Figure 2.5(a)), which is high by global standards (Figure 2.5(b)). Year-to-year variation is considerable (Figure 2.5(c)). Two-thirds of the total precipitation falls between June and October (Figure 2.5(d)). This rainy season is referred to as "Changma" in Korea, "Meiju" in China, and "Baiu" in Japan

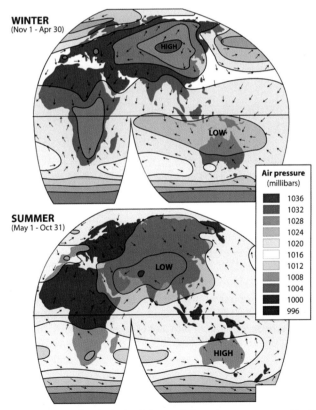

Figure 2.4 Seasonal (monsoonal) winds in Asia, and the air-pressure gradients that drive them. *Modified from Livingstone and Warren (1996)*. The monsoon is driven by the difference in heat capacity of the continent and ocean. In winter, the ocean cools slower than the continent, causing dry continental wind to blow southward across Korea, whereas in summer the continent heats up faster than the ocean, causing wet ocean wind to blow northward across Korea.

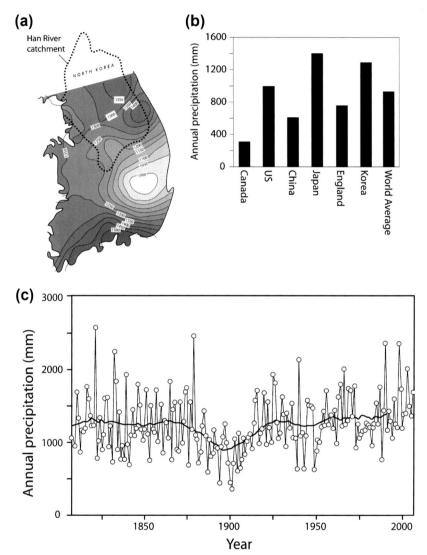

Figure 2.5 Precipitation in Korea. (a) Spatial distribution of precipitation in South Korea, averaged between 1959 and 1988 (Yoon and Woo, 2000). Numbers refer to total annual precipitation in millimeters. (b) Comparison of Korean precipitation with that of other countries (Korean Water Resources Website). (c) Annual precipitation (dots) in Seoul from 1807 to 2006 (Kim et al., 2009). Solid line denotes the 30 year moving average.

(Byun and Lee, 2002; Jin et al., 2005). Precipitation events are extremely intense—upward of 400 mm of rain can fall in one day (Figure 2.5(e))—and tend to be associated either with the episodic passage of tropical cyclones (Woo and Kim, 1997; Kim et al., 2004) or with convective cells that form along the Changma–Meiju–Baiu polar front, which oscillates northward and southward over Korea in the summer (Lee, 1976; Jeong et al., 2012).

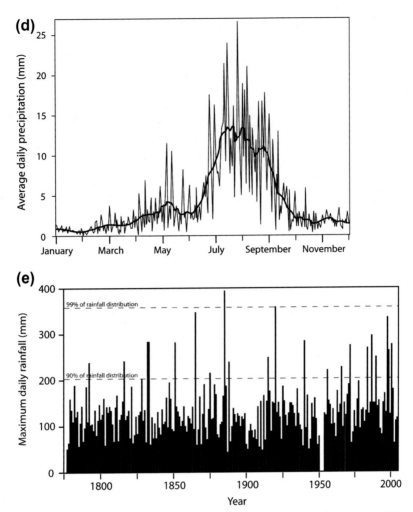

Figure 2.5—cont'd (d) Seasonal cycle of precipitation in Seoul based on data averaged from 1977 to 2006 (Kim et al., 2009). Solid line denotes the 15 day moving average. (e) Annual maximum daily rainfall in Seoul from 1777 to 2005 (Kim et al., 2007).

2.3 THE HAN RIVER

Seventy percent of the Korean peninsula is mountainous (Figure 2.6). Major rivers flow westward into the Yellow Sea from the eastern, elevated half of the peninsula. The Han is the largest of these (Figures 2.6 and 2.7), although its 26,000 km² catchment is moderate-sized and steep by global standards (Milliman and Syvitski, 1992). Comprised of the Nam, Puksan, and Imjin tributaries, the Han straddles the border between North and South Korea, and flows through Seoul before emptying into Gyeonggi Bay. Metasedimentary

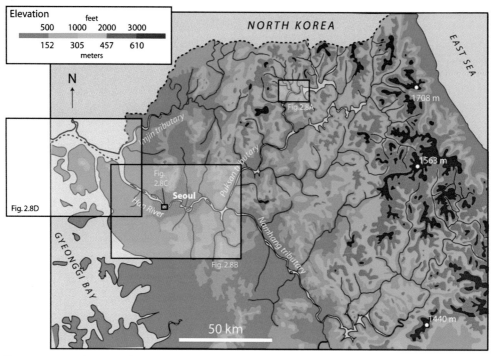

Figure 2.6 Han River catchment. The catchment is approximately 26,000 km² (23,000 km² in South Korea and 3000 km² in North Korea). Note that steep bedrock mountains commonly confine the channel and limit the extent of the floodplain.

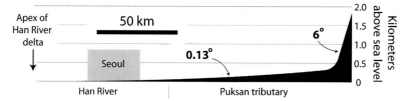

Figure 2.7 Longitudinal profile of Han River. Profile measured straight down valley from highest elevation in Han River catchment (1708 m), which occurs at the head of the Puksan tributary, to the delta apex, where the trunk channel splits into distributaries around several bedrock islands.

and granite rocks of the Gyeonggi massif outcrop throughout the catchment, forming small, steep, forested mountains (Lee et al., 2002). The upper Han River is meandering to straight, with meander scroll bars visible locally. The channel is commonly confined by bedrock, and floodplains are of limited extent (Figure 2.8(a)). The lower (tide-influenced) Han River near Seoul, by contrast, is wide, straight, and shallow. The channel is less commonly confined by bedrock, and floodplains/tidal flats are more extensive but still limited in size (Figure 2.8(b)). Sandy mid-channel bars, some inhabited, were ubiquitous in the

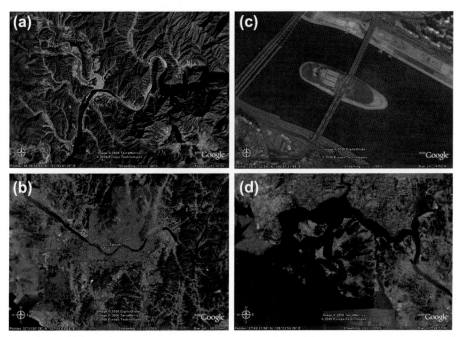

Figure 2.8 Han River planform images (for image locations see Figure 2.6). (a) Upper Han River (Puksan tributary). Image is 20 km wide. Note that steep bedrock topography commonly confines the channel and limits floodplain development, and that floodplains are commonly urbanized. Despite the presence of dams (e.g., the dam in center-right of photo with associated upstream reservoir), these factors, in addition to the intensity of summer rainfalls, decrease the ability of the Han River to dampen discharge fluctuations. (b) Lower Han River near Seoul. Image is 80 km wide. Channel is approximately 1 km wide near Seoul, and is tidally influenced (the upstream tidal limit is near the Paldang Dam in the east end of Seoul; Yoon and Woo, 2013). Note that bedrock less commonly confines the channel than in the Upper Han River, and that floodplains (and reclaimed tidal flats and salt marshes) are more extensive but are highly urbanized. Note also rare, sandy mid-channel bars and islands, which were ubiquitous prior to the start of aggressive aggregate mining in the 1960s. (c) Mid-channel island near Seoul. Island is 0.7 km long. (d) Mouth of the Han River. Long dimension of largest island (Ganghwa Island) is 25 km. The Imjin River is visible in the top right corner of the image. It joins the Han River just upstream of Ganghwa Island, where the channel network changes from tributary to distributary. *All images courtesy of Google Earth©.*

Han River near Seoul prior to the 1960s, but have been mined almost completely for aggregate during the post-1960 economic boom (Lankov, 2005) (Figure 2.8(c)). Downstream of Seoul, the Imjin River joins the Han River before it empties into Gyeonggi Bay (Figure 2.8(d)).

Water discharge from the Han River (Figure 2.9) parallels the monsoonal climate, with its wet summers and dry winters (Park et al., 2002; Kim et al., 2010). Total annual discharge is 2300 million m^3/year (Choi et al., 2004a). Over 70% of the total annual discharge occurs between June and September, which is typical for Chinese and Korean rivers than drain into

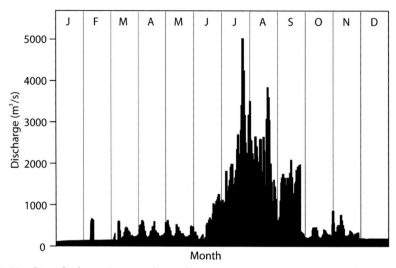

Figure 2.9 Han River discharge (averaged over the 1980–1997 period) based on daily water measurements made at the Henju bridge, Seoul (Park et al., 2002). Total annual discharge is approximately 23 km³/year. Note pronounced seasonality. Summer discharge can increase by an order of magnitude during floods (max. 30,000 m³/s).

the Yellow Sea (Meade, 1996; Chang and Isobe, 2005). Approximately three quarters of the Han's discharge in summer is associated with precipitation from convective cells along the polar front, with the remaining one quarter from tropical cyclones (Kim et al., 2010). Average summer discharge (2326 m³/s) is an order of magnitude greater than average winter discharge (150–400 m³/s), and may increase by an additional order of magnitude during extreme floods. Maximum flood discharge can exceed 30,000 cubic meters per second (Woo and Kim, 1997). This rate is close to the 1993 flood discharge of the Mississippi River at St. Louis, even though the Han River catchment is 40 times smaller than the Mississippi upstream of St. Louis. Several factors reduce the capacity of the Han River system to dampen discharge spikes, including the small size and steepness of the catchment, the rocky and impermeable nature of much of the drainage basin, the limited surface area of floodplains, the ubiquity of concrete levees, urbanization of watersheds, and the intensity of the summer rainfalls themselves. Floods are therefore a socioeconomic problem, local riverbed aggradation/degradation is common (Figure 2.10), and water and sediment are transferred readily to Gyeonggi Bay.

Few studies have investigated the sediment discharge of the Han River. The total annual sediment discharge is estimated to be 10–12 million tons (Chough and Kim, 1981; Hong et al., 2002), which is moderate compared to rivers worldwide (Figure 2.11). For example, the Han discharges in a year the same quantity of sediment that the Mississippi discharges in 20 days, or what the Huanghe and Changjiang discharge in 1–2 days. The sediment load consists approximately of 10% sand and 90% clay and silt (Oh, 1995). Suspended-sediment concentrations are roughly 10–20 mg/l during low discharge periods in winter, but may

Figure 2.10 Riverside walkway in Seoul excavated following burial by several meters of sediment during the 1990 flood (Woo and Kim, 1997). *Photo courtesy of Hyoseop Woo, Korea Institute of Construction Technology.*

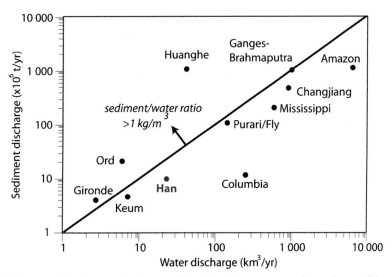

Figure 2.11 Water and sediment discharge of the Han River compared to other well-known rivers (Wright and Nittrouer, 1995), including the Keum (Korea), Huanghe (China), and Changjiang (China), all of which empty into the Yellow Sea or the adjacent East China Sea.

reach 5000 mg/l during summer floods (Yoon and Woo, 2000). In the main deltaic distributary channel (Sukmo Channel; see next paragraph), gravel and coarse sand are present in the thalweg, medium and fine sand and sandy silt cover the shallow subtidal zone, and sandy silt to fine silt cover the intertidal zone (Choi et al., 2004a).

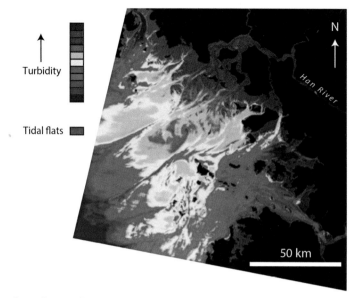

Figure 2.12 False-color satellite image of water turbidity in Gyeonggi Bay during a low-discharge period (February 1999). Note turbidity maximum (red colors) near Ganghwa Island in the distributary-channel zone; this turbidity maximum is displaced seaward during summer floods. Another area of high suspended-sediment concentration (red) occurs on the crest of the large tidal bar, in the area seaward of the diagonal swatchways, presumably as a result of wave-generated resuspension. Note also the low turbidity in the southern part of Gyeonggi Bay (blue colors). *Image courtesy of KIGAM (Korean Institute of Geoscience and Mineral Resources).*

Given the extreme seasonality of the Han River discharge, it is likely that most river-borne sediment enters Gyeonggi Bay in summer. During low discharge months, a *turbidity maximum* develops in distributary channels near Ganghwa Island (Figure 2.12). During high discharge months following precipitation events, satellite images show that the turbidity maximum is displaced outward into Gyeonggi Bay. A freshwater signal does not typically reach south-central Gyeonggi Bay, even during summer (Figure 2.13). This is in part a reflection of intense tidal mixing of river and marine water (Lee et al., 2002) and in part because the river plume is deflected slightly to the right (northward), as is typical for mid-latitude, northern-hemisphere river plumes due to the Coriolis effect (Garvine, 1999; Geyer et al., 2004). Satellite data in Choi et al. (2013) suggest that suspended-sediment concentrations are higher over a tidal cycle in the north part of Gyeonggi Bay than in the south part.

The macrobenthic community is affected by the salinity gradient at the mouth of the Han River (Figure 2.14). On average, the transition from fresh to brackish water occurs near the delta apex (Yu et al., 2012). The freshwater community upflow of this (stations 1–10 in Figure 2.14) is dominated by annelid worms and has a relatively low density and diversity. The brackish water community downflow of this (stations 11–16 in Figure 2.14) has a greater density and diversity. It consists primarily of annelid worms and crustaceans with lesser numbers of echinoderms and molluscs.

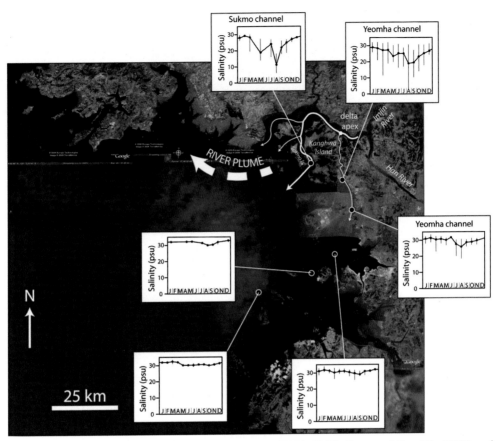

Figure 2.13 Average monthly salinity measurements (1986–1998), Gyeonggi Bay (Oh, 1995) and hypothesized trajectory of the river plume. Note that pronounced salinity lows occur in both Sukmo and Yeomha distributary channels during the summer, but not in south-central Gyeonggi Bay. This likely reflects strong tidal mixing (note pronounced downstream salinity increase in Yeomha Channel), but also suggests that the river plume is deflected to the north, as is to be expected for mid-latitude, northern-hemisphere river plumes because of the Coriolis effect. The lowest salinity readings occur in Sukmo Channel, the deepest distributary, suggesting that it is the main conduit for freshwater (and sediment) entering Gyeonggi Bay. (Note that southward reworking by winter waves and a southward-flowing current in the Yellow Sea may also be important in governing the ultimate fate of Han River sediments, as is the case for other Yellow Sea dispersal systems such as the Keum (Korea) (Chough et al., 2004), Changjiang (China) (Wright and Nittrouer, 1995), and Huanghe (China) (Liu et al., 2004) rivers.)

2.4 PHYSICAL OCEANOGRAPHY

Very large tides define the physical oceanography of Gyeonggi Bay: at the head of the bay, near the mouth of the Han River, tidal range is approximately 5 m during neap tide and 9 m during spring tide (Figure 2.15). The tides that affect Gyeonggi Bay originate in the Pacific Ocean and move northward into the Yellow Sea. As the tidal wave shoals

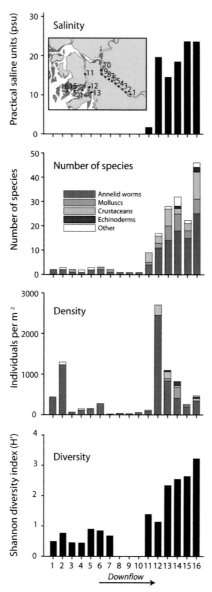

Figure 2.14 Macrobenthic community as a function of salinity across the Han River fluvial–marine transition in July 2006. *Modified from Yu et al. (2012)*. Numbers 1–16 refer to measurement stations, as shown in the inset map.

across the shelf edge, it is amplified and shortens in wavelength, but retains its period and becomes pushed up against the east coast of the Yellow Sea due to the Coriolis effect (Figure 2.16). Tidal range along the western Korean coast therefore varies from mesotidal (2–4 m) along the open coast to *macrotidal* (>4 m) in embayments such as Gyeonggi

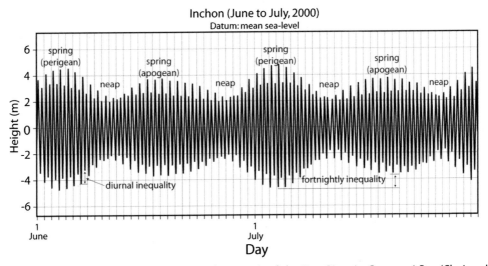

Figure 2.15 Tidal chart for Incheon, near the mouth of the Han River in Gyeonggi Bay (Choi and Dalrymple, 2004). Note neap–spring cycles, diurnal inequality (minor), and fortnightly inequality (minor).

Bay and Korea Bay (Figure 2.17). As the tidal wave continues to pass northward along the Korean coast, it moves in and out of coastal embayments, generating rectilinear currents in these regions (Figure 2.18). In Gyeonggi Bay during spring tide, the tidal range increases landward from 5 m at the bay mouth to 9 m at the head of the bay. Tidal currents in Gyeonggi Bay are fast (max. 1–2 m/s), semidiurnal, and almost rectilinear, but retain a slight counterclockwise rotation due to the relatively large width of the embayment (Choi, 1991; Lee et al., 2001; Kang et al., 2002) (Figure 2.19). As is typical of macrotidal settings, Gyeonggi Bay is hypersynchronous: tidal range increases landward due to convergence, reaches a maximum, then decreases landward due to bottom friction (Figure 2.20). The location of the largest tidal range occurs just downstream of the delta apex, the point where the river splits around bedrock islands into three distributary channels. By the time the tidal wave is funneled into the Han River, it is distinctly asymmetric (Figure 2.21), and may evolve further into a 1–2 m high tidal bore (Kim et al., 2004). The tidal limit is located approximately 90 km upstream of delta apex, near the Paldang Dam, just east of Seoul (Yoon and Woo, 2013). During high spring tide (tidal range ~9 m), the tidal flux of water in and out of the lower Han River is estimated to be 1600 m³/s (peak flood) and 1500 m³/s (peak ebb) (Oh, 1995). When these figures are compared with the average fluvial discharge during winter (150–400 m³/s) and summer (2326 m³/s), it seems likely that there is a seasonal change in process domination: the mouth and lower reaches of the Han are likely tide dominated in winter, but become river dominated during summer, especially during summer floods.

Wind patterns in Korea are a function of two separate systems, the East Asian monsoon, which causes regional, seasonal variation in wind intensity and direction, and

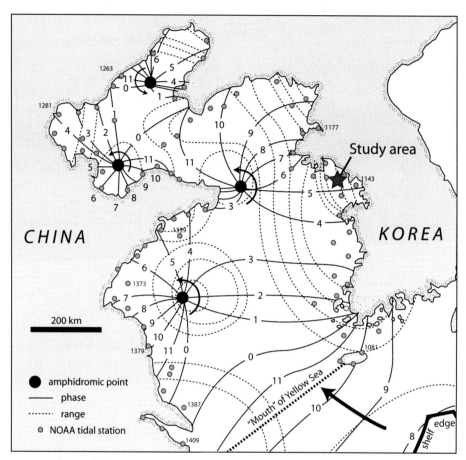

Figure 2.16 Observed tides in the Yellow Sea (M$_2$ tide only (Guo and Yanagi, 1998). The tidal wave moves northward into the Yellow Sea from the Pacific Ocean. As it shoals across the shelf edge, it shortens in wavelength and is amplified, and then banks up against the right-hand (eastern) side of the Yellow Sea basin due to the Coriolis effect. This generates meso- to macrotidal conditions along the west Korean coast (see Figure 2.17). The reflected tidal wave moves out of the Yellow Sea along the Chinese coast. Interference of incoming and reflected waves produces four amphidromic systems in the Yellow and Bohai seas. Because of tidal friction and energy dissipation, the reflected wave is weaker than the incoming wave, causing displacement of the amphidromic points toward the Chinese coast. Although the Yellow Sea is large enough to support rotary tides, Gyeonggi Bay is not. Tidal currents are almost (but not quite) rectilinear in Gyeonggi Bay, retaining a slight component of counterclockwise rotation due to the large width of the bay (see Figure 2.19).

tropical cyclones, which function to transport heat poleward from the equator, punctuating the summer monsoon. During the winter monsoon, cold fronts sweep out of Asia and cross the Yellow Sea with a frequency of once per week and a duration of 1–2 days (Graber et al., 1989). Winds during these events generally blow toward the south to southeast (i.e., obliquely onshore) across the west Korean coast, with wind speeds at times exceeding

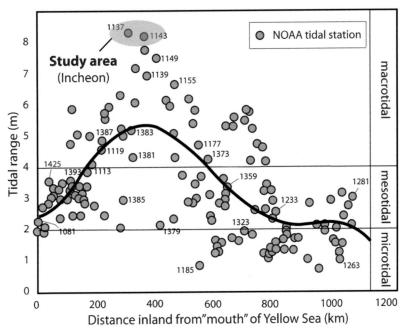

Figure 2.17 Tidal range along the Yellow Sea coast (Archer and Hubbard, 2003). Tidal ranges greater than the average (black line) occur on the eastern (Korean) side, whereas ranges less than the average occur on the western (Chinese) side because of the Coriolis effect. Note that the tidal range in the Yellow Sea reaches a maximum in Gyeonggi Bay. Numbers refer to National Oceanic and Atmospheric Administration (NOAA) tidal stations (see Figure 2.16 for their positions).

35 km/h and reaching 90 km/h (Figure 2.22). By contrast, the summer monsoon generates gentle winds that blow toward the north across the region with an average speed of 10 km/h (US Navy Website www.nrlmry.navy.mil/port_studies/thh-nc/korea/incheon/text/sect5.htm). Approximately one tropical cyclone per summer disrupts the monsoonal wind pattern in Gyeonggi Bay (Figure 2.23), most commonly in August (Figure 2.24). Tropical cyclones are convection-driven, counterclockwise-spinning storms that affect any given area for 1–2 days, diminish in intensity upon landfall, and commonly cause up to several hundred millimeters of rainfall. *Storm surges* of several meters have been generated along the west coast of Korea by tropical cyclones that pass to the north (Moon et al., 2003), with wind speeds exceeding 150 km/h near the storm center (Table 2.1). However, such storm surges and wind speeds have been very rare in Gyeonggi Bay over the past 50 years because tropical cyclones typically pass east of the Korean peninsula (Figures 2.25 and 2.26). A typical tropical cyclone, therefore, will generate winds in Gyeonggi Bay that blow toward the west or south (i.e., generally offshore directed) with an average wind speed of 40 km/h (US Navy Website www.nrlmry.navy.mil/port_studies/thh-nc/korea/incheon/text/sect5.htm). Because of the Coriolis effect and Ekman transport, such winds likely push surface water away from the coast,

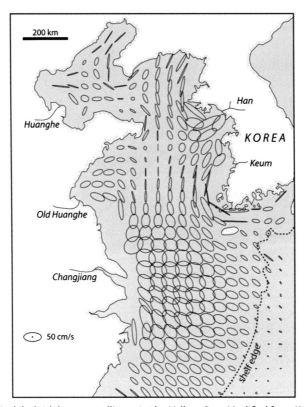

Figure 2.18 Modeled tidal-current ellipses in the Yellow Sea. *Modified from Kang et al. (2002).*

thereby inhibiting storm-surge development while promoting upwelling (Swift and Niederoda, 1985).

Maximum wave height values from 2004 onward are available for two wave buoys offshore Korea, Duk-Juk in southern Gyeonggi Bay and Chil-Bal in the southeastern Yellow Sea (Figure 2.27) (Kumar et al., 2003). In 2004, winter storms generated waves with heights that reached 3 m in southern Gyeonggi Bay and 5 m in the southeastern Yellow Sea, whereas summer waves rarely exceeded 1 m in southern Gyeonggi Bay and 2 m in the southeastern Yellow Sea. Three tropical cyclones occurred in 2004, all in late August–early September. All three passed east of the Korean peninsula, as is typical. These storms generated 3 m high waves in the southeast Yellow Sea and very small (1 m high) waves in southern Gyeonggi Bay. Theoretically, therefore, winter storms are therefore more likely to resuspend sediment in Gyeonggi Bay than summer storms (Box 2.1).

Water properties and general circulation in the Yellow Sea are strongly influenced by several factors, including seasonal variations in wind strength and insolation, freshwater input from rivers, and current activity in the neighboring Pacific Ocean (Figure 2.28).

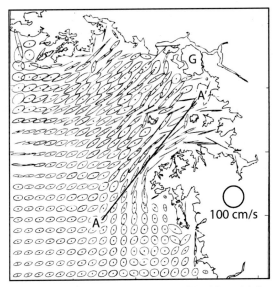

Figure 2.19 Modeled tidal-current ellipses in Gyeonggi Bay (M$_2$ tide only) (Lee et al., 2001). Note that tidal currents tend to be rotary outside of Gyeonggi Bay, but become more rectilinear in Gyeonggi Bay, although they retain a distinct rotary component due to the relatively large width of the embayment. Data for points over bar crests are much more rotary than the data for points in the channels, which reflects the local bathymetric control on the currents. For scale, Ganghwa Island (G), the largest island at the mouth of the Han River, is 25 km long. A–A′ is the location of the cross-section shown in Figure 2.29.

In winter, the Yellow Sea becomes vertically well mixed as a result of surface cooling, tidally generated turbulence, and strong winds. Come spring, surface heating from increased insolation, in addition to greater freshwater input from rivers (of which the Changjiang River contributes over 80%), causes stratification in the central Yellow Sea, although coastal waters remain well mixed by tides (e.g., Gyeonggi Bay; Figure 2.29). Near-bottom circulation patterns are similar year round, although the Kuroshio Current penetrates into the Yellow Sea during winter but not in summer. Residual currents rarely exceed several cm/s (Jung et al., 2001). It is notable that the South Korean Coastal Current appears to flow southward year round (Figure 2.28), but becomes stronger in the winter (up to 10 cm/s locally at the sea surface) in response to wind forcing (Wells, 1988).

2.5 HAN RIVER DELTA: THE BROADER ENVIRONMENTAL CONTEXT

The continental shelf between Gyeonggi Bay and the shelf edge (130–140 m water depth) is wide (800 km), shallow (average ~55 m), and nearly horizontal (Figure 2.30). Sand covers the outer shelf and mud occurs immediately offshore from major rivers and in a north–south belt in the center of the Yellow Sea (Figure 2.31). The sediment distribution

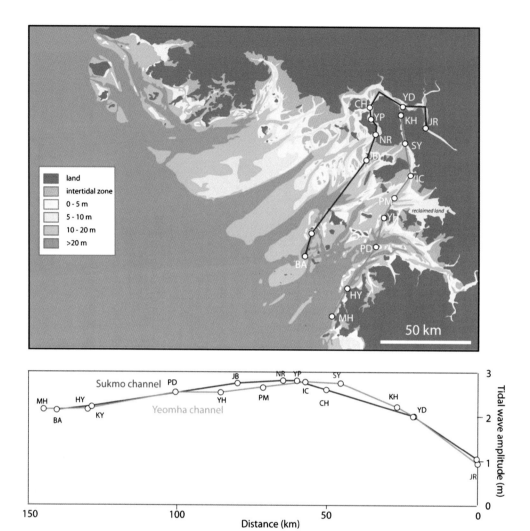

Figure 2.20 Observed tidal wave amplitude in Gyeonggi Bay. *Modified from Woo and Yoon (2011)*. The increasing-then-decreasing trend indicates that the bay is hypersynchronous (*sensu* Salomon and Allen, 1983): convergence initially causes amplitude increase, the effects of which are eventually overcome by bottom friction, causing the tidal wave to decrease in amplitude landward of the tidal maximum. Letters refer to tidal stations.

and seafloor morphology are believed to reflect the combined influence of the physical processes that operated during the LGM lowstand (~18 ka BP), the ensuing eustatic transgression (~18–7 ka BP), and the Holocene highstand (~7 ka BP to present) (Figure 2.32).

During the LGM, global sea level was 120 m below the present level and the continental shelf was subaerially exposed, although sea level apparently did not fall below the shelf break (130–140 m isobath; Figure 2.30). In seismic data, buried LGM fluvial

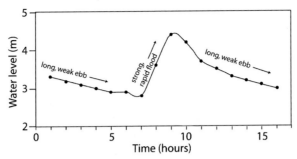

Figure 2.21 Asymmetric tidal–wave profile revealed from measurements of the water surface eleva-tion in the Han River downstream of Seoul (Kim et al., 2002). Note short, steep leading wave-face (flood tide) and long, gentle trailing wave-face (ebb tide). This type of asymmetry is a common feature of tidal waves (and wind waves) as they shoal because the greater frictional retardation of the wave trough, due to interaction with the seafloor, slows down the trough relative to the crest. This causes the crest to catch up with the trough, which in turn leads to a steepening of the leading face of the wave (flood tide) and a lengthening of the trailing wave-face (ebb tide). Because bedload transport increases exponen-tially with current velocity, bedload is commonly transported landward over time in tide-dominated coastal systems due to the overall flood dominance, generating a landward fining trend in gravel- and sand-sized sediment supplied from the sea; by contrast, river-supplied sediment fines seaward.

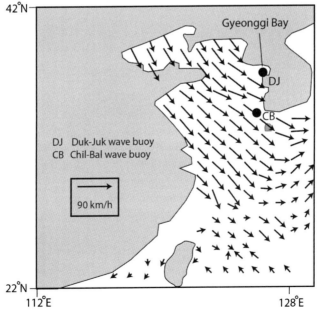

Figure 2.22 Wind pattern during an extreme winter storm, January 1, 1997 (Kumar et al., 2003). Winter storms affect the Yellow Sea approximately once per week, and last for 1–2 days. This figure is based on satellite data and wave buoy data from Duk-Juk (DK) and Chil-Bal (CB). Note that wind speeds during this extreme event approached 90 km/h over much of the Yellow Sea. Typical winter storm wind speeds are closer to 35 km/h.

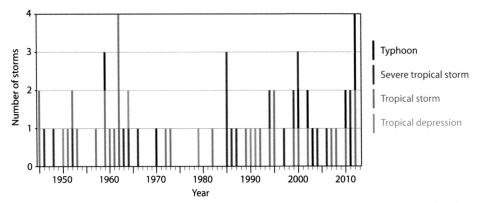

Figure 2.23 Chronology of 62 tropical cyclones that passed within 333 km (180 nautical miles) of Incheon between 1945 and 2012. *Data from Table 2.1.* Only nine of these were classified as typhoons (maximum sustained wind speed >118 km/h). Because most tropical cyclones pass east of the Korean peninsula (Figure 2.23), average wind speed in Gyeonggi Bay tends to be much lower, averaging about 40 km/h (less for cyclones that pass outside of the 333 km radius).

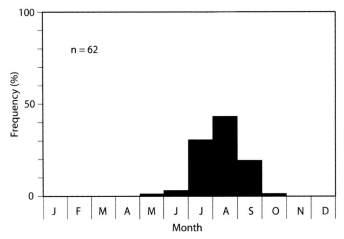

Figure 2.24 Histogram showing monthly distribution of the 62 tropical cyclones that passed within 333 km of Incheon between 1945 and 2012. *Data from Table 2.1.*

systems have been identified in several locations on the outer continental shelf (Figure 2.33). The base of these fluvial units can be traced laterally to interfluves where paleosols are locally developed (Jin and Chough, 2002; Lim et al., 2004; Choi, 2005)—this surface defines the LGM sequence boundary. Offshore of China, LGM fluvial deposits are interpreted to form a <15 m thick, >350 km wide sheet that consists primarily of Changjiang River sediments, with possible contributions from the Huanghe (Wellner and Bartek, 2003). The absence of large shelf-margin deltas at the terminal end of this system may be symptomatic of the low sediment supply that resulted from a dry LGM climate in

Table 2.1 Description of the 62 tropical cyclones that passed within 333 km of Incheon over the 67 year period between 1945 and 2012 (Korean Meteorological Administration Website)

Storm number	Storm name	Year	Month	Day	Maximum wind speed at storm center (km/h)*	Closest point of approach to Incheon	Class§
1	Eva	1945	August	4	74	31	TS
2	Ursula	1945	September	13	81	132	TS
3	Lilly	1946	August	20	93	32	STS
4	Pearl	1948	July	7	93	32	STS
5	Grace	1950	July	21	56	145	TD
6	Marge	1951	August	23	82	39	TS
7	Karen	1952	August	18	98	72	STS
8	Mary	1952	September	3	83	45	TS
9	Kit	1953	July	6	83	89	TS
10	Agnes	1957	August	21	83	130	TS
11	Billie	1959	July	17	83	82	TS
12	Louise	1959	September	7	68	90	TS
13	Sarah	1959	September	17	178	269	TY
14	Carmen	1960	August	23	83	26	TS
15	Betty	1961	May	28	67	266	TS
16	Helen	1962	August	3	50	53	TD
17	Joan	1962	July	10	74	61	TS
18	Nora	1962	August	2	72	34	TS
19	Amy	1962	September	7	62	14	TS
20	Shirley	1963	June	19	98	20	STS
21	Flossie	1964	July	29	111	183	STS
22	Helen	1964	August	2	115	228	TD
23	Betty	1966	August	30	56	175	STS
24	Billie	1970	August	31	89	114	TY
25	Rita	1972	July	26	130	225	TS
26	Iris	1973	August	17	63	27	TS
27	Irving	1979	August	17	80	140	TD
28	Cecil	1982	August	14	60	158	TS

Continued

Table 2.1 Description of the 62 tropical cyclones that passed within 333 km of Incheon over the 67 year period between 1945 and 2012 (Korean Meteorological Administration Website)—cont'd

Storm number	Storm name	Year	Month	Day	Maximum wind speed at storm center (km/h)*	Closest point of approach to Incheon	Class§
29	Jeff	1985	August	2	83	285	STS
30	Kit	1985	August	10	93	161	STS
31	Lee	1985	August	14	93	142	STS
32	Vera	1986	August	28	91	72	STS
33	Thelma	1987	July	15	102	138	STS
34	Judy	1989	July	28	65	72	TS
35	Abe	1990	September	2	59	42	TD
36	Gladys	1991	August	23	57	255	TD
37	Ted	1992	September	24	74	173	TS
38	Brendan	1994	July	31	76	75	TS
39	Seth	1994	October	11	94	276	STS
40	Faye	1995	July	23	83	182	TS
41	Janis	1995	August	26	65	40	TS
42	Tina	1997	August	9	94	296	STS
43	Neil	1999	July	27	76	21	TS
44	Olga	1999	August	3	101	30	STS
45	Bolaven	2000	July	30	65	303	TS
46	Prapiroon	2000	September	1	130	116	TY
47	Saomai	2000	September	15	101	220	STS
48	Rammasun	2002	July	5	83	45	TS
49	Rusa	2002	August	31	119	112	TY

50	Maemi	2003	September	12	137	241	TY
51	Megi	2004	August	19	112	307	STS
52	Ewiniar	2006	July	10	94	12	STS
53	Nari	2007	September	16	68	289	TS
54	Kalmaegi	2008	July	20	61	263	TS
55	Dianmu	2010	August	10	86	311	TS
56	Kompasu	2010	September	1	137	47	TY
57	Meari	2011	June	26	90	270	STS
58	Muifa	2011	August	7	122	230	TY
59	Khanun	2012	July	18	79	7	TS
60	Tembin	2012	August	30	83	121	TS
61	Bolaven	2012	August	28	144	103	TY
62	Sanba	2012	September	17	137	169	TY

* Maximum wind speeds correspond to those measured in the center of the storm, and do not necessarily correspond to those measured at Incheon.
§ World Meteorolgial Organization classification. TD (tropical depression), maximum wind speed <61 km/h TS (tropical storm), maximum wind speed range 62–88 km/h STS (severe tropical storm), maximum wind speed range 89–117 km/h TY (typhoon), maximum speed greater than 118 km/h.

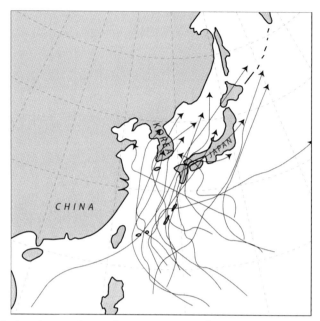

Figure 2.25 Tropical cyclone pathways between 1993 and 1996. *Data from Korean Meteorological Association website.* Note that most tropical cyclones pass east of the Korean peninsula, generating winds that blow toward the west or south for 1–2 days in Gyeonggi Bay.

central Asia (Figure 2.32). Offshore Korea near Cheju Island, in the central axis of the Yellow Sea, LGM fluvial deposits are interpreted to fill discrete channels that are several kilometers wide and tens of meters deep (Jin and Chough, 1998; Chough et al., 2004). These channels may record the position of one or more Korean rivers during the LGM. Although the above examples provide us with glimpses of what the LGM lowstand drainage network might have looked like, the regional picture is largely incomplete, and the courses shown in Figure 2.33 should not be considered exact.

Between the LGM and the present day, global sea level rose 120 m, shorelines in the Yellow Sea translated up to 1500 km landward across the low-gradient shelf at a rate of up to 80–100 m per year, and a thin transgressive sand layer (typically <3 m thick) was deposited throughout the Yellow Sea. Numerical modeling suggests that the changing basin configuration promoted the development of strong tides early in the transgression, and that these were responsible for sculpting the seafloor locally into elongate tidal ridges as transgression proceeded (Uehara and Saito, 2003). Tidal ridges occur in six fields (Figure 2.34): (1) offshore South Korea (Jin and Chough, 1998, 2002; Chough et al., 2002; Park et al., 2006), in the eastern half of the Yellow Sea; (2) SE of the modern Changjiang mouth (Berné et al., 2002; Chen et al., 2003); (3) NW of the modern Changjiang mouth (Yang, 1989); (4) in Gyeonggi Bay (Jung et al., 1998; Jin, 2001); (5) in Korea Bay (Off, 1963); and (6) in the Bohai Sea (Liu et al., 1998). These tidal ridges

Figure 2.26 Satellite image of Typhoon Rusa, August 30, 2002—the most damaging tropical cyclone to hit Korea since 1959. Note that Rusa passed east of Gyeonggi Bay, which is the typical path for tropical cyclones that affect Korea (Figure 2.25). Wind speed reached 120 km/h on the southern tip of the Korean peninsula during Typhoon Rusa, but only for several hours, and only within 250 km of the eye of the storm (Lee and Niler, 2003). Winds preceding and following passage of the eye had relatively low speeds. Despite high wind speeds associated with tropical storms, the short duration of the winds, their general offshore direction, restricted fetch and loss of energy upon landfall likely renders such typhoons ineffective at generating big waves in Gyeonggi Bay. *Photo courtesy of NASA.*

are all elongate, sand covered, and oriented roughly parallel to the present-day tidal ellipses (Figure 2.35). They differ, however, in their present level of activity (active vs moribund), morphology (degree of elongation and width), surface ornamentation (smooth vs dune-covered), internal sedimentology (sandy vs mud-dominated hetero-lithic), cross-sectional profile (symmetric vs asymmetric), seismic architecture (lateral-accretion vs complex), and, therefore, presumably, in their origin (largely erosional vs predominantly depositional). We limit our review here to the tidal ridges offshore of South Korea (field 1 in Figure 2.34), for which long cores are available, in addition to short piston cores and extensive seismic profiles.

Tidal ridges offshore of South Korea occur in an arcuate swath between Gyeonggi Bay and Cheju Island in water depths ranging from 20 to 80 m (Figure 2.36). They are 15–25 m high, 3–10 km wide, 25–100 km long, and tend to be rounded and symmetric in cross–section, with low–angle (<1°) flanks. Most are sand covered. Their surfaces are smooth or are covered by *dunes* up to 5 m in height. Some of the ridges closest to the coast are buried by prodelta mud supplied by the Keum River (Figure 2.36).

The tidal ridges offshore of South Korea display three different stratigraphic architectures, presumably reflecting three different origins (Figure 2.37): (1) erosional sculpting of the seafloor (Type 1 ridges), (2) reworking of the transgressive lag (Type 2 ridges), and (3) deltaic–like deposition (Type 3 ridges). Ridges offshore of central Korea described by Jin and Chough (1998) are Type 1 ridges, the product of tidal sculpting of the seafloor that have been referred to as "erosional ridges" (Dalrymple, 2010). These ridges consist of a core of older heterolithic deposits overlain by a thin (<3 m) sandy carapace that represents a transgressive lag. The sediment removed from

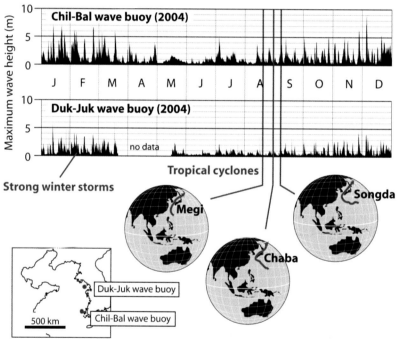

Figure 2.27 Maximum wave height data for 2004 from wave buoys located at Duk-Juk (southern Gyeonggi Bay) and Chil-Bal (southeastern Yellow Sea). Red lines indicate time of closest point of approach to Korea of three tropical cyclones. Note that winter waves are larger than summer waves at both locations. Waves are also larger at Chil-Bal than at Duk-Juk, likely due to the fetch-limited, shallow-water conditions inside Gyeonggi Bay. Note also that all three tropical cyclones, which are considered typical because they passed east of the Korean peninsula, had little effect at Duk-Juk, but generated waves at Chil-Bal. Duk-Juk is located in the southern half of Gyeonggi Bay, which is exposed to open-ocean waves entering the bay from the Yellow Sea during winter. It is likely that waves in the sheltered northern half of Gyeonggi Bay, where the study area is located, are smaller than at Duk-Juk. *Data from the Korea Meteorological Association.*

the seafloor to form such erosional ridges might have moved onshore because of the flood dominance discussed earlier. If this is the case, then the existence of such erosional ridges suggests that a significant volume of sediment might have been supplied to Gyeonggi Bay during the transgression. Type 2 ridges are also present offshore of southwest Korea. They are constructional in origin and consist of sandy sediment that represents a local thickening of the transgressive lag that lies above a flat *ravinement surface* (Jin and Chough, 2002). Jung et al. (1998) identified a Type 3 ridge in the southern part of Gyeonggi Bay. The ridge consists of an elongate nucleus (possibly a Type 1 or 2 tidal ridge) that is overlain on one side by a mud-rich clinothem package that

---

Box 2.1 Predicting near-bed fluid motion and sand transport beneath wind waves

Previous workers have developed equations that allow the characteristics of wind-generated waves (wave height and period) to be calculated based on three input variables: wind speed, fetch, and water depth (see later). The calculated wave heights and periods can in turn be used to predict the intensity of near-bed fluid motion and sediment transport.

Winter storm waves

The 3-m high waves recorded in southern Gyeonggi Bay during the winter storms of 2004 (Figure 2.25) should generate near-bed oscillatory velocities of several tens of centimeters per second on the top of the large tidal bars (water depth ~20 m). Such velocities are capable of resuspending fine sand. Even 2-m high waves—a more likely maximum wave size in the central part of Gyeonggi Bay, given the limited fetch (~50 km)—would still resuspend fine sand. Because of their once-per-week frequency, winter storms are therefore suspected to affect sedimentation on the top of the large tidal bars, possibly considerably. Evidence of significant sediment resuspension by winter waves is visible in satellite images showing elevated suspended-sediment concentrations in the shallow water on the bar tops (Figure 2.12).

Summer storm waves

Summer storms, by contrast, are much less likely to influence sedimentation on the top of the large tidal bars. As discussed previously, the average summer tropical cyclone passes to the south of Gyeonggi Bay and generates winds that blow offshore in the bay at a speed of ~40 km/h (11 m/s). These winds would generate ~1 m high waves in central Gyeonggi Bay. On the top of the large tidal bars, near-bed oscillatory velocity would be several centimeters per second, which is almost an order of magnitude lower than that required to move fine sand.

Input variables
Wind speed, U_A (m/s)
Fetch, f (m)
Water depth, h (m)
Grain size, D (m)

Continued

Box 2.1 Predicting near-bed fluid motion and sand transport beneath wind waves—cont'd

(1) Wave height $(H) = 0.283 \tanh\left[0.53\left(\dfrac{gh}{U_A^2}\right)^{3/4}\right]\tanh\left\{\dfrac{0.00565\left(\dfrac{gf}{U_A^2}\right)^{1/2}}{\tanh\left[0.53\left(\dfrac{gh}{U_A^2}\right)^{3/4}\right]}\right\}U_A \div g$

(2) Wave period $(T) = 7.54\tanh\left[0.833\left(\dfrac{gh}{U_A}\right)^{3/8}\right]\tanh\left\{\dfrac{0.0379\left(\dfrac{gf}{U_A^2}\right)^{1/3}}{\tanh\left[0.833\left(\dfrac{gh}{U_A^2}\right)^{3/8}\right]}\right\}U_A \div g$

(3) Wavelength $(L) = T\sqrt{\dfrac{gh}{F}}$

where

$$F = G + \dfrac{1}{1 + 0.6522G + 0.4622G^2 + 0.0864G^4 + 0.0675G^5}$$

and

$$G = \left(\dfrac{2\pi}{T}\right)^2\dfrac{h}{g}$$

(4) Bottom oscillatory velocity $(u_b) = \dfrac{H\pi}{T\sinh(2\pi h/L)}$

(5) Bottom orbital diameter $(d_o) = \dfrac{H}{\sinh(2\pi h/L)}$

(6) Threshold bottom oscillatory velocity $(u_{b\,thresh}) = 0.337(g^2 TD)^{1/3}$ if $D < 0.5$ mm

$= 1.395(g^4 TD^3)^{1/7}$ if $D > 0.5$ mm

Equations 1 and 2 from US Army Corps (1984), Equation 3 from Hunt (1979) as quoted in US Army Corps (1985), Equation 4 from Wiberg and Sherwood (2008), Equation 5 from Komar and Miller (1973) and Equation 6 from Komar and Miller (1973, 1975) as quoted in Clifton and Dingler (1984). Equation 6 applies for quartz density sand (0.063 mm < D < 2 mm) in seawater.

records outbuilding from the nucleus as a result of mud supplied by the Han River. The entire ridge is then mantled by a thin (<3 m) carapace of sand that is thought to represent reworking of the ridge after the supply of mud diminished. Because the ridge form is composed predominantly of muddy sediment, Jung et al. (1998) referred

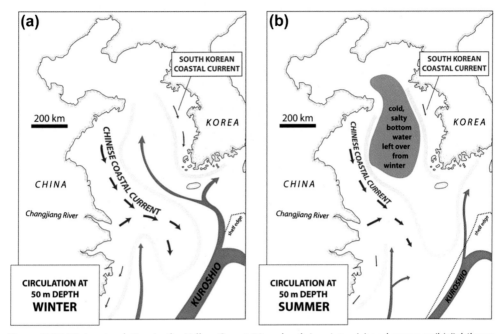

Figure 2.28 Water circulation in the Yellow Sea at 50 m depth in winter (a) and summer (b) (Ichikawa and Beardsley, 2002). Figures are based on hydrographic data collected between 1953 and 1970 from 50 m depth (shallower near coast), a depth that avoids the direct influence of atmospheric (wind) forcing and river discharge on the distribution of sea-surface temperature and salinity. Note that the warm, salty Kuroshio Current—the East Asian equivalent of the Gulf Stream—does not branch significantly into the Yellow Sea in summer, and that strong southwestward winds strengthen both the Chinese and South Korean coastal currents in winter.

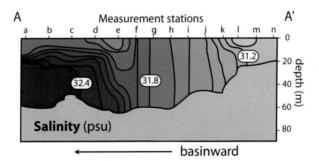

Figure 2.29 Salinity cross-section from southern Gyeonggi Bay based on measurements made at stations "a" to "n" between July 26 and 29, 1998 (i.e., during the period of high river discharge; Lee et al., 2002). See Figure 2.19 for cross-section location. Stations a–e are in the Yellow Sea, whereas stations f–n are in Gyeonggi Bay. The difference between the central and coastal regions in the Yellow Sea is exemplified in this figure: stratified conditions exist outside Gyeonggi Bay (between measurement stations a and e), whereas the water column is well mixed (i.e., the salinity contours are vertical) in Gyeonggi Bay (between measurement stations f and n). Weak stratification does occur in the innermost part of Gyeonggi Bay (between k and n), probably due to freshwater outflow from the Han River. The acronym psu stands for practical salinity units, a unit based on the conductivity of water.

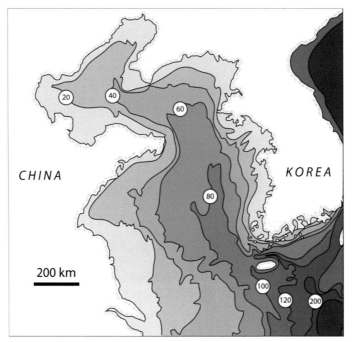

Figure 2.30 Bathymetry of the Yellow Sea (Chough et al., 2004). Contours are in meters. Gyeonggi Bay is approximately 800 km from the Late Pleistocene lowstand shoreline (120 m isobath). The shelf between Gyeonggi Bay and the shelf edge (130–140 m isobath) is almost horizontal (~0.007° dip to the south).

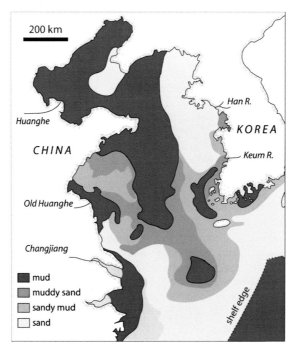

Figure 2.31 Present-day distribution of sediments in the Yellow Sea and adjacent East China Sea (Chough et al., 2004). Note that the outer shelf tends to be sandy, whereas mud occurs directly offshore from the mouth of major rivers and in a north–south belt in the middle of the Yellow Sea.

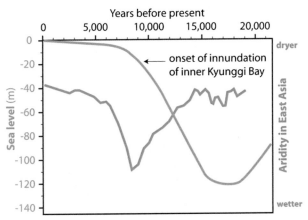

Figure 2.32 Sea-level curve (blue) for the west Korean coast (Choi and Dalrymple, 2004) and aridity index (green) for East Asia (Prins and Postma, 2000). Note that climate was more humid at the start of the Holocene than during either the last glacial maximum (18,000 years BP) or today.

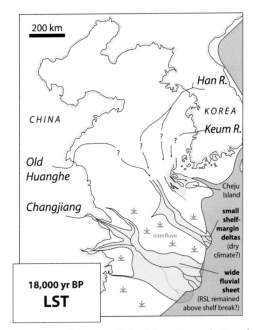

Figure 2.33 Approximate positions of lowstand fluvial systems during the last glacial maximum (18,000 years BP) interpreted from seismic data and a limited number of cores (Jin and Chough, 1998; Wellner and Bartek, 2003). (LST refers to lowstand systems tract. RSL refers to relative sea level.) Some workers believe that the Changjiang River deposited a <15 m thick, >350 km wide sheet of sand on the outer shelf during the last glacial maximum (LGM), but this interpretation is controversial as other research-ers believe that the sequence boundary lies at the top of the sand sheet rather than at its base (Berné et al., 2002). In the former interpretation, the sheet-like geometry is thought to be the result of sea level not falling below the shelf edge, which prevented the formation of a knickpoint and channel incision, thus allowing frequent channel avulsion. Shelf-margin deltas associated with the Changjiang River are volu-metrically insignificant, presumably because of substantial accommodation on the shelf and reduced sediment supply resulting from a dry climate (Wellner and Bartek, 2003). On the Korean side of the Yellow Sea, discrete channels interpreted to be incised valleys occur near Cheju Island (Jin and Chough, 1998). They may record the position of one (or more) Korean rivers during the LGM. It is unclear whether the Han River drained toward Cheju Island or joined the Huanghe and fed sediment to the central shelf edge. The former situation appears more likely, given the bathymetry of the Yellow Sea (Figure 2.30).

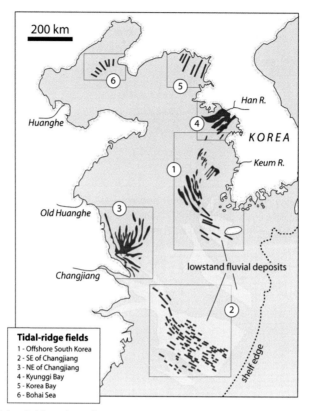

Figure 2.34 Tidal-ridge fields in the Yellow Sea and adjacent parts of the East China Sea (Chough et al., 2002; Jin and Chough, 1998, 2002; Chen et al., 2003; Berné et al., 2002; Yang, 1989; Jin, 2001; Jung et al., 1998; Liu et al., 1998). Most tidal ridges are interpreted to have formed during the transgression that followed the last glacial maximum (i.e., within the last 18,000 years).

to the ridge as a "pseudo tidal sand ridge." They hypothesize that it formed at the mouth of the Han River during a short progradational pulse that punctuated the post-LGM transgression at a time when sea level was lower than today. Subsequently, the ridge became detached from the Han River mouth and was reworked by tides, generating the sandy carapace. Of the three types of ridges, this depositional ridge is most similar in size, shape, and facies architecture to the large tidal bars that make up the subaqueous portion of the Han River delta.

The rate of relative sea-level rise is thought to have slowed soon after the start of the Holocene, a time when the East Asian climate may have been more humid than today, and more humid than during the LGM (Figure 2.32) (Box 2.2). In the last few thousand years, major rivers in China (Changjiang, Huanghe) and Korea (Han, Keum) have deposited deltas that have built out over the sandy transgressive lag. Locally, pro-delta mud buries the transgressive tidal ridges (Figure 2.38) and large dunes on the

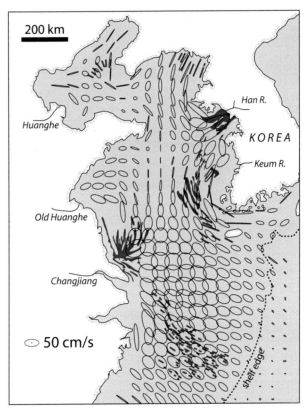

Figure 2.35 Tidal-ridge fields in the Yellow and East China seas, with superimposed tidal-current ellipses for the present-day tides (Kang et al., 2002). Tidal-current ellipses represent depth-averaged values. They are derived from a numerical model and agree with current measurements made at several locations offshore of Korea and China. Note that the long axes of the tidal ridges are commonly parallel or close to parallel with the long axis of the tidal ellipses, and that the currents approach or exceed 50 cm/s in almost all locations where tidal ridges are present.

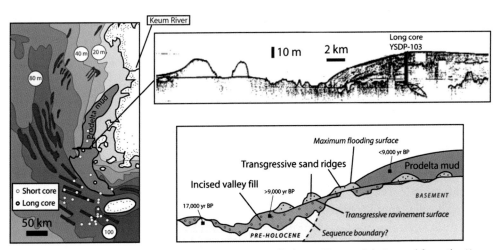

Figure 2.36 Tidal ridges offshore of South Korea, partially buried by prodeltaic mud from the Keum River. These are Type 2 ridges (cf. Figure 2.37). *Modified from Jin and Chough (1998).* Note that, unlike typical prodeltaic deposits, the mud belt is detached from the shoreline.

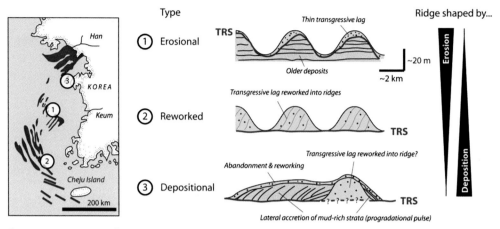

Figure 2.37 Stratigraphic architecture of tidal ridges, offshore Korea. Type 1 ridges owe their form to the erosive removal of sediment from the intervening low areas (i.e., erosional ridges), whereas Type 2 and 3 ridges are formed by the build-up of sediment and are termed "constructional" ridges. Type 1 and 2 ridges, as well as the sandy nucleus of the Type 3 ridge formed during transgressive reworking of the sea floor, whereas the muddy lateral-accretion deposits of the Type 3 ridge are thought to represent a progradational episode. The Type 1 and 2 ridges are based on Jin and Chough (2002) and Jin and Chough (1998), respectively. The Type 3 ridges are based on a ridge studied in southern Gyeonggi Bay by Jung et al. (1998). These authors refer to the ridge as a "pseudo tidal sand ridge" because it consists of mud-rich deposits underlying a sandy carapace. It and the Type 1 ridge could easily be misidentified as a tidal *sand* ridge based on grab samples alone. TRS refers to transgressive ravinement surface.

Box 2.2 Holocene climate

Since the mid-Holocene, climate in East Asia, and possibly Korea, appears to have become increasingly arid. During the mid- to early Holocene, pollen samples suggest moisture-demanding vegetation occupied regions that are now arid (Prins and Postma, 2000; Wellner and Bartek, 2003; Liu et al., 2004). The rate of progradation of the Ganges Brahmaputra delta was high in the mid-Holocene (Goodbred and Kuehl, 2000) and Indian rivers incised their valleys due to the increased water discharge (Gibling et al., 2005).

shelf (Figure 2.39). Even though the volume and areal extent of mud deposited by Chinese rivers is enormous compared to Korean rivers, the architecture and fate of the muddy deposits are similar. It is of note that the deposits of both Chinese and Korean rivers extend southward of the river mouths from which they originated, suggesting that the ultimate fate of sediment in these systems is more a function of winter resuspension and reworking than of summer river plume processes.

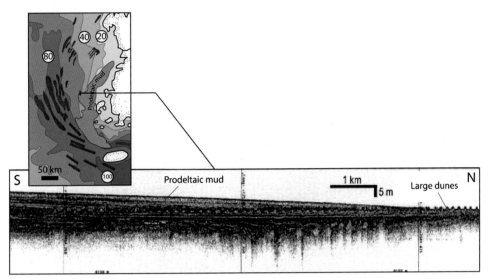

Figure 2.38 Large dunes that form part of the transgressive lag buried by prodeltaic mud from the Keum River (Shinn et al., 2004).

2.6 PHYSICAL PROCESS REGIME: SUMMARY

Based on the depositional parameters outlined earlier, the total hydraulic energy, as well as the contribution of energy from various sources (e.g., waves, tidal currents, river out-flow, and oceanic circulation) to the total hydraulic energy, is hypothesized to vary in an onshore–offshore (i.e., proximal–distal) transect through Gyeonggi Bay (cf. Dalrymple et al., 1992, 2012; Dalrymple and Choi, 2007, Figure 2.39, top). In addition, pronounced along-strike (i.e., coast parallel) variations in hydraulic energy also occur in Gyeonggi Bay (Figure 2.39, left side). This is typical of tide-dominated coastal environments because of their high degree of channelization. Furthermore, compared to narrow-mouthed estuaries, wide-mouthed, tide-dominated coastal embayments such as Gyeonggi Bay are more likely to experience along-strike variations in wave energy, in addition to experiencing along-strike variations in tidal energy. For example, the south-ern half of Gyeonggi Bay is exposed to large winter storm waves incoming from the Yellow Sea, whereas the northern half of Gyeonggi Bay typically remains sheltered. These along-strike energy variations, in conjunction with proximal–distal energy varia-tions, control the spatial distribution of sedimentary facies accumulating on the seafloor. Overall, tide-dominated facies should characterize the axial portion of the deltaic embayment and much of the subaqueous delta plain, especially in any tidal channels that dissect this region. By contrast, wave-dominated facies should become more abundant as one moves seaward. They will also become more abundant on the shallowly submerged top of the elongate tidal bars, and will predominate in coastal deposits adjacent to the

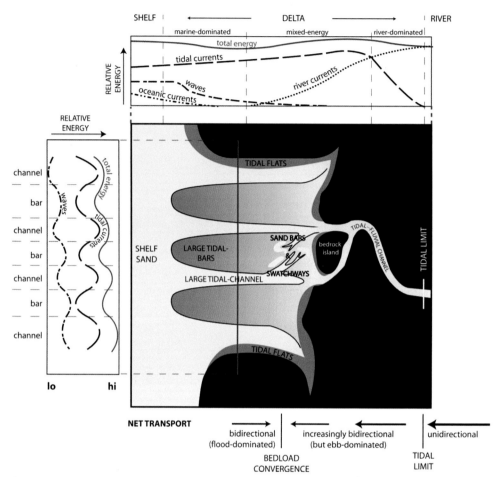

Figure 2.39 Spatial variation of physical processes in Gyeonggi Bay, in proximal–distal (at top) and along-strike (left-hand side) directions (highly simplified). Spatial variation of energy, in conjunction with the spatial variation in the rate and caliber of sediment supply, are believed to control the spatial distribution of sedimentary facies in the study area (see Chapter 5). Note that (1) any sediment moving landward from the shelf will likely be impoverished in modern woody organic matter relative to river sediments; (2) the water column is vertically well mixed due to tides except during river floods, although salinity increases steadily offshore; (3) the turbidity maximum may on average lie somewhere near, or slightly landward of, the bedload convergence (see bottom of figure), but likely moves basinward in summer and landward in winter due to significant seasonality in fluvial discharge; and (4) wave energy during strong winter storms likely increases southward across the bay due to greater sheltering along its northern flank. It is important to note that considerable seasonality exists at the basinward end of the proximal–distal energy transect with respect to waves and the South Korean Coastal Current. In particular, as discussed in Chapter 9, significant wave energy in winter combined with shear stress from tides has the potential to substantially increase energy levels in shallow water over the tops of the large tidal bars.

mouth of the delta. River-dominated facies will be present only in the inner part of the tidal–fluvial transition, including within the inner part of the distributary channels. Other first-order controls on the nature of the facies include the location, rate and caliber of sediment supply, the vertical and lateral salinity structure, and the location of the turbidity maximum.

CHAPTER 3

Data

Contents

Chapter Points

- The study focuses on a large tidal bar, one of three that make up the subaqueous delta platform, and the large tidal channels adjacent to the large tidal bar.
- The data collected consist of 66 short cores, two long cores, and 1800 km of seismic data extending from the intertidal zone to the open continental shelf.
- Previous work on tidal flats and distributary channels is integrated to cover all of the delta's subenvironments.

The data used in this study were collected primarily in May 2004. Their collection was funded and carried out by the Korean Institute of Geoscience and Mineral Resources (KIGAM), a South Korean government research institute, with supplementary financial support from a consortium of six international petroleum companies. The study focused on a large tidal bar, one of three that make up the subaqueous portion of the Han River delta, and the adjacent large tidal channels (Figure 3.1). A previous KIGAM campaign in the 1990s had collected data from the inner part of the same large tidal bar (see KIGAM, 1997; Jin, 2001; Choi et al., 2004b). Both datasets are used in this study, in addition to a set of hammer cores collected in July 2005. In total, 66 short cores, two long drill cores, 105 grab samples, and 1800 km of two-dimensional reflection seismic data were analyzed (Figure 3.1). Core and seismic are described in detail in Appendices 1 and 2 (1C to 1E refer online – http://booksite.elsevier.com/9780128007686/), respectively.

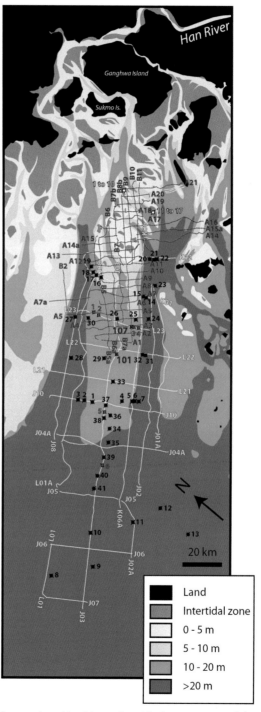

Figure 3.1 Geographic distribution of the data analyzed in this study, with the exception of the grab samples (see Figure 3.13). Drill cores 107 and 101 refer to Yellow Sea Drilling Project core 107 (YSDP 107) and New Yellow Sea Drilling Project core 101 (NYSDP 101), respectively.

3.1 CRUISE SCHEDULES

Two vessels were used to collect the 1990s dataset. In 1993, the Tamhae, a small vessel (173 grosse tonnage international) capable of navigating in shallow water collected seismic data from the inner part of the large tidal bar (Figure 3.1). Following this, the MV Kan 407 was commissioned to collect the long YSDP-101 drill core (Figure 3.1). Few additional details are available on the 1990s data collection.

The May 2004 data collection was undertaken using two vessels, the RV Tamhae II (Figure 3.2(a)) and the CV Royal (Figure 3.2(b)). The RV Tamhae II (64.4 m length; 2085 grosse tonnage international) is a large vessel operated by the KIGAM that is suitable for operation in deeper water and the handling of heavy equipment. Work in the outer part of the study area was carried out with this vessel. The CV Royal (~22 m length; 80 metric tones dead weight) is a chartered vessel with a smaller draft, more suitable for work in shallower water. It was assigned the work on the inner part of the study area. The work performed on each vessel is described later in the chapter. Personnel on the cruises are summarized in Table 3.1. A third vessel, the MV Kan 407 (Figure 3.2(c)), was used later in 2004 to collect the long NYSDP-101 drill core and several vibrocores.

Figure 3.2 The three main research vessels used in the study. (a) The RV Tamhae II, which collected seismic data and piston cores in deeper water. (b) The CV Royal, which collected seismic data in shallower water. (c) The MV Kan 407, which collected the two long drill cores (YSDP-107 and NYSDP-101) and the vibrocores.

Table 3.1 Personnel, May 2004 cruise

Crew member	Organization	Position
Jin, Jae Hwa	KIGAM	Chief scientist
Kim, Won Sik	KIGAM	Co-chief scientist
Dalrymple, Robert	Queen's University	Consultant
Kim, Sung Pil	KIGAM	Scientist and navigator
Kim, Ji-Hoon	KIGAM	Scientist and navigator
Kwon, Yi Kyun	KIGAM	Scientist and sample coordinator
Kim, Jung Ki	KIGAM	Technician (seismic and Chirp)
Choi, Joung Gyu	KIGAM	Technician (seismic and Chirp)
Kim, Hag Ju	KIGAM	Technician (mechanic)
Seo, Kap-Seog	KIGAM	Technician (mechanic)
Kim, Hyeon Sig	KIGAM	Captain, RV Tamhae II
Lee, Heui Baeg	KIGAM	Chief officer, RV Tamhae II
Ban, Chi Myong	KIGAM	Second officer, RV Tamhae II
Kwon, Tae Hyeon	KIGAM	Third officer, RV Tamhae II
Hwang, Young-Sin	KIGAM	Chief engineer, RV Tamhae II
Seo, Jeong Deok	KIGAM	First engineer, RV Tamhae II
So, Chang Won	KIGAM	Second engineer, RV Tamhae II
Bag, Yeong Gyoon	KIGAM	Third engineer, RV Tamhae II
Kim, Jae Seong	KIGAM	Boatswain, RV Tamhae II
Jeon, Seok Yong	KIGAM	Able-bodied seaman A, RV Tamhae II
Im, Hyo Min	KIGAM	Able-bodied seaman B, RV Tamhae II
Kwon, Yeong Jin	KIGAM	First oiler, RV Tamhae II
Ko, Young Ho	KIGAM	Oiler A, RV Tamhae II
Hong, Gyu Ig	KIGAM	Oiler B, RV Tamhae II
Go, Bong su	KIGAM	Cook, RV Tamhae II
Jung, Jang Kweon	KIGAM	Cook, RV Tamhae II
Kim, Jeong Ryeol	Hyundai Playing Boat	Captain, CV Royal
Kim, In su	Hyundai Playing Boat	First engineer, CV Royal
Lee, Jeong Nam	Hyundai Playing Boat	Cook, CV Royal

KIGAM, Korean Institute of Geoscience and Mineral Resources.

To determine surface-sediment characteristics and the nature of sedimentary structures and stratigraphy in the surficial portion of the sea bed, the first half of the May 2004 survey period on the RV Tamhae II was devoted to the collection of grab samples and piston cores (Table 3.2). A two-dimensional seismic survey and a swath-bathymetry survey were undertaken on the RV Tamhae II during the second half of the survey period, simultaneously using an air-gun, a Chirp, a Simrad multibeam system, and an echo sparker. The air-gun seismic records were generally of good quality, except for the first line, and the swath bathymetry was uniformly useful at showing the presence of bedforms. By contrast, the Chirp and sparker data were of variable quality, ranging from reasonable penetration to almost useless.

On the CV Royal, a two-dimensional seismic survey was undertaken simultaneously using a sparker, a 3.5 kHz sub-bottom profiler, and an echo sounder (Table 3.2). Eleven seismic records were acquired along 10 track lines, nine of which are of good quality.

Table 3.2 Schedule for the 2004 cruise

Date	RV Tamhae II	CV Royal
May 3, 2004	Mobilization	Equipment transportation
May 4, 2004	Mobilization	Equipment transportation and harbor test
May 5, 2004	Mobilization	Mobilization and scouting of survey area
May 6, 2004	Day lost to bad weather	Geophysical survey
May 7, 2004	Piston coring	Geophysical survey
May 8, 2004	Piston coring	Geophysical survey
May 9, 2004	Day lost to bad weather	Geophysical survey
May 10, 2004	Day lost to bad weather	Geophysical survey
May 11, 2004	Day lost to bad weather	Geophysical survey
May 12, 2004	Grab sampling	Geophysical survey
May 13, 2004	Grab sampling	Geophysical survey
May 14, 2004	Grab sampling	Geophysical survey
May 15, 2004	Day lost to bad weather	Geophysical survey
May 16, 2004	Piston coring	Geophysical survey
May 17, 2004	Piston coring	Geophysical survey
May 18, 2004	Piston coring	Geophysical survey
May 19, 2004	Day lost to bad weather	Geophysical survey
May 20, 2004	Prep for geophysics survey	Demobilization
May 21, 2004	Prep for geophysics survey	
May 22, 2004	Geophysical survey	
May 23, 2004	Geophysical survey	
May 24, 2004	Geophysical survey	
May 25, 2004	Geophysical survey	
May 26, 2004	Geophysical survey	
May 27, 2004	Sampling	
May 28, 2004	Sampling	
May 29, 2004	Demobilization	
May 30, 2004	Demobilization	
May 31, 2004	Demobilization	

3.2 TRACK LINE GRID

All geophysical surveys were conducted on a grid that was designed to provide regional coverage of the large tidal bar (Figure 3.1). Lines were laid out transverse (NW–SE) and parallel (NE–SW) to the axis of the large tidal bar, as determined from bathymetric data provided by bathymetric charts. Due to time lost to unusually bad weather early in the survey period, not all planed lines could be completed; as a result, the survey grid was modified as the survey continued to maintain representative coverage of the study area.

Longitudinal (shore normal) track lines were run along the crest of the large tidal bar, along its southern flank, where much of the modern deposition appears to occur (see subsequent chapters), and down the axes of the adjacent large tidal channels. The longest continuous line was along the large tidal bar crest and was 150 km in length. Longitudinal lines were not run in the inner part of the large tidal bar because of bad weather. Transverse (shore parallel) lines were more or less evenly distributed, extending from areas that lie seaward of the outer end of the large tidal bar to its shallow inner end (but still seaward of where it becomes dissected by a series of discontinuous "*swatchway*" channels that cut obliquely across the large tidal bar). In addition, three lines were run across the landward extension of the southern large tidal channel.

An attempt was made to run all lines with the same ship speed. However, because of variable current speeds and directions, the speed over ground varied between approximately 4.9 and 5.7 knots (i.e., 2.5–3 m/s). Due to the presence of many poorly marked fishing nets (Figure 3.3), surveys could only be conducted during daylight hours (i.e., from approximately 7:00 until 20:30).

3.3 NAVIGATION

The position and time synchronization on both the RV Tamhae II and CV Royal were controlled by a global positioning system (GPS), which was differentially corrected using shore control stations in real time. A Trimble 4000DS GPS was used on RV Tamhae II and a Trimble NT300D GPS was used on the CV Royal. Positional accuracy is estimated to have been ±1–2 m for the RV Tamhae II and ±5 m for the CV Royal. The position of the GPS satellite antenna projected to sea surface was referenced with respect to all the installed sensors to measure the along- and cross-track offsets (Figure 3.4). Vessel maneuvering and data logging were performed using Hypack Max software (v.005a; Coastal Oceanographics) and a Simrad ECDIS (electronic chart display and

Figure 3.3 Fishing net in shallow water, exposed at low tide. Other types of nets, marked by buoys, were encountered throughout the study area and were especially common in certain channel and ridge-flank locations. They posed a serious hazard to the vessels and any towed geophysical gear and were avoided accordingly.

information system). Positioning was maintained within 10 m of the predetermined track lines, unless the course was obstructed by fishing nets (Figure 3.3), in which case a deviation was required.

3.4 BATHYMETRY

The bathymetry was measured continuously throughout the surveys at approximately 1 s intervals using an Elac HydroStar 4300 echosounder (200 kHz) onboard the CV Royal and Simrad EA500 echosounder (12 kHz) onboard the RV Tamhae II (Figure 3.4). The depth data were logged simultaneously with the navigation data and were corrected to a fixed (non-tidal) reference frame using tidal-elevation data as recorded at Deokjeokdo ("do"="island" in Korean), the nearest tidal station to the survey area. The datum for all bathymetry data was the approximate lowest low water level.

3.5 SWATH BATHYMETRY

During all geophysical surveys on board the RV Tamhae II, a Simrad EM950 multi-beam instrument (95 kHz; pulse rate 0.2 ms) was used to obtain an image of bedforms and other sea-bed features beneath the ship (Figure 3.4(a)). Due to time constraints, a

(a) RV Tamhae II

Sensors	Along	Across	Above
Echosounder	28.15 m	0.25 m	-5.05 m
Multi-beam and SBP	N/A (approximately installed at the same location)		
GPS	-0.3 m	-0.17 m	-23.30 m

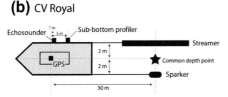

(b) CV Royal

Figure 3.4 Instrument layouts and offsets from global positioning system (GPS) antennae on (a) the RV Tamhae II and (b) the CV Royal.

standard vertical acoustic–velocity structure was employed throughout the entire cruise (i.e., vertical profiles were not obtained to allow the development of corrections for salinity/temperature/density stratification). In total, 520 line kilometers of swath bathymetry were collected. The beam footprint (i.e., the width of the swath imaged) is 7.4 times the water depth and therefore varied between approximately 100 and 400 m. Horizontal resolution is therefore different in the cross–track and along–track directions. Lateral resolution (i.e., one pixel) is 1/60 of the total width. All data, including positional data, were recorded using Simrad MERMAID data storage software.

The swath bathymetry data form the basis for a master's thesis (Kim, 2009). They were not analyzed in detail in this study. The results of Kim (2009) will be commented on briefly in subsequent chapters.

3.6 SPARKER AND AIR-GUN SEISMIC SURVEYS

The source and receiver used for the sparker seismic system on the CV Royal (Figures 3.5 and 3.6; Table 3.3) were towed ca. 30 m behind the stern to avoid vessel noise (Figure 3.4(b)). The constituent units were grounded to remove the potential for electrical interference. In total, the CV Royal collected 220 line kilometers of seismic data from the inner portion of the large tidal bar and adjacent large tidal channels (Figure 3.1; Table 3.4). All lines of good quality were transverse to the large tidal bar. The one line that was run along the crest of the large tidal bar was of very poor quality because of interference by wave noise that obscured reflections within the sediment; the other records contain variable but much less significant levels of wave noise. In general, the good-quality records had a penetration of approximately 150 ms two-way travel time (~100 m depth). However, sea-bed and other multiples limit the highest-quality record to a depth equal to the two-way travel time in the water column. Strong reflections

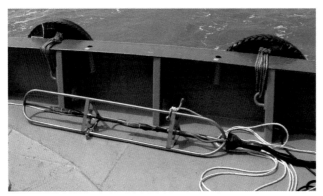

Figure 3.5 Korean Institute of Geoscience and Mineral Resources-constructed sparker array used to collect high-resolution seismic data on the CV Royal. The red wires are the sparker electrodes.

can be seen below this but the degree of certainty of the architectural interpretations is lower. The sparker data are archived in Appendix 2.

The source and receiver for the air-gun seismic system used on the RV Tamhae II (Figure 3.7) were towed ca. 50 m behind the stern to avoid vessel noise (Figure 3.4). An attempt was made to use the sparker system that had been used onboard the CV Royal, but interference from unknown sound sources hindered its use. In total, the RV Tamhae

Figure 3.6 Capacitor bank, power supply, and data-logger units for the sparker array on the CV Royal.

Table 3.3 Air-gun and sparker seismic systems

	RV Tamhae II	**CV Royal**
Source	30 in³ air gun (bolt; Figure 3.7) with 4 s firing interval, Geco-Prakla gun controller, and Hamworthy air compressor	EG&G 231-triggered capacitor bank (Figure 3.6), EG&G power supply (max. 3399V), and KIGAM*-constructed multielectrode sparker array (100 electrodes; Figure 3.5). The firing interval and energy level were set to 1 pps and 400 J, respectively.
Receiver	Benthos 50/249 hydrophone array (24 elements)	Benthos hydrophone array (10 elements)
Recorder	EPC Labs 9802 thermal graphic recorder (sweep speed: 300 ms at full scale) (Figure 3.8)	EPC Labs GSP 1086 thermal graphic recorder (sweep speed: 250 ms at full scale)
Filter		Krohn-hite 3700 band pass filter (set to 0.5–1.2 kHz)
Logging		Inkel minidisk recorder (maximum 74 min recording time)
Real-time digital recording and postprocessing	Geoacoustics SES 881 Sonar Enhancement System	

* KIGAM, Korean Institute of Geoscience and Mineral Resources.

Table 3.4 Seismic lines

Line number	Orientation§	Ship used	Seismic energy source	Length (km)	Geomorphic zone	Date collected	Reference
04DCJ01A	Dip	RV Tamhae II	Air gun	44	Channel (and bar) south of LTB*	May 2004	Dalrymple and Jin (2004)
04DCJ02	Dip	RV Tamhae II	Air gun	58	Channel south of LTB	May 2004	Dalrymple and Jin (2004)
04DCJ02A	Dip	RV Tamhae II	Air gun	19	Shelf	May 2004	Dalrymple and Jin (2004)
04DCJ03	Dip	RV Tamhae II	Air gun	96	LTB to shelf	May 2004	Dalrymple and Jin (2004)
04DCJ04	Strike	RV Tamhae II	Air gun	7	Distal LTB clinoform	May 2004	Dalrymple and Jin (2004)
04DCJ04A	Strike	RV Tamhae II	Air gun	34	Distal LTB clinoform	May 2004	Dalrymple and Jin (2004)
04DCJ05	Strike	RV Tamhae II	Air gun	18	Shelf	May 2004	Dalrymple and Jin (2004)
04DCJ06	Strike	RV Tamhae II	Air gun	24	Shelf	May 2004	Dalrymple and Jin (2004)
04DCJ07	Strike	RV Tamhae II	Air gun	7	Shelf	May 2004	Dalrymple and Jin (2004)
04DCJ08	Dip	RV Tamhae II	Air gun	48	Channel north of LTB	May 2004	Dalrymple and Jin (2004)
04DCJ09	Dip	RV Tamhae II	Air gun	59	North flank of LTB	May 2004	Dalrymple and Jin (2004)
04DCJ10	Strike	RV Tamhae II	Chirp	25	LTB top, flanks, and adjacent channels	May 2004	Dalrymple and Jin (2004)
04DCJ10	Strike	RV Tamhae II	Air gun	35	LTB top, flanks, and adjacent channels	May 2004	Dalrymple and Jin (2004)
04DCK02	Strike	CV Royal	3.5 kHz sub–bottom profiler	29	LTB top, flanks, and channel south of LTB	May 2004	Dalrymple and Jin (2004)

04DCK02	Strike	CV Royal	Sparker	29	LTB top, flanks, and channel south of LTB	May 2004	Poor quality due to high seas during data collection; Dalrymple and Jin (2004)
04DCK05	Strike	CV Royal	Sparker	6	Small "bar" in channel north of LTB	May 2004	Dalrymple and Jin (2004)
04DCK06	Dip	CV Royal	Sparker	48	South flank of LTB	May 2004	Shows apparent LTB-downlap onto channel-base dunes; Dalrymple and Jin (2004)
04DCK06A	Dip	RV Tamhae II	Chirp	41	South flank of LTB to shelf	May 2004	Dalrymple and Jin (2004)
04DCK06A	Dip	RV Tamhae II	Air gun	41	South flank of LTB to shelf	May 2004	Dalrymple and Jin (2004)
04DCL01	Dip	RV Tamhae II	Air gun	15	Shelf	May 2004	Dalrymple and Jin (2004)
04DCL01A	Dip	RV Tamhae II	Chirp	16	Shelf	May 2004	Dalrymple and Jin (2004)
04DCL01A	Dip	RV Tamhae II	Air gun	16	Shelf	May 2004	Dalrymple and Jin (2004)
04DCL21	Strike	CV Royal	3.5 kHz sub-bottom profiler	39	LTB top and flanks	May 2004	Dalrymple and Jin (2004)
04DCL21	Strike	CV Royal	Sparker	39	LTB top, flanks, and adjacent channels	May 2004	Dalrymple and Jin (2004)
04DCL22	Strike	CV Royal	3.5 kHz sub-bottom profiler	31	LTB top, flanks, and channel south of LTB	May 2004	Dalrymple and Jin (2004)

Continued

Table 3.4 Seismic lines—cont'd

Line number	Orientation[§]	Ship used	Seismic energy source	Length (km)	Geomorphic zone	Date collected	Reference
04DCL22	Strike	CV Royal	Sparker	31	LTB top, flanks, and channel south of LTB	May 2004	Dalrymple and Jin (2004)
04DCL23	Strike	CV Royal	3.5 kHz sub-bottom profiler	22	LTB top and flanks	May 2004	Dalrymple and Jin (2004)
04DCL23	Strike	CV Royal	Sparker	22	LTB top and flanks	May 2004	Dalrymple and Jin (2004)
A1	Strike	Tamhae	Sparker	15	LTB top	1993	Jin (2001), Choi et al. (2004b)
A2	Strike	Tamhae	Sparker	26	LTB top	1993	Jin (2001), Choi et al. (2004b)
A3	Strike	Tamhae	Sparker	32	LTB top	1993	Jin (2001), Choi et al. (2004b)
A4	Strike	Tamhae	Sparker	33	LTB top	1993	Jin (2001), Choi et al. (2004b)
A4a	Strike	Tamhae	Sparker			1993	Jin (2001), Choi et al. (2004b)
A5	Strike	Tamhae	Sparker	38	LTB top and N flank	1993	Jin (2001), Choi et al. (2004b)
A6	Strike	Tamhae	Sparker	12	LTB top	1993	Jin (2001), Choi et al. (2004b)
A7	Strike	Tamhae	Sparker	30	LTB top	1993	Jin (2001), Choi et al. (2004b)
A7a	Strike	Tamhae	Sparker	20	N Channel	1993	Jin (2001), Choi et al. (2004b)
A8	Strike	Tamhae	Sparker	20	LTB top	1993	Jin (2001), Choi et al. (2004b)

A10	Strike	Tamhae	Sparker	33	LTB top and flanks	1993	Jin (2001), Choi et al. (2004b)
A11	Strike	Tamhae	Sparker	33	LTB top and flanks	1993	Jin (2001), Choi et al. (2004b)
A12	Strike	Tamhae	Sparker	30	LTB top and N flank	1993	Jin (2001), Choi et al. (2004b)
A13	Strike	Tamhae	Sparker	42	LTB top to N channel	1993	Jin (2001), Choi et al. (2004b)
A13a	Strike	Tamhae	Sparker			1993	Jin (2001), Choi et al. (2004)
A14	Strike	Tamhae	Sparker	34	S Channel and swatchway	1993	Jin (2001), Choi et al. (2004b)
A14a	Strike	Tamhae	Sparker	30	LTB top to N channel	1993	Jin (2001), Choi et al. (2004b)
A15a	Strike	Tamhae	Sparker	34	S Channel and swatchway	1993	Jin (2001), Choi et al. (2004b)
A16	Strike	Tamhae	Sparker	40	S Channel, swatchways, and sand bars	1993	Jin (2001), Choi et al. (2004b)
A17	Strike	Tamhae	Sparker	20	Sandbars and swatch-ways	1993	Jin (2001), Choi et al. (2004b)
A18	Strike	Tamhae	Sparker	12	Sandbars and swatch-ways	1993	Jin (2001), Choi et al. (2004b)
A19	Strike	Tamhae	Sparker	18	Sandbars and swatch-ways	1993	Jin (2001), Choi et al. (2004b)
A20	Strike	Tamhae	Sparker	14	Sandbars and swatch-ways	1993	Jin (2001), Choi et al. (2004b)
A21	Strike	Tamhae	Sparker	30	S channel, sand bars, and swatchways	1993	Jin (2001), Choi et al. (2004b)
B2	Dip	Tamhae	Sparker	20	N of N channel	1993	Jin (2001), Choi et al. (2004b)
B5	Dip	Tamhae	Sparker	15	Sandbars and swatch-ways	1993	Jin (2001), Choi et al. (2004b)

Continued

Table 3.4 Seismic lines—cont'd

Line number	Orientation[§]	Ship used	Seismic energy source	Length (km)	Geomorphic zone	Date collected	Reference
B5a	Dip	Tamhae	Sparker	40	Sandbars and swatch-ways	1993	Jin (2001), Choi et al. (2004)
B6	Dip	Tamhae	Sparker	17	Sandbars and swatch-ways	1993	Jin (2001), Choi et al. (2004)
B6a	Dip	Tamhae	Sparker	35	Sandbars and swatch-ways	1993	Jin (2001), Choi et al. (2004)
B7a	Dip	Tamhae	Sparker	30	Sandbars and swatch-ways	1993	Jin (2001), Choi et al. (2004)
B7b	Dip	Tamhae	Sparker	15	Sandbars and swatch-ways	1993	Jin (2001), Choi et al. (2004)
B8a	Dip	Tamhae	Sparker	38	LTB top	1993	Jin (2001), Choi et al. (2004)
B9	Dip	Tamhae	Sparker	36	Sandbars and swatch-ways	1993	Jin (2001), Choi et al. (2004)
B10	Dip	Tamhae	Sparker	28	Sandbars and swatch-ways	1993	Jin (2001), Choi et al. (2004)
B11	Dip	Tamhae	Sparker	35	Sandbars and swatch-ways	1993	Jin (2001), Choi et al. (2004)

* LTB, large tidal bar.
§ Strike = line oriented approximately northwest–southeast, transverse to the large tidal bars and large tidal channels dip = line oriented approximately northeast–southwest, parallel to the length of the large tidal bars and large tidal channels.

Figure 3.7 Air-gun source unit and hydrophone array (attached to end of black cable at right) being towed behind the RV Tamhae II.

Figure 3.8 EPC Labs recorder, printing hard copy of air-gun seismic data, RV Tamhae II.

II collected 520 line kilometers of seismic data from the outer part of the large tidal bar and the adjacent floor of the Yellow Sea (Table 3.4). Overall, sea conditions were very good during this phase of the geophysical survey, so wave noise is generally minimal. In general, the good-quality records had a penetration of about 200 ms two-way travel time (~135 m). However, as with the CV Royal sparker data, sea-bed and other multiples cause the same limitations on data quality. The air-gun data are archived in Appendix 2.

3.7 CHIRP AND 3.5 kHz SEISMIC SURVEYS

Ultrahigh resolution seismic data were also collected on both CV Royal and RV Tamhae II. On the CV Royal, a sub-bottom profiler was used (Table 3.5). On the RV Tamhae II, a Chirp unit was used.

The source and receiver for the sub-bottom profiler used by the CV Royal were towed alongside the vessel (Figures 3.4 and 3.9). The same track lines (520 km) as those occupied for the sparker seismic surveys were used. Record quality was mixed and depended significantly on the sea state. During calm weather, good penetration was achieved (ca. 20 ms), producing high-resolution records.

On the RV Tamhae II, a CAP-6600 Chirp II 3.5 kHz acoustic-profiling system (Benthos Inc.) was used. It utilized a 16-element, hull-mounted array with a firing rate of 2 pulses per second. The track lines were the same as those on which air-gun records were obtained. As with the sub-bottom profiler used on the CV Royal, record quality was variable, but was generally of poor quality for unknown reasons; the system was being used for nearly its first time and unknown problems appeared to degrade performance. For most of the survey, it performed essentially as a depth sounder, with no penetration. During the very brief periods of optimal penetration, good-quality records with penetration to ca. 80 ms two–way travel time were obtained. Because of the poor record quality and the limitation on the number of data feeds (2) from the GPS navigation system (the two feeds from the GPS navigation systems were devoted to the air-gun and swath-bathymetry systems), Chirp data were not logged digitally until the last two days of the geophysical survey. Almost all of the track-line data were printed in paper format on an EPC Labs thermal graphic recorder (Figure 3.8).

Table 3.5 Sub-bottom profiler used on CV Royal

Source	Geoacoustics Geopulse transducer: a four-element array on the CV Royal
Transmitter	Geoacoustics 5430A transmitter (2 pps trigger interval)
Receiver	Geoacoustics Geopulse 5210A receiver
Recorder	EPC Labs GSP 1086 thermal graphic recorder (sweep speed: 150 ms at full scale)

Figure 3.9 3.5 kHz source and receiver being towed alongside the CV Royal.

3.8 CORE COLLECTION

The RV Tamhae II collected 41 piston cores (maximum possible length 6 m; inside diameter 8.5 cm) from the outer portion of the large tidal bar (Table 3.6). A Ewing-type piston corer was used (Figure 3.10), which weighs ca. 500 kg. The piston corer had been significantly improved by the KIGAM from its original design through the addition of a textile core catcher, which deforms the sediment less, and a swivel-type piston that operates more smoothly. The piston cores were collected on transect lines oriented transverse to the axis of the large tidal bar (Figure 3.1). They sample the flank and top of the large tidal bar, the adjacent large tidal channels, and the shelf basinward of the subaqueous delta.

Following the May 2004 campaign, the MV Kan 407, the same Chinese drill vessel that had collected the long YSDP-107 core in 1997, was commissioned to collect another long core (NYSDP-107) in order to ground-truth the new seismic data (Figures 3.1 and 3.2(c)). A mud rotary drilling system mounted over the center well of the vessel was used. During this time, the MV Kan 407 also collected six vibrocores (maximum length 1.6 m; inside diameter 8.5 cm).

In July 2005, 17 hammer cores were collected from the exposed, intertidal portions of two sandbars that ornament the inner part of the large tidal bar (Figure 3.1). The cores were collected by hammering PVC tubes into the sandy, dune-covered surfaces of the sandbars during low tide.

Detailed core logs are archived in Appendix 1 (1C to 1E refer online - http://booksite.elsevier.com/9780128007686/), along with photographs and X-radiographs.

3.9 CORE LOGGING AND SAMPLING

With the exception of several piston cores that were sectioned on the RV Tamhae II, all cores were shipped to the KIGAM's climate-controlled core storage facility in Daejeon, South Korea, where they were sectioned, X-radiographed, sampled, and logged. The cores were sectioned by cutting the plastic core liner lengthwise on two opposite sides using a skill saw (Figure 3.11). A thin stainless-steel wire was then pulled lengthwise through the sediment to separate the core into two halves. A rectangular plastic box was pushed into one half of the core and a wire was passed under the box to obtain a slab with a thickness of ca. 1 cm (Figure 3.12). The resulting sediment slabs were X-radiographed. Selected shell fragments were retained for ^{14}C dating. Subsamples were collected from the sample half for textural and other analyses. The other half of the core was photographed digitally and then described in lamina-scale detail, generally with the aid of the X-radiographs, noting sedimentary textures and structures, body fossils, trace fossils, macroscopic diagenetic features, and nature of bed contacts.

The cores experienced varying amounts of deformation during collection (Table 3.7). The hammer cores and vibrocores experienced little deformation and therefore provide "high-resolution" insight into the sedimentology of the uppermost 2 m of the large tidal bar.

Table 3.6 Summary of information about the piston cores collected from the RV Tamhae II (datum for water depths is the approximate lowest low water level)

Piston core name	Latitude		Longitude		Water depth (m)	Date	Time	Core length (m)	Subsample
	Deg	Min	Deg	Min					
04DC–P01	37	13.349	125	34.291	20.0	07–05–2004	13:00	3.20	2
04DC–P02	37	14.422	125	33.473	24.8	07–05–2004	13:45	3.77	2
04DC–P03	37	15.214	125	32.746	29.6	07–05–2004	14:15	3.54	2
04DC–P04	37	10.060	125	38.268	19.5	07–05–2004	15:00	4.64	2
04DC–P05	37	8.855	125	39.583	24.0	07–05–2004	15:40	2.14	2
04DC–P06	37	8.582	125	39.921	41.0	07–05–2004	16:15	5.81	2
04DC–P07	37	8.089	125	40.462	57.0	07–05–2004	16:40	0.92	2
04DC–P08	37	0.024	125	4.962	50.0	08–05–2004	10:35	4.30	2
04DC–P09	36	56.222	125	11.662	53.0	08–05–2004	11:30	2.70	2
04DC–P10	37	0.002	125	16.001	48.0	08–05–2004	13:00	4.85	1
04DC–P11	36	56.259	125	22.993	53.0	08–05–2004	14:05	4.19	1
04DC–P12	36	54.561	125	28.555	53.0	08–05–2004	14:55	5.18	1
04DC–P13	36	48.670	125	28.589	56.0	08–05–2004	15:50	5.12	1
04DC–P14	37	17.924	125	55.827	45.0	16–05–2004	14:50	2.37	1
04DC–P15	37	18.362	125	55.657	26.0	16–05–2004	18:00	3.20	1
04DC–P16	37	25.200	125	52.741	15.0	17–05–2004	8:30	1.48	1
04DC–P17	37	25.972	125	52.432	26.0	17–05–2004	9:00	2.12	3
04DC–P18	37	26.653	125	52.429	35.0	17–05–2004	9:25	0.77	1
04DC–P19	37	27.448	125	52.963	40.0	17–05–2004	9:50	2.29	2
04DC–P20	37	21.487	126	1.718	28.0	17–05–2004	12:25	1.45	2

04DC-P21	37	20.924	126	2.162	34.0	17-05-2004	12:45	1.95	2
04DC-P22	37	20.620	126	2.649	44.0	17-05-2004	13:00	1.59	1
04DC-P23	37	18.114	125	58.709	53.0	17-05-2004	13:50	1.24	2
04DC-P24	37	15.432	125	53.141	50.0	17-05-2004	14:40	3.73	2
04DC-P25	37	16.876	125	51.419	18.0	17-05-2004	15:30	1.70	2
04DC-P26	37	19.080	125	49.059	18.0	17-05-2004	15:55	1.70	2
04DC-P27	37	24.083	125	43.858	48.0	17-05-2004	17:00	1.33	2
04DC-P28	37	20.307	125	37.665	49.0	17-05-2004	17:35	2.20	2
04DC-P29	37	16.513	125	41.874	18.0	17-05-2004	18:20	1.83	1
04DC-P30	37	22.486	125	45.378	28.0	17-05-2004	16:30	0.62	1
04DC-P31	37	12.055	125	47.755	50.8	18-05-2004	8:05	1.38	1
04DC-P32	37	12.449	125	47.480	44.0	18-05-2004	8:45	2.92	1
04DC-P33	37	13.080	125	39.789	17.6	18-05-2004	9:30	3.76	2
04DC-P34	37	8.623	125	32.861	19.2	18-05-2004	10:25	3.54	2
04DC-P35	37	7.328	125	30.779	23.6	18-05-2004	10:50	3.49	3
04DC-P36	37	9.910	125	34.782	19.6	18-05-2004	11:20	3.36	2
04DC-P37	37	11.461	125	35.567	19.6	27-05-2004	17:10	3.20	1
04DC-P38	37	10.376	125	33.580	20.0	27-05-2004	17:30	1.05	1
04DC-P39	37	6.313	125	28.123	33.0	27-05-2004	18:20	2.68	2
04DC-P40	37	5.138	125	24.749	46.0	27-05-2004	18:50	1.70	2
04DC-P41	37	4.226	125	22.557	46.0	27-05-2004	19:20	2.90	1

Figure 3.10 Piston corer used on the RV Tamhae II.

Figure 3.11 Device used to cut core tubes lengthwise, in preparation for splitting the core into sample and archive halves. Photographs of cores are shown in Appendix 1 (1C to 1E refer online - http://booksite. elsevier.com/9780128007686/).

Figure 3.12 Sample half of core showing emplacement X-radiograph slab holder. The sides of the holder extend 1 cm into the sediment. A wire is passed under the holder and the holder and sediment are removed and wrapped in plastic for later X-radiography. X-radiographs for cores that survived significant coring induced deformation are presented in Appendix 1 (1C to 1E refer online - http://booksite.elsevier.com/9780128007686/).

Table 3.7 Summary of cores collected

Core type	Number of cores	Length	Geomorphic zone	Date collected	Comments and references	Data quality
Hammer cores[‡]	17	0.65–1 m	Sand bars on proximal half of LTB	July 2005	Typically no coring-induced deformation.	Highest
Vibrocores[§]	8	1.2–3.2 m	Center and distal half of LTB	May 2004	Only a slight amount of coring-induced deformation.	
Drill core NYSDP-101[†]	1	75.3 m	Center of LTB	2004	Only a small amount of coring-induced deformation. Missing intervals are common, especially lower in core. Purple–gray colored drilling mud is injected locally, especially near base of core. This has commonly filled the pore space in sand beds and caused soft-sediment deformation.	
Drill core YSDP-107*	1	60.55 m	Center of LTB	1997	Only a slight amount of coring-induced deformation. Missing intervals common, especially lower in core. Purple–gray colored drilling mud is injected locally, especially near base of core. This has commonly filled the pore space of sand beds and caused soft-sediment deformation. Protective plexiglass seal could not be removed, so cores had to be logged through clear cover, although core photos taken previously by KIGAM members with plexiglass removed helped in making final logs. YSDP-107 was used by J.H. Jin in his PhD thesis (Jin, 2001).	
Piston cores¶	41	0.65–5.6 m	LTB (top, flanks, distal clinoform), adjacent channels, and shelf	May 2004	Moderate to substantial coring-induced deformation. In particular, sand beds >10 cm thick were liquefied and rendered structureless during core collection.	Lowest

KIGAM, Korean Institute of Geoscience and Mineral Resources; LTB, large tidal bar.
* See (Appendix 1A).
[†] See (Appendix 1B).
[‡] 05JB series hammer cores (Appendix 1C) (http://booksite.elsevier.com/9780128007686/).
[§] NYSVC series vibrocores (Appendix 1D) (http://booksite.elsevier.com/9780128007686/).
¶ 04DC series piston cores (Appendix 1E) (http://booksite.elsevier.com/9780128007686/).

The two long cores (YSDP-107 and NYSDP-101) also experienced little coring-induced deformation, missing segments and local injections of drilling mud notwithstanding. By contrast, most of the piston cores were deformed during collection. They provide relatively a "low-resolution" insight into the deposits covering the large tidal bar.

3.10 CORE–SEISMIC TIE

The short cores (piston cores, hammer cores, or vibrocores) fall entirely within the *bubble-pulse* reflections, which masks stratigraphic detail in the uppermost 6 m of sediment below the seafloor (~8 ms). The two long cores (YSDP-107 and NYSDP-101) were tied to seismic data by assuming the speed of sound in water and sediment was 1500 m/s.

3.11 GRAB SAMPLES

A Korean-made Van Veen-type grab sampler was used on the CV Royal to obtain samples of the surficial sediments. This sampler was 50 cm wide by 70 cm long, and could penetrate to a depth of 30 cm. Grab sample locations are shown in Figure 3.13.

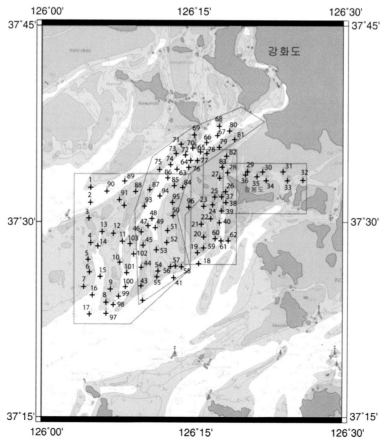

Figure 3.13 Grab samples collected by the CV Royal from the inner portion of the large tidal bar.

Immediately upon retrieval, the undisturbed central portion of each grab sample was subsampled using a plastic sleeve similar to those used for the X-radiograph samples from the cores. The slabs from the grab samples were 20 cm deep, 10 cm wide, by 3 cm thick. They were X-radiographed to document the near-surface structures. Additional material was retained for textural analysis.

CHAPTER 4

Geomorphology

Contents

Chapter Points

- The delta can be subdivided into three geomorphic zones: (1) distributary channels, (2) large tidal bars, and (3) large tidal channels. Almost all of the delta's sediment is contained within the large tidal bars, which together make up the subaqueous delta. A fourth geomorphic zone, the open shelf, occurs basinward of the delta.

- The position of the delta apex, where the fluvial channel splits into multiple distributaries, is bedrock controlled, and the distributaries themselves bifurcate around bedrock islands. However, the distributaries are alluvial in the sense that sediment has aggraded in them, the distributaries meander through this sediment, and tidal–fluvial point bars are present locally. The distributary channels are 15 to 40 m deep. Small to medium dunes are ubiquitous subtidally. Steep, narrow tidal flats form the intertidal portions of the point bars. There has been extensive reclamation of tidal flats in the tidal–fluvial zone.

- The large tidal bars that make up the subaqueous delta are 15–40 km wide and extend approximately 100 km offshore. There are at least three of them, possibly four. They rise 20–40 m above the base of the large tidal channels that separate them.

- The inner halves of the large tidal bars are flat topped, slope gently seaward, and are relatively symmetric in cross-section. Open-coast tidal flats are developed where the heads of the large tidal bars attach to the land. Immediately offshore of this, partially intertidal sandbars and swatchways are present. Small to medium dunes are ubiquitous in both the intertidal and subtidal zones here.

- The outer halves of the large tidal bars are fully subtidal, largely dune-free, devoid of sandbars and swatchways, and asymmetric in cross-section. Their distal faces make up the deltaic clinoform, the rollover of which occurs about 100 km offshore, in 20 m of water.

- The trumpet shaped large tidal channels that separate the large tidal bars link landward to either a fluvial distributary or a coastal embayment, suggesting an enhanced tidal prism controls their position. The channels extend basinward through the subaqueous delta, widening and deepening in the process, eventually bottoming out on the shelf. Their bases are lined with dunes, some of which are several meters in height.

- The shelf immediately offshore of the delta is gently undulating, relatively dune-free, and devoid of the tidal ridges so common to other parts of the Yellow Sea.

The Han River debouches into Gyeonggi Bay, a wide-mouthed embayment along the west coast of Korea (Figure 4.1). The bay is 125 km wide (north-south) and is indented 75 km relative to the rest of the Korean coast. Water depth is variable (0–50 m; average 15 m), and increases to 50–60 m outside the bay on the open Yellow Sea continental shelf. The northern part of the bay is muddier and shallower than the southern part of the bay.

The coastline in Gyeonggi Bay is typical of west Korea in that it is rocky, dotted with small bedrock islands, and fringed by broad open-coast *tidal flats*. No delta-like shoreline protruberance exists at the Han River mouth. However, if one could see beneath the water, huge sediment bars would be observed that extend 100 km offshore of the river like the fingers of a giant hand. These bars parallel the rectilinear tidal currents. As mentioned in Chapter 1, we refer to them as large tidal bars. They are the dominant

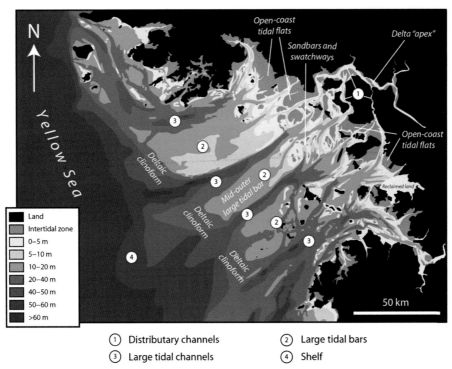

Legend	
■	Land
	Intertidal zone
	0–5 m
	5–10 m
	10–20 m
	20–40 m
	40–50 m
	50–60 m
	>60 m

① Distributary channels   ② Large tidal bars
③ Large tidal channels   ④ Shelf

Figure 4.1 Major morphological elements in the Han River delta. Bathymetric data from standard charts (Korea Hydrographical and Oceanographic Administration, 2011). The datum for these data is the mean spring-tide low water level.

geomorphic feature in the bay and contain most of the delta's sediment. Together, they make up the subaqueous delta platform (Figure 1.1).

In a very simple way, the Han River delta consists of three geomorphic zones: (1) deltaic distributary channels, (2) large tidal bars, and (3) large tidal channels. A fourth geomorphic zone, the open shelf, is present basinward of the delta (Figure 4.1).

4.1 DISTRIBUTARY CHANNELS

The Han River splits around bedrock islands before emptying into Gyeonggi Bay (Figure 4.2). The location of the *delta apex*—the position of the initial channel bifurcation—as well as the location and stability of the distributary channels, is controlled largely by this antecedent topography. The initial bifurcation occurs upstream of Gang-hwa Island, creating two first-order distributaries. The southern branch (Yeomha Chan-nel; Figure 4.2) empties directly into Gyeonggi Bay, whereas the northern branch bifurcates around several more bedrock islands, forming three channels. A total of four distinct distributaries therefore empty into Gyeonggi Bay. Sukmo Channel, located between Sukmo and Ganghwa islands, is the deepest and has the lowest salinity during the discharge peak in summer (see Figure 2.13). This suggests it is the main conduit through which fresh water and sediment pass into Gyeonggi Bay.

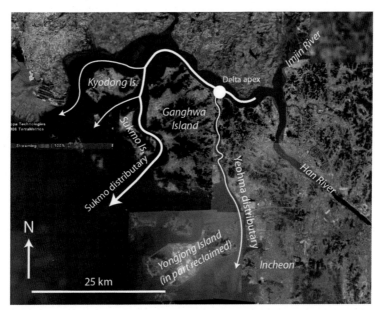

Figure 4.2 Distributary channels of the Han River delta. The position of the delta apex and the stability of the distributary-channel network are controlled by several bedrock islands around which the Han River splits before entering Gyeonggi Bay. The major distributary is believed to be the Sukmo Channel: it is the deepest and has the lowest salinity in summer. *Image courtesy of Google Earth©.*

Although fixed in space by bedrock islands, the distributary channels are not barren of sediment. Rather, they have aggraded their beds and banks so that the width of the active channel is typically less than the space between the bedrock islands. Tide-influenced point bars occur locally (Choi et al., 2004a; Choi, 2011). These have well developed intertidal flats along their sides, lateral-accretion slope angles between 3–14°, and rare step-like terraces. Point bars reach 40 m in relief, but only in scoured bedrock constrictions. Elsewhere, there is 15 to 30 m of relief between the channel thalweg and the top of the flanking tidal flats. Blind (headward-terminating) channels termed *flood barbs* (Robinson, 1960; van Veen et al., 2005), which partially to completely dissect point bars from their banks, are identifiable in bathymetric charts at the seaward end of some distributary-channel point bars.

Lobate mouth bars, such as those that characterize river-dominated deltas (Olariu and Bhattacharya, 2006; Bhattacharya, 2010), are absent at the terminations of Han River distributary channels. Rather, once past the zone of bedrock influence, distributaries become straighter, continue to widen and deepen, and merge with large tidal channels in Gyeonggi Bay, which in turn continue to widen and deepen until their bases pass outward without interruption onto the floor of the open shelf (Figure 4.3). The main distributary, Sukmo Channel, also bifurcates up onto the large tidal bar to the south and feeds into a zone of ebb-dominated sandbars and swatchways (see below).

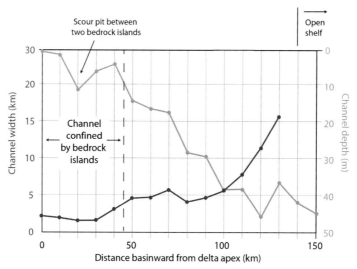

Figure 4.3 Width and thalweg depth of the Sukmo Channel downstream of the delta apex and its extension seaward to the floor of the Yellow Sea. Note that, unlike the distributaries in most deltaic systems, including most other tide-dominated deltas, channel depth and width increase progressively basinward. Between 50–90 km downstream of the apex, the Sukmo distributary merges with a large tidal channel in Gyeonggi Bay, which in turn continues to widen and deepen basinward, eventually merging with the continental shelf outside Gyeonggi Bay.

4.2 LARGE TIDAL BARS

Three large shore-attached bars—the large tidal bars mentioned previously—extend subtidally offshore of the Han River mouth (Figure 4.1). A fourth, narrower bar—the pseudo-tidal sand ridge described by Jung et al (1998)—is also present in southern Gyeonggi Bay, along with several smaller sandbars (Choi and Park, 1992; Kum et al., 2010). Unlike the other large tidal bars, the pseudo-tidal sand ridge of Jung et al (1998) is completely detached from shore. However, it has a similar internal stratigraphy as the other large tidal bars. It may constitute a fourth large tidal bar, possibly formed during an earlier progradational pulse before being abandoned as transgression continued (Jung et al., 1998). The three (or four) large tidal bars in concert make up the subaqueous delta. We focus here on the central large tidal bar, where most of our data were collected.

The central large tidal bar in the subaqueous delta platform is 110 km long, 20 km wide and 35 m high relative to the bases of the adjacent channels (Figure 4.2). It contains roughly 1 billion cubic meters of sediment. For perspective, this is equal to 50 million dump trucks worth of sediment, or half the mineable volume of the Athabasca Oil Sands in Canada. The proximal end of the large tidal bar extends into the intertidal zone and attaches to Ganghwa Island, where open-coast tidal flats are developed (see Figures 4.5a, 4.6). Its top surface, the delta "topset", slopes basinward at 0.01° for 100 km to a break in slope—the clinoform "rollover"—at the 20 m isobath, then steepens into a 10-km long deltaic clinoform/prodelta that slopes basinward at 0.2° before merging with the gently undulating continental shelf in about 50 m of

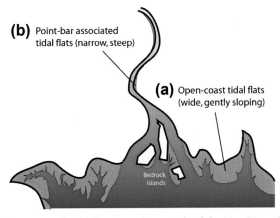

Figure 4.4 Simplified depiction of tidal flats near the mouth of the Han River, Gyeonggi Bay. In terms of morphology (Figure 4.6), two end-member types exist. (a) Open-coast tidal flats. Most tidal flats in Gyeonggi Bay fringe the rocky coastline in "interfluve" areas between rivers and tidal channels. These tidal flats are broad, gently sloping, and dissected by tidal creeks and channels that broaden and deepen basinward. Many form the shore-attached portions of the large tidal bars that make up the subaqueous delta. (b) Point-bar associated tidal flats. Tidal flats also fringe distributary channels in the tidal–fluvial transition, commonly on point bars. These tidal flats are much narrower and steeper than the tidal flats that fringe the open coast.

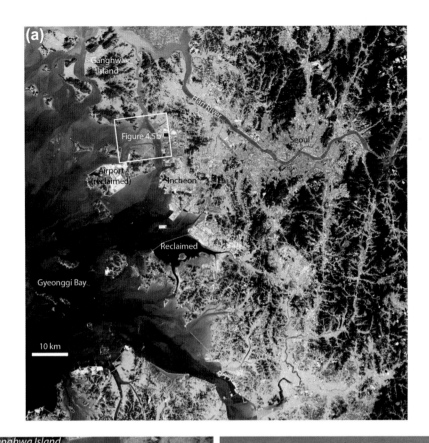

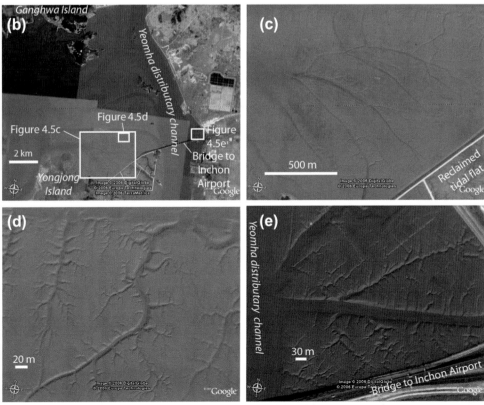

water (Figures 4.9, 4.12). In strike (cross-bar) section, the surface of the large tidal bar is relatively symmetric and flat-topped close to shore (e.g., seismic line L23 in Figure 4.13), irregularities associated with swatchways and sandbars notwithstanding (see below). However, it becomes asymmetric basinward (seismic line J10 in Figure 4.13), with a gently sloping (~0.25°) northern flank and a steeper (~1°) southern flank. This asymmetry is common to the outer portions of the three northernmost large tidal bars in the Han River delta, although their relief becomes progressively more subdued towards the south of Gyeonggi Bay (Figures 4.7, 4.8).

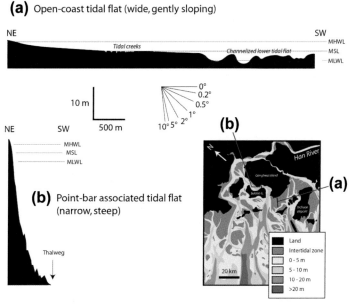

Figure 4.6 Representative topographic profiles of (a) an open-coast tidal flat in Gyeonggi Bay (Yeochari tidal flat; Choi et al., 2011), and (b) a point-bar associated tidal flat in a Han River distributary channel (Oepori tidal flat in the Sukmo Channel; Choi et al., 2004a). The Yeochari tidal flat, also known as the Ganghwa tidal flat (Kim, 2006; Lee et al., 2011), is located along the open, unbarred coast of Gyeonggi Bay, but it is somewhat sheltered from wave energy by several small bedrock islands and the broad, shallow and dissipative subaqueous delta platform. MSL stands for mean sea level. MHWL and MLWL stand for mean high water level and mean low water level, respectively.

Figure 4.5 Tidal flats near the mouth of the Han River, viewed at various scales. (a) Regional view of northeastern Gyeonggi Bay, showing channelization of both the intertidal flats and the shallow subtidal zone. (b) Open-coast tidal flat near Ganghwa Island. This particular tidal flat (Lee et al., 2011) is easily visible on the drive from Seoul to the Incheon airport. (c) The same tidal flat viewed at a higher magnification, showing the well-developed tidal-creek networks. (d) A close up of (c), showing the almost fractal-like geometry of the tidal creeks. (e) Close up of the mainland-attached tidal flats on the other side of the Yeomha distributary channel. Note the absence of supratidal vegetation, likely due to land reclamation, a trait common to most Korean tidal flats. *Image (a) courtesy of the NASA's Earth Observatory. Images (b) to (e) courtesy of Google Earth©.*

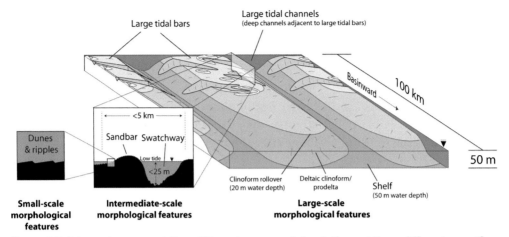

Figure 4.7 Schematic representation of the subaqueous delta platform at three different magnifications. The platform consists of large tidal bars separated by large tidal channels. At the next smaller scale, mounds (*sandbars*) and channels (*swatchways*) ornament the inner–mid part of the large tidal bar. At the smallest scale, dunes and ripples mantle parts of the seafloor, especially in the sandbar–swatchway zone and at the base of the large tidal channels.

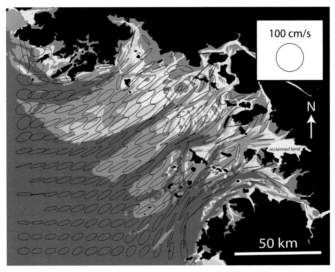

Figure 4.8 Tidal current ellipses in Gyeonggi Bay (Lee et al., 2001) superimposed on the seafloor bathymetry. Note that the long axes of the tidal ellipses parallel the large tidal bars and tidal channels. *Bathymetric data from standard charts (Korea Hydrographical and Oceanographic Administration, 2011).*

To discuss the geomorphology of the large tidal bar in detail, it is useful to subdivide it into the following zones: open-coast tidal flats, sandbars and swatchways, the asymmetric outer-mid large tidal bar, and the deltaic clinoform/prodelta. These subdivisions are depicted in Figure 4.1.

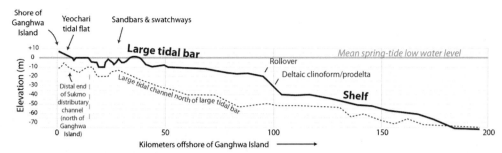

Figure 4.9 Low-resolution onshore–offshore profile of the seafloor down the axis of the large tidal bar (solid line) and down the axis of the adjacent large tidal channel to the north (dashed line). *Data traced from bathymetric maps (Korea Hydrographical and Oceanographic Administration, 2011), except the Yeochari tidal flat profile, which is simplified from Choi et al. (2011).*

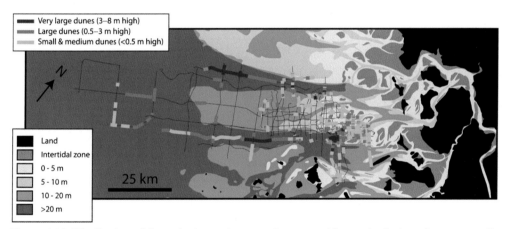

Figure 4.10 Distribution of dunes in the study area as interpreted from seismic data, the most areally extensive dataset available. The largest dunes tend to occur at the base of large tidal channels on either side of the large tidal bar. Dunes also occur locally on the large tidal bar, but not on the distal clinoform or on the southern flank. Dunes are much more common on the proximal half of the large tidal bar, where they ornament the swatchways and sand bars, than on the distal half of the large tidal bar. Dunes are also present locally on the shelf. Small dunes (0.05–0.5 m high) and medium dunes (0.25–0.5 m high) are grouped together in this figure because in most cases they are difficult to differentiate in seismic data. Portions of seismic tracklines depicted as lacking dunes are either featureless, or are ornamented by bedforms that are too small to resolve in the seismic data (e.g., ripples or small dunes). Dune nomenclature (small, medium, etc.) and height ranges from Ashley (1990), as modified by Dalrymple and Rhodes (1995). *Bathymetric data from standard charts (Korea Hydrographical and Oceanographic Administration, 2011).*

4.2.1 Open-Coast Tidal Flats

Broad (3–8 km), low-gradient (0.1–0.5°) *tidal flats* fringe most of the open rocky coastline in Gyeonggi Bay, as well as the seaward-facing sides of bedrock islands around which the Han River splits (Figures 4.4 to 4.6) (Frey et al., 1989; Alexander et al., 1991; Kim,

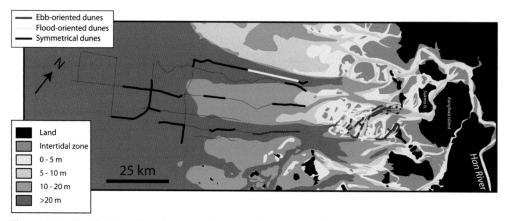

Figure 4.11 Distribution of flood- versus ebb-oriented asymmetric dunes, as interpreted from multibeam and sidescan sonar data (Kim, 2009). Note there are two sets of tracklines: the widely spaced tracklines from outer Gyeonggi Bay correspond to the multibeam (swath bathymetry) data, whereas the closely spaced tracklines from inner Gyeonggi Bay correspond to the sidescan sonar data. Because symmetrical dunes only occur in areas with almost zero net sediment transport (i.e., where flood and ebb currents are nearly equal, a situation that is not common in other, similar tide-dominated depositional settings), it is possible that some of the dunes mapped as "symmetrical" are actually asymmetrical, but that the degree of asymmetry was too small to be distinguished clearly, or that they were imaged after the subordinate tide when the asymmetry is reduced. *Bathymetric data from standard charts (Korea Hydrographical and Oceanographic Administration, 2011).*

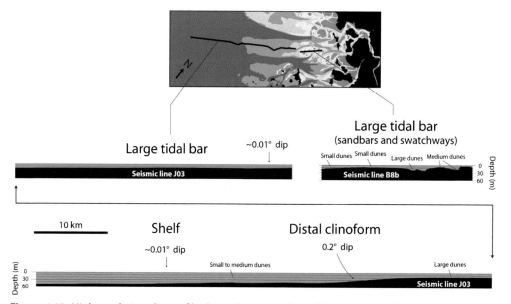

Figure 4.12 High-resolution dip profile down the center-line of the large tidal bar and out onto the shelf. The profile is traced from seismic lines, the locations of which are given on the inset map. Note that the inner large tidal bar is ornamented by cross-bar channels (*swatchways*), as well as mounds adjacent to the swatchways that protrude into the intertidal zone (*sandbars*). Dunes are abundant on the inner large tidal bar, where they ornament the sandbars and swatchways, whereas they are rare on the outer large tidal bar. For details, see seismic lines J03 and B8b in Appendix 2.

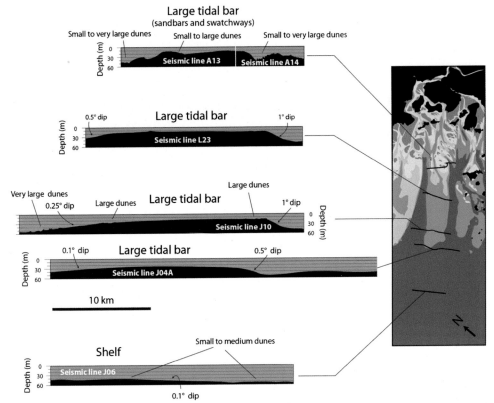

Figure 4.13 Strike profiles across the large tidal bar and parts of the adjacent large tidal channels (seismic lines A13/A14 to J04A) and a profile from the Yellow Sea shelf just outside Gyeonggi Bay (seismic line J06). The proximal half of the large tidal bar is relatively symmetric, and is ornamented by intermediate-scale channels (*swatchways*) and intervening sand-covered topographic highs (*sandbars*). By contrast, the distal half of the large tidal bar lacks swatchways and sand bars, and is markedly asymmetric, with a steep south flank and gentle north flank. Dunes occur locally on both proximal and distal halves of the large tidal bar, but they are more common in the proximal half. See individual seismic lines in Appendix 2 for details.

2006; Choi et al., 2011; Lee et al., 2011). These tidal flats occupy "interfluve" locations between distributary channels, and form the inner portions of the large tidal bars.

Because these tidal flats do not occur behind wave-formed barriers, we refer to them as open-coast tidal flats. However, they are exposed to considerably less wave energy than many previously studied open-coast tidal flats along the Korean and Chinese coasts (e.g., Yang et al., 2005; Dalrymple, 2010; Fan, 2012). This is due to the fetch-limited nature of Gyeonggi Bay, local shelter provided by bedrock islands, and dampening of incoming waves by the broad subaqueous delta platform. Dendritic tidal creek networks are well developed (Figure 4.5). These creeks are generally no more than several meters deep and 50 meters wide. They become progressively deeper and wider downslope towards the low-water line and their confluence with broader, deeper tidal channels that are typically lined by *dunes*. Supratidal salt marshes, common to temperate-region tidal flats (Amos, 1995; Flemming, 2012), are largely absent in Gyeonggi Bay due to extensive land reclamation.

4.2.2 Sandbars and Swatchways

Seaward of the open-coast tidal flats is a zone of swatchways and sandbars (Figure 4.14). These features are an order of magnitude smaller than the large tidal bar itself (Figure 4.15), and are more comparable in size to the elongate bars typically found in tide-dominated deltas, estuaries and shelves (cf. Off, 1963; Dyer and Huntley, 1999; Wood, 2003).

Following Robinson (1960), the term swatchway is used to refer to any channel that cuts diagonally across the large tidal bar (Figure 4.12). The swatchways are short (<20 km long), narrow (<5 km wide) and shallow (<25 m relief relative to the adjacent bar crests). They become narrower and shallower when traced upslope onto the large tidal bar, commonly bifurcating in the process. Only several cross the large tidal bar completely. Swatchways on the north flank of the large tidal bar become shallower basinward (southward), whereas those on the south flank become shallower landward (northward).

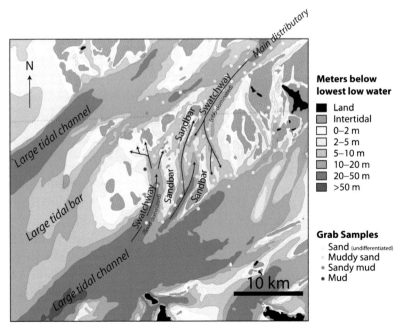

Figure 4.14 Swatchways and sandbars on the proximal half of the large tidal bar. Grab-samples are from the 05JB series collected in July 2005. *Bathymetric data from standard charts (Korea Hydrographical and Oceanographic Administration, 2011).*

The term sandbar is used here to refer to any sand-covered bathymetric high adjacent to, or at the end of, a swatchway. Unlike swatchways, which are subtidal, sandbars commonly protrude into the intertidal zone. In plan view, they tend to be oblong or horseshoe shaped. Oblong sandbars occur adjacent to swatchways, trend parallel to them, and have average heights, widths and lengths of approximately 15 m, 5 km, and 15 km, respectively. Horseshoe-shaped sandbars are located at the terminations of

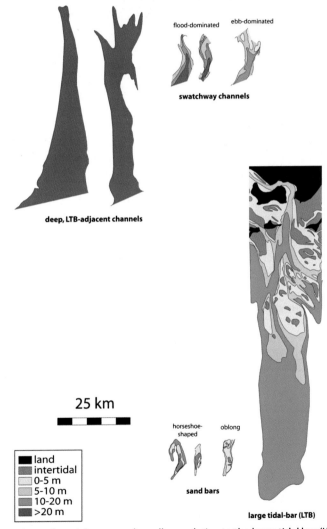

Figure 4.15 Dimensions of swatchways and sandbars relative to the large tidal bar (LTB) and large tidal channels. Note the order of magnitude size difference between swatchways and the large tidal channels, and between sand bars and the large tidal bar. *Bathymetric data from standard charts (Korea Hydrographical and Oceanographic Administration, 2011).*

swatchways. Some appear to consist of two oblong sandbars joined at the tip, suggesting oblong and horseshoe forms are inter-gradational.

Dunes cover the sandbars (Figure 4.16) and swatchways (Figure 4.17). A rough cor-relation between dune size and water depth is observed: dunes on the intertidal parts of sandbars tend to be <0.5 m high, whereas those at the base of swatchways are larger, and can reach heights of 3 m. However, reverse trends are common. In sidescan sonar data,

Figure 4.16 A slightly sinuous, continuous-crested 2-D dune (50 cm high) on the intertidal surface of a sandbar in the Han River delta. Flood currents move to the right of the photo and ebb currents to the left. Ebb reworking of this flood-dominant dune has been minimal.

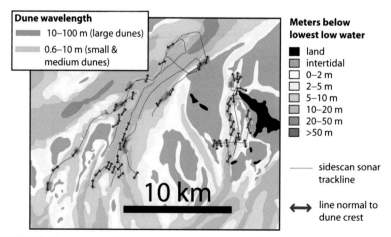

Figure 4.17 Distribution and crestline orientation of dunes within the swatchways that cut the inner part of the large tidal bar, as interpreted from sidescan-sonar data. Dune sizes follow classification scheme in Dalrymple and Rhodes (1995). *Bathymetric data from standard charts (Korea Hydrographical and Oceanographic Administration, 2011).*

dune crestlines in general are relatively continuous, only slightly sinuous, and oriented roughly perpendicular to the tidal currents (Figure 4.17).

4.2.3 Mid–Outer Large Tidal Bar

The mid–outer portion of the large tidal bar extends from the sandbar and swatchway zone to the clinoform rollover. The zone is fully subtidal, devoid of sandbars and swatchways, and generally devoid of dunes. In the inner part of this zone, the large tidal bar is symmetric in cross section, with flanks that dip at 0.5 to 1° and a relatively flat top (Figure 4.13). Moving basinward, the cross-section of the large tidal bar becomes progressively more asymmetric, and the bar develops a gently sloping (0.25° dip) north flank and steeper (1° dip) south flank (Figure 4.13). This asymmetry is a trait shared by all the large tidal bars in the subaqueous delta (Figure 4.1), and is also reflected in the intervening large tidal channels (see below). Rare dunes near the distal end of this zone (seismic line J10 in Figure 4.13) may potentially be migrating southward over top of the large tidal bar given their apparent sense of asymmetry.

4.2.4 Deltaic Clinoform/Prodelta

The distal tips of the large tidal bars make up the basinward-dipping face of the deltaic clinoform (also referred to here as the *prodelta*). For the centremost large tidal bar investigated herein, the clinoform "rollover"—the break in slope between topet and clinoform—is located 100 km from shore, submerged beneath 20 m of water. The clinoform itself is relatively planar, 10 km long, and 20 m high. It slopes basinward at 0.2° before flattening out onto the continental shelf in about 40 m of water (Figures 4.9, 4.12). Dunes are not observed on the clinoform in seismic data (Figure 4.10), but Kim (2009) reports the presence of small (<25 cm), symmetric dunes at this location based on an analysis of the multibeam data (Figure 4.11), which provide a higher resolution view than the seismic data.

4.3 LARGE TIDAL CHANNELS

Large tidal channels separate the large tidal bars (Figure 4.1). At least four are present, spaced 15 to 40 km apart (average 30 km). They extend basinward through the entire subaqueous delta platform, from a location where there is a larger *tidal prism* and enhanced tidal flow, which can be either a deltaic distributary channel or a coastal embayment, to the open shelf, becoming wider and deeper in the process (Figure 4.3). Their reliefs range from 15 to 40 meters (Figure 4.9). Their cross-sections change basinward in tandem with those of the adjacent large tidal bars: they are relatively symmetric close to shore, but become progressively asymmetric basinward, eventually developing gently sloping (~0.25°) southern flanks and steep (~1°) northern flanks.

Dunes are common at the base of the large tidal channels (Figures 4.10, 4.11). They tend to occur in groups of similar sized individuals. Some reach heights of 8 m, the largest observed in the dataset. Based on swath bathymetry, Kim (2009) concluded that the dunes in the large tidal channels vary from being asymmetric to near-symmetric in profile, and that their crests are straight to sinuous and generally aligned perpendicular to the local tidal currents (Figure 4.11). He observed flood-oriented dunes in the large tidal channel that links to the main distributary (Sukmo Channel) and ebb-oriented dunes in the next large tidal channel south of this (Figure 4.11), and therefore argued that flood- and ebb-dominated bedload transport was occurring in these two locations, respectively.

4.4 SHELF

The continental shelf offshore of the Han River delta has an average water depth of 50 meters. The seafloor is devoid of tidal ridges common to other parts of the Yellow Sea, the nearest of which are located 100 km to the south (Figure 2.34). Rather, the shelf seafloor undulates gently and dips 0.01° southwestward toward the 80-meter deep central axis of the Yellow Sea. Rare fields of dunes are present on the shelf, with individual dunes reaching 3 m in height (Figure 4.10).

CHAPTER 5

Near-Surface Sediment

Contents

Chapter Points

- Much of the near-surface sediment in the delta is heterolithic: interlaminated mud and sand intervals were recovered in at least some of the short (1–2 m average) cores from most subenvironments, including distributary channels, large tidal bars, and large tidal channels. Only parts of these subenvironments consist exclusively of sand or mud.

- The four geomorphic zones that make up the large tidal bar investigated herein have distinct near-surface facies. (1) The open-coast tidal flats that make up the inner part of the large tidal bar are covered by dune cross-stratified sand subtidally, heterolithic sediment intertidally, and bioturbated mud near the high tide mark. Most stratification is generated by tidal currents, with waves having a subordinate influence. (2) Sandbars and swatchways on the inner-mid large tidal bar are covered by dune cross-stratified sand, both intertidally and subtidally. The largest dunes occur at the base of channels, a common theme in channels throughout the delta. (3) The mid–outer large tidal bar is largely dune-free, and is covered by interlaminated mud and sand, which is variably bioturbated. (4) The outer end of the large tidal bar, which forms part of the prograding deltaic clinoform, is covered by a veneer of fine sand. This sand is anomalous: the underlying older deltaic clinoform deposits consist of variably bioturbated interlaminated mud and sand, as revealed by long cores (see Chapters 7 and 8).

- Dune cross-stratified sand is present in the base of most channels in the delta, including distributary channels in the tidal–fluvial transition, the large tidal channels that separate the large tidal bars, the swatchways, and the larger of the tidal creeks that drain the open-coast tidal flats. Beneath the turbidity maximum, fluid mud deposits (~1 cm thick) are locally present in the bottomsets of dune cross-stratification.

- The shelf is covered by a thin (2–250 cm) sandy transgressive lag. The underlying transgressive ravinement surface is flat, is ornamented by firmground burrows, and truncates older, firmer back-barrier lagoonal mud. The ravinement surface is interpreted to be wave-generated.

The Tide-Dominated Han River Delta, Korea
http://dx.doi.org/10.1016/B978-0-12-800768-6.00005-5
87

- "Cyclic" tidal rhythmites—those that contain neap–spring or other tidal cyclicities—are rare. They are observed only in point bars of (1) distributary channels in the tidal–fluvial transition and (2) small tidal creeks that drain the open-coast tidal flats.

All strongly tide-influenced deltas and estuaries studied to date are pervasively hetero-lithic in that they contain abundant interlaminated sand and mud (Box 5.1). For example, heterolithic sediment is actively accumulating in parts of the Amazon River delta (Jaeger and Nittrouer, 1995), Fly River delta (Dalrymple et al., 2003), Changjiang River delta (Hori et al., 2001), Bay of Fundy estuary (Dalrymple et al., 1991), Bay of Mont Saint Michel estuary (Tessier, 2012), Qiantang estuary (Zhang et al., 2014), and the Gironde estuary (Fenies and Tastet, 1998) (Box 5.2). Heterolithic sediment also abounds in ancient equivalents of such systems, including the McMurray Formation, Canada (Hein et al., 2013); the Tilje Formation, Norway (Martinius et al., 2001; Ichaso and

Box 5.1 Heterolithic deposits, Part 1: Processes and products

Heterolithic deposits, such as those that are present throughout the Han River delta, require three things to form: mud, sand, and an unsteady flow that repeatedly superimposes the two sediment types. Several factors help to furnish these conditions in tide-dominated estuaries and deltas.

1. Semidiurnal tides (~12.4 h period)

Tidal currents in macrotidal and even upper mesotidal river mouths are extremely unsteady: they stop and start rhythmically over a period of several hours. To understand why, one must first understand how tides are generated. (See Kvale (2012) for a more detailed explanation than what is possible here.) Under equilibrium tidal theory, tides are caused by rotation of the Earth beneath two tidal bulges, one generated by gravitational attraction between the Moon and the ocean, the other by the inertia of the ocean in a spinning Earth–Moon system. As a given region passes beneath a tidal bulge, water levels rise and then fall as the bulge migrates past the location of interest. This causes the ocean to flood rapidly into low lying coastal areas, including river mouths. This incoming flow (the "flood tide") stops at high water, then drains out rapidly (the "ebb tide"), stopping again at low water. Most coastal areas experience two of these tidal cycles every ~25 h (i.e., semidiurnal tides). Semidiurnal tides would ideally deposit two **sand–mud couplets** during each 12.4-h tidal cycle, provided there is sufficient current and sediment to record the tidal variations. More commonly, however, ebb and flood currents are not equal in strength, and the subordinate of the two is not strong enough to deposit a sand layer. This leads to amalgamation of the two slackwater mud drapes and the development of only one sand–mud couplet for each complete tidal cycle.

2. Neap–spring tides (~14 day period)

Neap–spring tides do not so much generate heterolithic sediment as they modulate the thickness of sand–mud couplets produced by semidiurnal tides (e.g., Dalrymple et al., 1991). Neap–spring tidal cycles have a fortnightly (~14 day) period: over the course of the first week,

Box 5.1 Heterolithic deposits, Part 1: Processes and products—cont'd

tidal range and current speeds increase, then decrease the following week. The neap–spring cycles arise because the Moon rotates about the Earth at a period of once per month (~28 days). When the Sun, Moon, and Earth are aligned, tide-generating forces are maximized, and maximum (spring) tides occur. When Sun and Moon are orthogonal to the Earth, tide-generating forces are minimized, and minimum (neap) tides occur. Similar variations can be generated by the monthly cycle in lunar declination; see Kvale (2006) for details. Neap–spring tidal variations cause gradual thickening and thinning of sequential sand–mud couplets. In the common situation where only one sand–mud couplet is generated per tidal cycle, neap–spring packages can consist of seven progressively thicker sand–mud couplets overlain by seven progressively thinner ones, for a total of 14 couplets. This ideal number rarely occurs. The number can be greater than this: the subordinate (ebb or flood) tide can be strong enough to deposit a sand layer, especially during spring tides, leading to somewhere between 14 and 28 couplets per neap–spring cycle. The number can also be less: muddy couplets deposited during neap tides can become amalgamated, and "missing beats" (and in some cases even "added beats") can occur due to noise in the system (see below). The number of couplets is also less than this ideal in the case where sedimentation occurs above the neap high tide level because the surface is not inundated every tide.

If tides were the only factor controlling sedimentation in tide-dominated estuaries and deltas, ideal neap–spring rhythmite successions might be ubiquitous. In reality, they are rare. One of the two main reasons for this is that "noise" in the system (e.g., waves) episodically overwhelms the tidal current signal, punctuating and obscuring neap–spring thickening and thinning trends (Green and Coco, 2014). The second reason why neap–spring cycles are rare is that only a few locations have sufficiently rapid sedimentation to record every tide. Perhaps not surprisingly, therefore, neap–spring cycles in tidal rhythmites are most commonly reported in settings sheltered from waves and in close proximity to channels that supply the sediment necessary for the rapid aggradation required to record the neap–spring cycle. Thus, some of the most common locations where neap–spring cycles occur are in dune cross-stratified channel-base sand (Boersma, 1969) and in inclined heterolithic stratification (IHS) produced by rapidly migrating tidally influenced point bars in proximal, semi-enclosed, and sheltered settings (Choi, 2011).

3. Seasonality of river discharge (1-year period)
As with many depositional environments, tide-dominated deltas and estuaries are generally subject to seasonal climatic changes that can affect sedimentation. Wind velocity and wave energy (e.g., Yang et al., 2005), microbial binding on upper tidal flats (Noffke and Krumbein, 1999), bioturbation intensity related to differences in temperature (van den Berg, 1981), precipitation-controlled tidal flat rilling (Choi, 2011), and temperature-controlled sediment settling, transport, and drying rates (Ettema and Daly, 2004; Chang et al., 2006) can all be affected. Seasonal changes in river discharge are particularly important because they can lead to seasonal variations in the delivery of bedload and suspended sediment to the basin. This in turn can generate annual sediment layers (e.g., Hovikoski et al., 2008; Jablonski, 2012) that can be difficult in some cases to differentiate from rhythmites generated by tidal currents.

Continued

Box 5.1 Heterolithic deposits, Part 1: Processes and products—cont'd

4. Dynamics of the turbidity maximum

Physical, chemical, and biological processes conspire to trap mud in tide-dominated deltas and estuaries, promoting the development of heterolithic deposits. Several key processes contribute to this.

- **Estuarine circulation**—the landward flow of dense, salty marine water beneath less dense, fresh river water—can occur during river floods in macrotidal deltas and estuaries, especially during neap tides, when less tidal mixing of the water column occurs (e.g., Allen et al., 1980; Harris et al., 2004). Mud that settles out from the river plume is transported back toward shore at depth, effectively trapping it in the system. During low-flow stages, however, such salinity stratification can be reduced or destroyed in macrotidal systems due to intense tidal mixing of the water column by strong tidal currents (e.g., Uncles et al., 2006a). Other processes such as tidal pumping (see below) therefore likely help trap mud in macrotidal systems.

- The concentration of suspended mud in the surficial waters of tide-dominated deltas and estuaries tends to be <10 g/L (Papenmeier et al., 2013). At these concentrations, individual mud particles settle through the water column relatively unimpeded by their neighbors (Mehta, 1991). They also tend to settle faster than predicted under Stokes Law because most clump together (i.e., they **"flocculate"**) into aggregates termed "flocs" (Figure A). Typical flocs are 0.1–1 mm in diameter and settle at approximately 1 mm/s (Geyer et al., 2004), nearly the same speed as very fine sand. Flocculation thus strips suspended mud from the water column and transfers it to greater depths, closer to the bed, especially during high- and low-tide slackwater periods when turbulent shear in the water column is reduced and flocculation is promoted (Uncles et al., 2006b). Classically, flocculation has been viewed as a chemical phenomenon caused by increased salinity (e.g., van Olphen, 1963). Indeed, in lab experiments, the addition of only a small amount of salt (<2‰) to freshwater can cause rapid flocculation of mud in suspension (Kranck, 1980). However, the effects of biological binding and increased sediment concentrations should not be discounted: in some cases, they may be equally if not more important than increased salinity in promoting flocculation (Droppo, 2001; Bale et al., 2002).

- Tidal waves become asymmetric as they shoal due to bottom friction. Water beneath the crest of the tidal wave is deeper than beneath the trough in front of it. The trough therefore experiences more frictional retardation and the front of the wave steepens, causing the crest to catch up to it. The result is a short, fast flood tide and a long, slow ebb tide. As seen previously in Chapter 2, this can lead to net landward transport of bedload because bedload moves preferentially in the direction of peak current speed (e.g., Figure 2.21). Such flood-dominated tidal asymmetry has also been hypothesized to transport mud landward, a process referred to as **tidal pumping** (Yu et al., 2014). Because of the asymmetry, more mud tends to be resuspended during the faster flood tide than during the slower ebb tide (Castaing and Allen, 1981). The high water slack also tends to lengthen relative to the low water slack (Dronkers, 1986), allowing for more of this mud to settle out and be deposited. Further accentuating landward mud transport is the fact that average water depths tend to be less at high water slack than low water slack because of extensive, low-gradient tidal flats (Dyer, 1995; Dalrymple et al., 2012). The mud settling from suspension during high water slack therefore has less distance to travel, on average, before it reaches the bottom.

Box 5.1 Heterolithic deposits, Part 1: Processes and products—cont'd

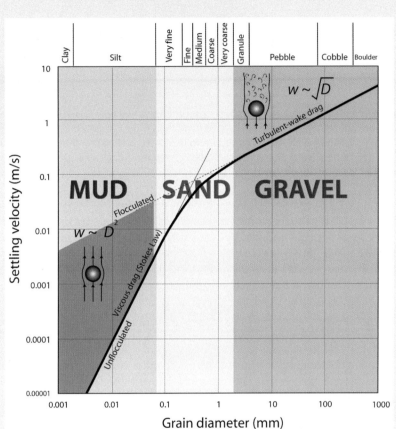

Figure A Settling velocity as a function of grain size, showing the effect of flocculation (dark gray zone) on the settling velocity of mud. *(Modified from Ferguson and Church, 2004.)* D is grain diameter and w is settling velocity. The settling velocities for mud in the graph only apply if suspended mud concentrations are less than 10 g/L. Above this, hindered settling starts to slow settling rates considerably.

The above processes generate a *turbidity maximum*, a feature that is particularly well developed in macrotidal deltas and estuaries due to enhanced resuspension by the strong tidal currents. Turbidity maxima are highly mobile reservoirs of suspended mud that develop at, or slightly landward of, the inner, fresher (1–5‰) end of the salinity front. Suspended sediment concentrations (SSCs) within them can range from 0.1 g/L (slightly murky water) to several 100 g/L (a yogurt-like muddy suspension) (Uncles et al., 2006c). Turbidity maxima tend to be tens of kilometers in horizontal dimension, with turbidity decreasing both upriver and seaward, and increasing downward toward the bed. The amount of mud in the turbidity maximum can be significant; in some systems, it is thought to equal or exceed the yearly supply of mud from the river (e.g., Nichols, 1985; Granboulan et al., 1989; Kineke and Sternberg, 1995; Manning et al., 2010).

Continued

Box 5.1 Heterolithic deposits, Part 1: Processes and products—cont'd

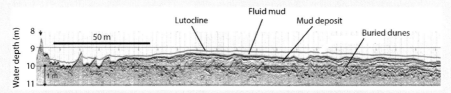

Figure B Fluid mud imaged acoustically in a high-energy channel-base setting in the Weser Estuary, Germany. *(After Schrottke et al., 2006.)* The fluid mud is separated from the overlying, dilute water column by a lutocline, a sharp jump in mud concentration that generates a coherent seismic reflection. Note that an acoustically layered mud deposit underlies the fluid mud, which in turn, overlies buried dunes.

Two layers are commonly definable in turbidity maxima (Figure B). Most of the water column tends to be occupied by an upper, turbid but relatively dilute (<10 g mud/L) layer. Flocs settle freely in this layer. Below this, there is commonly a layer of fluid mud, which forms a thin layer (centimeters to several meters thick) over the bed. A lutocline—a sharp jump in mud concentration—separates it from the overlying water column: "lutum" is Latin for "mud" (Kirby and Parker, 1983).

Fluid muds are observed in numerous marine-influenced river mouths, including the Fly (Wolanski et al., 1995), Weser (Schrottke et al., 2006; Papenmeier et al., 2013), Amazon (Kineke et al., 1995), Mississippi (Rotondo, 2004), Jiao (Jiang and Mehta, 2000), Severn (Kirby and Parker, 1983), and Gironde (Allen et al., 1980). Fluid mud is defined as a water–mud mixture with SSCs >10 g/L (Figure C). It forms because of hindered settling, a phenomenon that starts to occur when SSCs exceed ~10 g/L due to increased particle–particle interactions and the lift force generated by escaping fluid. Fluid mud layers consist predominantly of clay and silt (McNally et al., 2007), but can also include non-negligible amounts of very fine to fine sand (Papenmeier et al., 2013), microbial slime (Wurpts, 2005), and terrestrial organic material (Dalrymple et al., 2003). They are commonly imaged using acoustic devices in bathymetric lows during slackwater, including dune-covered channel bases (Figure B; Schrottke et al., 2006; Becker et al., 2013). In rare cases, they occur throughout the turbidity maximum zone, blanketing tens of square kilometers of the subtidal seafloor (e.g., Ems Estuary; Papenmeier et al., 2013).

Although 10 g/L is well accepted as the transition from a dilute suspension to fluid mud (McNally et al., 2007), the transition from fluid mud to consolidated bed has proven more difficult to define. Many take it to be the gelling concentration, the point at which the fluid mud suspension collapses sufficiently to become a particle-supported mud "deposit" with a small yet definable yield strength (McNally et al., 2007). This concentration can vary widely, from less than 70 g/L (Becker et al., 2013) to at least 250 g/L (McNally et al., 2007). Papenmeier et al. (2013) argue for an even higher limit, 500 g/L, and suggest that two types of fluid mud exist, low-viscosity mobile fluid mud (10–200 g/L) and high-viscosity stationary fluid mud deposits (200–500 g/L). Previous authors also observe mobile and stationary fluid mud in macrotidal settings, though the layers vary in concentration from setting to setting (e.g., Kirby and Parker, 1983; Mehta, 1991; Manning et al., 2010). This suggests that other factors influence the point at which fluid mud becomes a stationary

Box 5.1 Heterolithic deposits, Part 1: Processes and products—cont'd

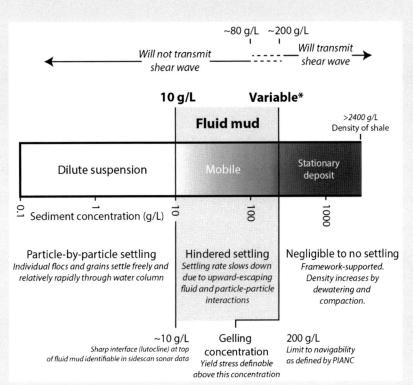

Figure C Fluid mud definition and properties. *Based on Ross and Mehta (1989), Mehta (1991), Dyer (1995), Wurpts (2005), Dankers and Winterwerp (2007) and Papenmeier et al. (2013).*

deposit, including time available for dewatering and consolidation, clay mineralogy (Mehta, 1989), microbial "slime" content (Wurpts, 2005), and sand content (Manning et al., 2010), in addition to downslope gravitational force and/or fluid shear from overriding currents and/or waves (Wright et al., 2001; Baas and Best, 2002; Friedrichs and Wright, 2004; Baas et al., 2009). The important point is that in adopting a two-tiered definition, Papenmeier et al. (2013) and previous authors are implying that fluid muds can consolidate and aggrade in high-energy, subtidal settings (cf. Mackay and Dalrymple, 2011; Becker et al., 2013), a hypothesis supported by the presence of acoustically layered mud sandwiched between fluid mud and sandy dunes in the Weser Estuary (Figure B) and by the recovery of thick (up to 10 cm), unbioturbated mud layers in the macrotidal Severn (Kirby, 1991) and Fly River (Dalrymple et al., 2003) systems (see also Choi and Jo, 2015).

The turbidity maximum, including any mobile fluid mud at depth, is the locus for mud deposition in tide-dominated deltas and estuaries. In and of itself, it does not promote deposition of heterolithic sediment. Left unfettered, it would promote deposition of mud. It is the **dynamic nature of the turbidity maximum** that promotes deposition of heterolithic deposits and helps generate tidal signatures. In essence, the turbidity maximum is the ink, the unsteady currents, the pen.

Continued

Box 5.1 Heterolithic deposits, Part 1: Processes and products—cont'd

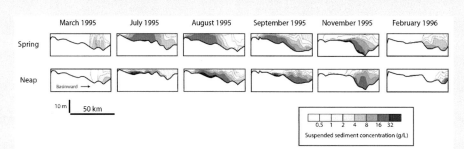

Figure D Dynamics of the turbidity maximum in the Humber Estuary, UK, over seasonal and neap–spring timescales. *(Redrawn from Uncles et al., 2006a.)* All profiles collected at high water. Note that river discharge is the main control on the time-averaged position of the turbidity maximum. Neap–spring cycles cause the turbidity maximum to collapse (neaps) and expand (springs), but have comparatively less effect on its time-averaged position. Basinward is to the right.

Semidiurnal tides cause the turbidity maximum to swell and collapse vertically on a ~12.4-h period. They also cause it to move horizontally somewhat (Ganju et al., 2004), but they do not alter its time-averaged horizontal location. Fluid muds can be impacted on these timescales (Schrottke et al., 2006). For example, during slacks, fluid muds in dune troughs in the Weser Estuary are stable for ~2 h, becoming partially or completely resuspended during floods and ebbs (Becker et al., 2013). Semidiurnal swelling and collapse of the turbidity maximum can accentuate the signature of the semidiurnal tides. Fluid mud deposition can occur during slacks in bathymetric lows (e.g., Becker et al., 2013) and slackwater mud can be deposited by flocculation settling throughout the turbidity maximum zone.

On weekly timescales, neap–spring tides perform a similar function: they cause the turbidity maximum to swell and collapse, while having little effect on its time-averaged horizontal position (Figure D). Turbidity throughout the water column is highest during springs, whereas it decreases during neaps. Again, from a sedimentological perspective, these dynamic variations accentuate the signature of the neap–spring tides. As spring leads to neap, mud collapses to the bed, promoting deposition of muddier couplets than might occur during the transition from neap to spring. Stationary fluid muds are particularly prone to develop during neap tides; those that form in the Severn Estuary during neap tide are not completely reentrained between slackwater periods (Kirby, 2010). This can lead to the deposition of anomalously thick (>1 cm) mud layers in the subtidal zone, especially in channel-base settings (e.g., Dalrymple et al., 2003; Ichaso and Dalrymple, 2009). Throughout the broader turbidity maximum, particularly thick (possibly 2–10 mm thick) mud drapes can be deposited by flocculation settling during neap-tide slacks (see Mackay and Dalrymple, 2011, and references therein).

On longer (e.g., seasonal or annual) timescales, river discharge is the key variable controlling the dynamics of the turbidity maximum. It dictates the time-averaged position of the turbidity maximum and the salinity front, with which the turbidity maximum tends to be closely associated. As discharge varies between wet and dry seasons, the turbidity maximum can be forced to

Box 5.1 Heterolithic deposits, Part 1: Processes and products—cont'd

move back and forth tens if not over a hundred kilometers horizontally (Figure D). Significant mud can become trapped upriver during the low flow season (e.g., Allen et al., 1980; Uncles et al., 2006a; Galler and Allison, 2008; Doxaran et al., 2009), and tidal signatures can form in entirely freshwater settings tens to hundreds of kilometers inland of the coast (Shanley et al., 1992; Archer, 2005). Brackish water can also intrude upriver during this time, in some systems up to several hundred kilometers, although the extent of brackish intrusion is typically a fraction of the tidal intrusion. This can lead to the generation of brackish water trace-fossil assemblages in fluvial deposits that are otherwise free of bioturbation (Shanley et al., 1992; Jablonski, 2012). By contrast, during high flow season, the mud depocenter and salinity front can be pushed basinward tens to over one hundred kilometers (Castaing and Allen, 1981; Kineke and Sternberg, 1995; Geyer et al., 2004). The sedimentary consequences of this latter point are discussed in Box 5.2.

In sum, tides and river currents are the primary controls on the dynamics of the turbidity maximum and salinity front in tide-dominated deltas and estuaries. Understanding the details of this relationship is a prerequisite to understanding the nature of sedimentary facies in these systems (Box 5.1), in addition to their geographic distribution (Box 5.2).

Box 5.2 Heterolithic deposits, Part 2: Controls on geographic distribution

No single "bed-scale" facies has yet proven diagnostic of a tide-dominated deltaic versus a tide-dominated estuarine environment. If one was handed a photo showing any of the "classic" tidally generated sedimentary structures—dune cross-stratified sand with double mud drapes, muddy tidal rhythmites with neap–spring cyclicities, or herringbone cross-stratification (e.g., Nio and Yang, 1991; Davis, 2012)—one would be hard pressed to say whether it came from an estuary or a delta. What is diagnostic, however, is the *distribution* of heterolithic facies in these systems (Dalrymple, 2010; Dalrymple et al., 2012).

Tide-dominated estuaries, like all estuaries, lack a muddy prodelta. Most mud is deposited intertidally and supratidally on mud flats located mainly in the inner part of the system. Offshore areas tend to be mud starved and typically consist of a lag of sand, gravel, and shell debris that is being slowly reworked onshore by flood-dominated tidal currents to create a complex of elongate tidal sandbars at the mouth of the estuary. There are exceptions. In particular, mud can occur offshore of estuaries located downcoast of major muddy rivers that supplies mud to the broader shelf, such as the Amazon River (Allison et al., 1995) or the Changjiang River (Zhang et al., 2014). Estuaries that do export small amounts of mud to the shelf, such as the Gironde (Castaing and Allen, 1981), might perhaps be best thought of as proto-deltas, or late-stage estuaries, perched at the transition between transgression and regression.

Like their estuarine counterparts, tide-dominated deltas also accumulate abundant mud on intertidal mud flats. However, these systems are net exporters of mud, not net importers. Over time, more mud moves basinward than moves landward. The offshore movement of mud likely commonly occurs in steps (Wright and Nittrouer, 1995) and may involve fluid mud underflows, possibly wave supported, as observed in the Amazon (Kineke and Sternberg, 1995) and Fly (Martin et al., 2008), and/or buoyant river plume overflows (Geyer et al., 2004). Most of this mud tends to be ultimately deposited on the deltaic clinoform (Pirmez et al., 1998), commonly forming heterolithic deposits (e.g., Hori et al., 2001) that can, in some cases, be difficult to distinguish from those generated elsewhere in the system, although there should be a greater abundance of wave-generated structures in such exposed locations.

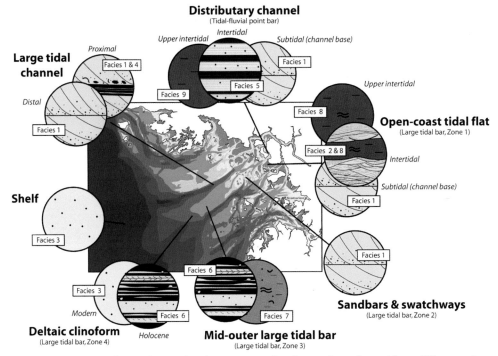

Figure 5.1 Dominant facies observed in short cores and grab samples collected from different suben-vironments of the Han River delta. *Data for the subaqueous delta collected during this study (Figure 5.2). Data for distributary channels and open-coast tidal flats from Choi et al. (2004a) and Kim (2006), respectively.* See Table 5.1 for summary of facies descriptions and interpretations.

Dalrymple, 2014); the Dir Abu Lifa Member, Egypt (Legler et al., 2013); and the Oficina Formation, Venezuela (Martinius et al., 2013).

The Han River delta is no exception to this generalization (Figure 5.1). Interlaminated mud and sand is present in many of the short cores and grab samples collected from most subenvironments, including tidal–fluvial point bars (Choi et al., 2004a), open–coast tidal flats (Kim, 2006), large tidal bars and large tidal channels in the subaqueous delta plain, and also the prodelta region (Figure 5.2), although the prodelta has recently switched to a sand-covered state (see long-core data in Chapter 7). In addition, sand dominates some subenvironments, most notably the base of channels of all types, whereas mud dominates parts of others, most notably the upper intertidal portions of tidal flats. A total of nine facies are identified in the short cores and grab samples (Table 5.1; Figure 5.3). We have opted to lump together facies with only slight visual differences instead of splitting them apart as many previous studies have done. The benefit of this is that each facies is believed to be diagnostic of a particular depositional process or set of processes, and in some cases, a particular subenvironment. The primary criteria used to differentiate facies are grain size, physical and biogenic structures, bed thickness, cyclicity or lack thereof, nature of bed contacts, and the abundance of terrestrial organic detritus (referred to as

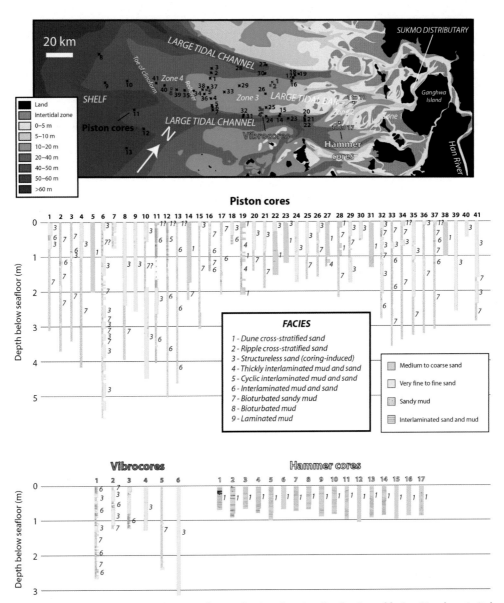

Figure 5.2 Short cores collected during this study, showing the distribution of facies. Numbers in italics refer to facies. See inset box for list of facies and Table 5.1 for details. Black lines in hammer cores depict dune cross-stratification. See Appendices 1C to 1E (http://booksite.elsevier.com/9780128007686/) for details.

"tea leaves" or "coffee grounds" in Appendix 1 (1C to 1E refer online – http://booksite.elsevier.com/9780128007686/), shell fragments, and mud pebbles. Facies associations, characteristic of each subenvironment, are described below, starting with the subaqueous realm, the focus of this study.

Table 5.1 Sediment facies observed in grab samples and short (<6m) cores

	FACIES	DOMINANT GRAIN SIZE	Dune cross-strat	Current ripple cross-strat	Small wave ripple cross-strat	Small combined-flow ripple x-strat	Undulating planar lamination	Gradational contacts	Neap–spring cycles	Slump structures	Bioturbation	Mud pebbles	Shell fragments	Root traces	Woody detritus	Dune-covered	"Smooth"[2]	Tidal–fluvial point bars	Large tidal channels	Zone 1 Open-coast tidal flats	Zone 2 Sandbars and swatchways	Zone 3 Mid-outer large tidal bar	Zone 4 Deltaic clinoform/prodelta	Shelf (above ravinement surface)	Shelf (below ravinement surface)	INTERPRETATION
SAND 1. Dune cross-stratified sand		Medium to coarse sand	X							X		x[3]	X		x	X		X	X	X[4]	X	X		x		Deposited by strong tidal currents.
2. Ripple cross-stratified sand		Fine sand		X	X	X	x										X	x		X						Deposited by relatively weak tidal currents. Deposition at times influenced by waves.
3. Structureless sand (coring induced)		Fine sand											X			x[5]	X					X	X[6]	X		Typically ripple cross stratified? Hummocky cross stratified in places? Impossible to tell because of coring-induced deformation.
HETEROLITHIC 4. Thickly inter-laminated mud and sand		Mud & very fine sand								x							X		X							Deposited rapidly from fluid-mud layers during slack water periods when the turbidity maximum was displaced basinward due to high river discharge.
5. Inter-laminated mud and sand with rare to common cyclic thinning–thinning trends		Mud & very fine sand		X		X		X	X		x				X		X	X		x[7]					Deposited rapidly by flocculation settling and traction from reversing tidal currents, in a setting sheltered from large waves, near the turbidity maximum.
6. Inter-laminated mud and sand		Mud & very fine sand		X		X					X						X	x				X	X[8]		Deposited by flocculation settling and traction, primarily from reversing tidal currents, in subtidal setting exposed to waves, at distal tip of time-averaged turbidity maximum position.
7. Bioturbated sandy mud		Mud & very fine sand								x	X		X				X		X			X			Bioturbated version of Facies 6.
MUD 8. Bioturbated mud		Silt									X			x[9]			X	X		X					Deposited slowly by flocculation settling during high water slacks in relatively unstressed, intertidal, nearly fully marine setting at distal tip of the turbidity maximum.
9. Laminated mud		Silt									X			x[7]			X	X						X	Deposited rapidly from flocculation settling during high water slacks in a stressed, sheltered, intertidal, brackish water setting beneath turbidity maximum.

X = Common. x = Rare. Blank = not observed.

[1] Assessment of the character of the seafloor for subtidal areas, including all parts of the subaqueous delta platform, is based on the surface morphology seen on seismic data. For intertidal areas (tidal-fluvial point bars, open-coast tidal flats, the intertidal portions of sandbars and swatchways), the assessment it is based on direct observation.

[2] Areas of the seafloor that appear smooth on seismic data may in fact be covered by small (<25 cm high) dunes or ripples. Such bedforms are likely at or below the resolution of the data.

[3] Mud pebbles in Facies 1 were only observed in cores collected from base of the proximal part of the large tidal channel that connects landward to the main deltaic distributary channel (Sukmo Channel). In these cores, Facies 1 is also interbedded with Facies 4.

[4] Only observed in tidal channels at the base of the open-coast tidal flats, not on the tidal flats sensu stricto.

[5] In seismic data, dunes were generally not observed on the seafloor in association with this facies, although in rare cases there were hints of small dunes that were at the limit of resolution of the seismic data.

[6] Forms a thin carapace over heterolithic (Facies 6) deposits.

[7] Observed in the deposits of rapidly migrating point bars of small, sheltered tributary tidal creeks on Ganghwa open coast tidal flat (K.S. Choi, unpublished data).

[8] Not intersected in the short (<2.5 m) cores, but it is the dominant facies in the progradational deltaic clinoform deposits imaged in seismic data and intersected by the long cores (see subsequent chapters, where these data are presented and discussed).

[9] Only close to and above the highest high water level.

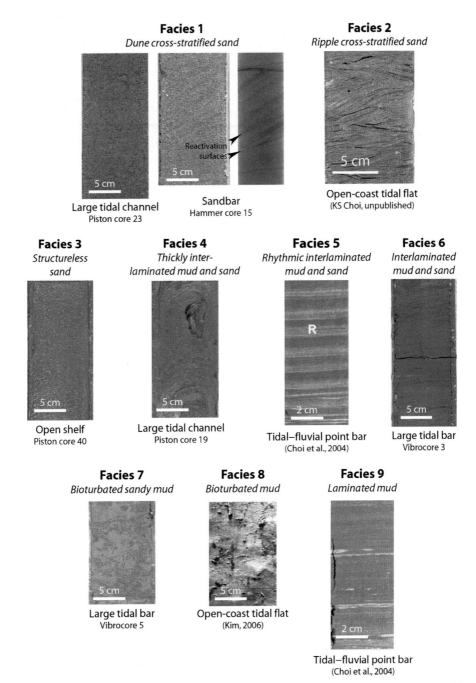

Figure 5.3 Sedimentary facies in short cores and grab samples from the Han River delta. See Table 5.1 for details. Note that piston cores 23 (Facies 1), 40 (Facies 3), and 19 (Facies 4) experienced coring-induced deformation: sand was liquefied and rendered structureless in cores 23 and 40, and the mud in piston core 19 was deformed into tight folds. *Images of open-coast tidal flat facies are from Kim (2006). Images of tidal–fluvial point bar facies are from Choi (unpublished).*

5.1 DISTRIBUTARY CHANNELS

Choi et al. (2004a) and Choi (2011) studied point bars in Sukmo Channel, the main distributary of the Han River delta. The following is based on this work.

The point bars in question fall in the tide-dominated part of the tidal–fluvial transition, near the tidal range maximum (Figure 2.20), and within the turbidity maximum and zone of brackish water (Figure 2.14). Choi et al. (2004a) recognized three main facies: dune cross-stratified sand (Facies 1), rhythmically interlaminated mud and sand (Facies 5), and laminated mud (Facies 9) (Table 5.1).

Grab samples collected from the base of the channel recovered medium to coarse sand, with gravel present locally. These deposits are likely dune cross-stratified (Facies 1) because channel-base dunes were imaged in echo-sounder data. No mud drapes or mud pebbles were observed, although such features are present nearby in the inner part of the large tidal channel to which the distributary connects (e.g., piston core 19; Figure 5.2).

In the upper subtidal to lower intertidal part of the point bar, rhythmically laminated mud and sand (Facies 5) is present. The rhythmites consist of normally graded layers, each typically <5 mm thick, which pass upward from sand to mud (Figure 5.3). Thicker sand layers are commonly current-ripple cross-laminated. Bioturbation is absent.

The intertidal zone is dominated by laminated silt-rich mud (Facies 9). Trace fossils are absent to rare, but increase in abundance toward the top of the tidal flat, especially above mean high water, where sedimentation rates decrease. Crabs are the main trace making organism. The lamination is commonly rhythmic. Packages consisting of ~30 mud–sand couplets that thicken then thin upward are common. These packages are interpreted to record neap–spring cycles. They are 1–3 cm thick near the base of the intertidal zone and less than 0.5 cm thick near the top.

The gradational contacts and neap–spring tidal cycles that occur in the tidal–fluvial point bar reflect rapid deposition from a combination of suspension (normally graded beds and laminated mud) and traction (current ripples) in a sheltered, stressed setting. Such cyclic rhythmites are exclusive to sheltered, tidally influenced channel margins in the proximal, intertidal part of the delta in samples collected to date.

5.2 LARGE TIDAL BARS

In addition to being geomorphologically distinct (Chapter 4), the four zones of the large tidal bar also have distinct near-surface facies characteristics (Figures 5.1 and 5.2).

5.2.1 Zone 1: Open-Coast Tidal Flats

Various workers have studied tidal flats in Gyeonggi Bay in the 1980s, including Frey et al. (1987, 1989), Wells et al. (1990), and Alexander et al. (1991). To date, the largest and best-studied tidal flat is the Ganghwa tidal flat, which is attached to Ganghwa Island where the large tidal bar studied herein attaches to land. It corresponds to Zone 1 of the large tidal bar in Figure 5.2. Because this tidal flat does not occur behind a wave-generated barrier island, it is referred to here as an "open-coast" tidal flat, even though it receives less wave energy than open-coast tidal flats farther south along the Korean coast (e.g., Yang et al., 2005) due to its partially sheltered location at the head of Gyeonggi Bay. The Ganghwa tidal flat has been investigated by Kim (2006), Lee et al. (2011), Choi et al. (2011), Choi et al. (2013) and Choi and Jo (2015). Of these studies, Kim (2006) contains the most detailed sedimentological descriptions. The following is a synthesis of this work, supplemented by our own observations (Choi, unpublished data).

Three main facies characterize the Ganghwa tidal flat: dune cross-stratified sand (Facies 1), ripple cross-stratified sand (Facies 2), and bioturbated mud (Facies 8). Cyclic tidal rhythmites (Facies 5) are accumulating on the flanks of the small tidal creeks that dissect the tidal flats (see below), but are volumetrically subordinate.

The lower intertidal portion of the Ganghwa tidal flat is cut by a network of sandy tidal channels. These are clearly visible in elevation profiles collected near Yeochari (see Figure 4.6A). During spring low water, the base of the shallower of these channels (<4 m deep) becomes exposed, revealing two- and three-dimensional dunes, which migrate actively during the alternating flood and ebb tides (Choi et al., 2013). No cores have been collected here, but it seems reasonable to suspect that dune cross-stratified sand (Facies 1) is the dominant facies being deposited here, as is the case at the base of most major channels in the delta.

The lower intertidal portion of the Ganghwa tidal flat between the tidal channels is also relatively sandy (Lee et al., 2013). Cores from this area (Kim, 2006) consist primarily of thinly bedded fine sand with little to no bioturbation (Facies 2). Beds are erosively-based and pass upward from parallel lamination to ripple cross-stratification. Parallel laminated beds are 1–10 cm thick. The parallel lamination is horizontal and commonly gently undulating. Undulations have millimeter-scale amplitudes and decimeter-scale wavelengths. Rare dewatering structures are observed. Ripple cross-stratified beds are 0.5–2 cm thick. Flat-based cross-sets are common. Rare scoop-shaped cross-sets are also present.

The ripple cross-stratification in Facies 2 is interpreted to have been deposited primarily by *current ripples* generated by relatively weak tidal currents. Rare scoop-based cross-sets are likely wave generated (e.g., Boersma, 1970; de Raaf et al., 1977). The gently undulating horizontal lamination is somewhat more enigmatic. It may be akin to a form of "micro *hummocky cross-stratification*" deposited by small, rapidly aggrading three-dimensional mounds ("hummocks") that were similar to (but smaller than) the meter-scale hummocks commonly generated beneath large-scale oscillatory flow in wave-tunnel experiments (e.g., Cummings et al., 2009). Yang et al. (2005, 2006) documented similar sedimentary structures on the open-coast tidal flats farther south along the Korean coast in areas that do not experience long-wavelength waves. Some of the low-angle cross-stratification in Facies 2 could also have been deposited by antidunes (e.g., Cartigny et al., 2014). Rainfall events can generate supercritical—and surprisingly dangerous—sheetflows on the Ganghwa tidal flat during low tide. It is suspected that intense, seasonal rainfall events are an under-appreciated geomorphic agent shaping Korean tidal flats (Choi, 2011).

Like most tidal flats (Amos, 1995), the upper intertidal part of the Ganghwa tidal flat is muddy. The main facies recovered in cores (Kim, 2006) is intensely bioturbated mud (Facies 8; Figure 5.2; Table 5.2). The mud is silt rich, with little clay. It is interpreted to have been deposited by particle-by-particle settling of flocs from suspension during high water slacks (Table 5.1). Given the intense bioturbation, sedimentation rates were likely low and the benthic community relatively unstressed. The silt-rich nature may suggest that the mud is largely fluvial in origin, given that the suspended load of the Han River is also silt rich (Lee et al., 2013). Moving down the tidal flat toward the low tide mark, the bioturbated mud becomes progressively interbedded at a decimeter scale with the erosion-based sand beds (Facies 2) described above, eventually disappearing close to the base of the intertidal zone. In general, flaser, lenticular, or wavy laminated beds (Reineck and Wunderlich, 1968)—styles of stratification typically associated with back-barrier tidal flats—are rare.

Additional facies are observed in the dendritic tidal-creek networks that ornament the Ganghwa tidal flat (Choi, unpublished data). Current-ripple cross-stratified sand (Facies 2) is observed in the point bars of the trunk channels of these small creeks, whereas cyclic tidal rhythmites (Facies 5) together with current-rippled sand (Facies 2) are observed in the point bars of the small tributaries. Preservation of cyclic rhythmites in the small tributaries may be due to rapid deposition rates (rapid lateral migration of channel) and sheltering from waves in these locations. These rhythmites are likely to be generated mainly during the summer when runoff and sedimentation rates are highest and wave energy is lowest.

Table 5.2 Terminology to describe the intensity of bioturbation

Adjective	Bioturbation index[*]	Description
None	0	No trace fossils observed.
Low	1–2	Rare trace fossils. Primary sedimentary structures clearly visible.
Moderate	3	Common trace fossils. Primary sedimentary structures obscured somewhat, but still easily recognizable.
High	4	Abundant trace fossils. Primary sedimentary structures and bedding surfaces obscured and commonly difficult to recognize.
Intense	5–6	Primary sedimentary structures and bedding obliterated.

[*] Taylor and Goldring (1993).

5.2.2 Zone 2: Sandbars and Swatchways

Sandbars and swatchways ornament the large tidal bar basinward of the open-coast tidal flats (Figure 5.2). The entire intertidal and subtidal area is sandy (Figure 4.14) and covered by dunes (Figure 4.10). Short cores collected from the intertidal portion of sandbars consist predominantly of brown, upper-fine to upper-medium sand (Figure 5.3; Facies 1). Beds are <1–15 cm thick (average ~5 cm), and typically contain high-angle (15°–30°) dune cross-stratification that is generally only visible in X-radiographs. Internal downlap surfaces within cross-sets, interpreted to be reactivation surfaces, are present locally (Figure 5.3). Mud layers and mud balls are absent. Shells (gastropods and bivalves), organic debris ("tea leaves"), and trace fossils (*Skolithos*) are present rarely. In most cases, the short cores lack obvious vertical tends in grain size, cross-set thickness, or sedimentary structures.

5.2.3 Zone 3: Mid–Outer Large Tidal Bar

The mid to outer part of the large tidal bar extends from the intertidal edge of the sandbar–swatchway zone basinward to the clinoform rollover at 20-m water depth, and includes both flanks of the large tidal bar as well as its flat top (Figure 5.2). The zone is fully subtidal. The large tidal bar at this location has a smooth surface: there are no swatchways or other channels, no sandbars, and only rare patches of dunes (Figure 4.8). The near-surface sediment cover is heterolithic and short cores are sedimentologically similar (Figure 5.2). However, two geotechnically distinct zones exist. Cores recovered from the steep, dune-free southern flank of the large tidal bar (e.g., piston cores 6 and 32; Figure 5.2) were highly water saturated. They consequently experienced considerable coring-induced deformation. By contrast, cores from the north flank and top of the large tidal bar were less water saturated, firmer, and experienced less coring-induced deformation.

Four facies are observed in the short cores from this zone. In decreasing order of volumetric importance, these are (1) thinly interlaminated mud and sand (Facies 6), (2) bioturbated sandy mud (Facies 7), (3) fine sand (Facies 3), and (4) medium sand (Facies 1) (Table 5.1).

Facies 6, which consists of interlaminated mud and sand, is the most common deposit type (Figure 5.3; Table 5.1). It is similar to the interlaminated mud and sand described below that is accumulating locally in the proximal large tidal channels (Facies 4), but it differs in that the mud laminae are thinner (<0.1–10 mm; average ~3 mm), the sand laminae are thicker (<0.1–4 cm; average ~1 cm), and the bioturbation levels are moderate to high (Table 5.2). Facies 6 units are 5–60 cm thick and they grade in and out of bioturbated sandy mud (Facies 7). The mud-to-sand ratio ranges from 60:40 to 40:60. Sand is very fine to fine grained. Rare medium-sand laminae occur locally. Sand and mud laminae are sharp-based. Sand laminae commonly pinch and swell across the width of the core, are commonly high-angle (15°–30°) cross-stratified, and are typically flat-based; scoop-based cross-sets are rare. Mud drapes on foresets are present locally. Nowhere were mud–sand couplets packaged into cyclic, thickening–thinning, or coarsening–fining units.

Facies 6 is interpreted to be predominantly a tidal current deposit. The flat-based geometry and high-angle cross-stratification of sand laminae are interpreted to reflect deposition by current ripples or possibly current-influenced combined-flow ripples; scoop-based cross-sets that might be interpreted as wave-generated (e.g., Boersma, 1970; de Raaf et al., 1977) are rare. This, along with the regular alternation between mud and sand, suggests that Facies 6 was deposited by periodic fluctuations of current energy, likely related to tidal currents acting on the seafloor. The absence of unequivocal tidal signatures (Box 5.1), however, may be symptomatic of low sedimentation rates basinward of the turbidity maximum, or possibly of interruption by wave action, especially given that waves generated during larger storms likely stir sediment on the seafloor at this loca-tion (see Box 2.1 and Figure 2.12). Similarly, Fan (2012) has invoked sporadic wave influ-ence as the reason for the lack of tidal rhythmicity in some Chinese open-coast tidal flats.

Bioturbated sandy mud (Facies 7) is nearly as common as interlaminated mud and sand (Facies 6) in the mid to outer part of the large tidal bar (Figure 5.2; Table 5.1). This facies consists of moderately to intensely bioturbated, medium- to fine-sandy mud. Beds range in thickness from several centimeters to at least 3 m. Bed contacts are bioturbated. Mud-to-sand ratios vary from ~60:40 to ~40:60, with sand typically occurring within burrows. Trace-fossil diversity is low, and trace-fossil assemblages commonly consist entirely of small "pin-prick" *Planolites* burrows (1–3 mm diameter). Vertical trace fossils are rare. Where primary stratification has not been obliterated by bioturbation, lamina-tion identical to that of Facies 6 is visible. As a result, Facies 7 is transitional to Facies 6 at lower levels of bioturbation. A similar origin is therefore proposed: Facies 7 is inter-preted to be a tidal current deposit generated when physical and/or chemical stress levels were lower than for Facies 6.

Fine sand (Facies 3) is common in cores from the middle part of the large tidal bar, where it forms sharp-based, decimeter-thick beds (Figure 5.3). Facies 3 beds were typi-cally rendered structureless in piston cores due to coring-induced deformation. However, in vibrocores, which experienced less deformation, crude horizontal stratification and rare subvertical burrows (*Skolithos?*) are visible locally (see vibrocore NYSVC-03, 40–50 cm, Appendix 1E (http://booksite.elsevier.com/9780128007686/)). Facies 3 beds are interpreted to reflect deposition at times when mud input was relatively low and hydraulic energy relatively high. Some of the structureless fine-sand beds could be lique-fied versions of Facies 2. Others could potentially be event beds generated during winter storms. Longer-term, multiyear changes in wave energy and storminess could have also conceivably led to deposition of the Facies 3 units.

Rarely, medium sand (Facies 1) was recovered in short cores from the top of the mid to outer large tidal bar. These cores invariably came from areas where dunes ornamented the seafloor. Given the orientation of the dunes (see Line J10 in Figure 4.9), one possible interpretation is that they migrated out of the northern channel and over the top of the large tidal bar toward the south, and will eventually deposit sand on the southern flank.

5.2.4 Zone 4: Deltaic Clinoform/Prodelta

Short cores from this zone consist entirely of fine sand (Facies 3) (Figure 5.2) that is indistinguishable from the thin sandy transgressive lag on the adjacent shelf. Shell fragments (bivalves) are rare, and terrestrial organic debris is absent. With the exception of several rare *Ophiomorpha* burrows, sedimentary structures are absent because they were destroyed by coring-induced liquefaction.

The fine sand on the deltaic clinoform is interpreted to have been reworked up the clinoform from the shelf by flood-dominated tidal currents. This sediment is anomalous: the long-core data presented in subsequent chapters show that the interior of the large tidal bar, including the older *clinoform* deposits, consists predominantly of muddy heterolithic sediment. A relatively recent switch in sedimentation has therefore occurred, the potential reasons for which will be discussed in subsequent chapters.

5.3 LARGE TIDAL CHANNELS

Three facies are identified in short cores collected from the large tidal channels. Medium–coarse sand (Facies 1) is volumetrically the most significant (Table 5.1). It is interbedded on a decimeter scale with thickly interlaminated mud and sand (Facies 4) in locations proximal to Sukmo Channel, the main distributary channel. In some distal locations, Facies 1 is interstratified with bioturbated sandy mud (Facies 7).

Brown medium–coarse sand (Facies 1)—the coarsest sediment in the study area—is present in all cores collected from the base of the large tidal channels (Figure 5.2). The sand was liquefied during coring and rendered structureless. Units are sharp-based and are a few decimeters to 1.5 m thick. Thin units tend to consist of well-sorted medium–coarse sand, whereas thicker units (>1 m) fine upward from coarse sand with mud pebbles and shell fragments to medium sand; this grading is likely a result of the coring-induced liquefaction. Mud laminae and woody organic debris are absent. Rare metasedimentary rock granules and pebbles are observed.

Facies 1 is interpreted to have been deposited by subaqueous *dunes*. Because the cores were rendered structureless, the main support for this comes from seismic records, in which 0.25–8 m high dunes (average ~3 m) are visible at the base of the large tidal channels in areas where Facies 1 was recovered (Figure 4.10). The occurrence of mud clasts and shell fragments exclusively in thicker (>1 m) units may reflect the tendency of larger dunes to scour more deeply into their substrates (Paola and Borgman, 1991; Best, 2005).

In short cores collected near the junction of the Sukmo Channel and the large tidal channel to which it links, sharp-based units of thickly interlaminated mud and sand, each 10–30 cm thick (Figure 5.3; Facies 4), are interbedded with medium–coarse sand units (Facies 1). Mud laminae in Facies 4 are 2–10 mm thick (average 8 mm), the thickest in the study area, whereas sand laminae are very fine grained and thin (1–5 mm; average 2 mm). Mud laminae are unbioturbated and lack sand or silt partings. The ratio of mud to sand in Facies 4 is typically high, usually above 80:20.

Given their substantial thickness and lack of bioturbation, mud laminae in Facies 4 are interpreted to be *fluid mud* deposits (cf. Ichaso and Dalrymple, 2009). Fluid mud, a dense, soupy substance with particle concentrations and physical properties somewhere between that of clear water and consolidated mud (Box 5.1), is commonly observed in modern high-energy, tidal coastal systems, commonly at the base of dune-covered channels in the turbidity maximum zone (Becker et al., 2013). A correlation between fluid mud deposition and elevated turbidity also appears to apply for the Han River system (Figure 5.4).

Farther basinward in the large tidal channels, bioturbated sandy mud (Facies 7; Figure 5.3) is interbedded with medium–coarse sand (Facies 1). Bioturbated units are sharp-based and 10–100 cm thick. Bioturbation levels are typically intense (Table 5.2), and discrete trace fossils are unidentifiable. The medium–coarse sand beds (Facies 1) are 20–175 cm thick. They were liquefied and rendered structureless during coring, but are suspected to be dune cross-stratified, given the presence of dunes on the seafloor in seismic records.

The bioturbated texture of Facies 7 likely reflects slow, particle-by-particle settling of mud from dilute flows seaward of the turbidity maximum where environmental stress was reduced. The mud at this distal location may have been flushed basinward during river floods, perhaps in conjunction with spring tides (Figure 5.4).

The medium–coarse dune cross-stratified sand (Facies 1) in the distal parts of the large tidal channels is suspected to have been derived from older underlying units. Some may have been reworked landward from the shelf. It is less likely to have been supplied by the river. Several lines of evidence support this. (1) The sand has an orangish tinge to it, whereas sand in the distributaries and swatchways that is expected to be supplied by the modern river tends to be brown to yellow. The orange color suggests derivation from an older, subaerially weathered deposit (see Chapter 8). (2) The asymmetry of the dunes on the floor of the large tidal channel basinward of the Sukmo Channel indicates that landward transport of *bedload* is occurring in this particular location due to *flood-dominated* tidal currents. (3) Mutually evasive sand transport pathways separated by a bathymetric high (5 m relief) exist at the junction between the Sukmo Channel and the large tidal channel to which it links (Figure 4.13). Moving basinward, the Sukmo Channel veers south at this location, passing up onto the adjacent large tidal bar where it feeds into a network of *ebb-dominated* swatchways (Figures 4.13 and 5.2). These are suspected to receive most of the fluvial bedload. Moving landward, the large

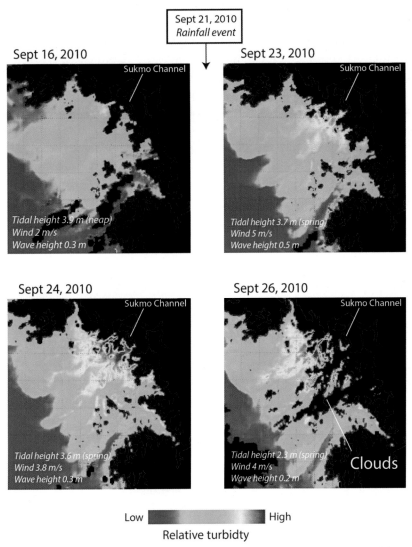

Figure 5.4 MODIS satellite image showing the behavior of a turbid plume that issued from the Han River following an intense late summer rainfall event. Daily precipitation on September 21 was 260 mm, an extreme amount only exceeded 10 or so times in the past 200 years (see Figure 2.5E). The turbid plume is deflected to the right, along the coast, due to the Coriolis effect. This type of plume behavior may explain in part why tidal flats are more extensive and why large tidal bars are wider and shallower on the north (North Korean) side of Gyeonggi Bay than on the south side. Note the spatial association of high turbidity values and the occurrence of interpreted fluid mud deposits (Facies 4; Figure 5.2).

tidal channel veers north at this location, passing up onto the adjacent large tidal bar where it feeds into a network of flood–dominated swatchways. These are suspected to receive most of the bedload that is moving landward in the northern large tidal channel.

5.4 SHELF

The gently undulating shelf outside Gyeonggi Bay is covered by a sharp-based veneer of fine to medium sand (Facies 1 and 3) ranging in thickness from 2 to 250 cm (Figure 5.2). Woody organic debris is rare and shell fragments (bivalves) are present but uncommon. Sedimentary structures are typically absent due to coring-induced liquefaction. Dunes (<3 m high) are visible locally in seismic data where the veneer coarsens from fine to medium sand.

Firm interlaminated mud and sand (Facies 5; Table 5.1) underlies the shelf sand veneer. Sand laminae are fine to very fine grained, flat-based, and current-ripple cross-stratified. Scoop-based, wave-rippled cross-sets (e.g., Boersma, 1970) are rare. Cyclic tidal rhythmites (Facies 5) are present in one piston core (piston core 12; see Figure 5.2). The mud–sand couplets are organized into thickening-then-thinning cycles that are 15–25 cm thick, each comprised of 10–25 mud–sand couplets. Sand-sized woody detritus is present locally. Bioturbation levels are low to intense (cf. Table 5.2). Horizontal burrows predominate, most of which are *Planolites*. Sand-filled *Thalassinoides* burrows subtend locally from the erosional base of the overlying sand veneer. Their robust and unlined nature suggests that the mud was firm during burrow excavation.

The sharp-based sand veneer on the shelf is interpreted to be a lag overlying a transgressive erosion (i.e., *ravinement*) surface. The underlying muddy heterolithics are likely older back-barrier lagoonal deposits given their firmness, abundant woody detritus, and the presence of cyclic rhythmites that may record neap–spring cycles (e.g., Fischer, 1961; Demarest and Kraft, 1987). Waves, not tides, are suspected to have caused the ravinement given the lack of tidal ridges on this part of the shelf and the lack of a paleo-embayment in bathymetric charts between the water depths of 50–80 m that could have amplified tides (Figure 2.30). Previous workers have argued that wave ravinement occurred along other parts of the Korean coast. For example, near Baksu and Dongho, wave-generated barrier islands are stranded locally at the highstand shoreline, and muddy lagoonal deposits underlie a transgressive sandy surface layer (Yang et al., 2006b). Further support for wave ravinement offshore of Gyeonggi Bay is presented in Chapter 6.

CHAPTER 6

Seismic Stratigraphy

Contents

Chapter Points

- The system is *progradational*. The large tidal bars of the subaqueous delta platform have prograded at least 75 km basinward, widening in the process. The widening occurred at the expense of the intervening large tidal channels, which are becoming narrower, while apparently remaining as deep as before.
- The delta is highly channelized, both intertidally and subtidally. As such, packages of seismic reflections produced by *lateral accretion* of channels abound.
- The large tidal bar focused on herein nucleated on two smaller elongate ridges that amalgamated by infilling of the intervening channel. The progradational clinoforms of the subaqueous delta platform downlap onto these older ridges.

As well as being sedimentologically distinct, the main morphological elements of the subaqueous delta—the large tidal bars and large tidal channels—are also seismically distinct, as is the shelf basinward of the delta. Key seismic reflections, seismic facies, and lapout relationships for each morphological element are described below. No seismic data exist for the open-coast tidal flats or point bars in the tidal–fluvial transition. Thus, they are not covered in this chapter.

6.1 LARGE TIDAL BARS

The large tidal bar investigated in this study consists of two seismic units, an outer carapace composed of clinoforms and two elongate ridges that form a nucleus underlying it (Figures 6.1 and 6.2). The former downlaps onto the latter. Sandbars and swatchways, which constitute smaller scale elements that ornament the inner-mid part of the large tidal bar, are discussed separately below.

The outer carapace of the large tidal bar (0–30 m thick) was generated by a combination of lateral widening and seaward progradation. The widening is apparent in shore-parallel seismic transects, which trend NW-SE (Figure 6.1). These reveal relatively continuous, moderately low-angle (~1°) clinoform reflections that build outward into the adjacent large tidal channels (Figures 6.3 and 6.4). Short cores from the areas of active lateral

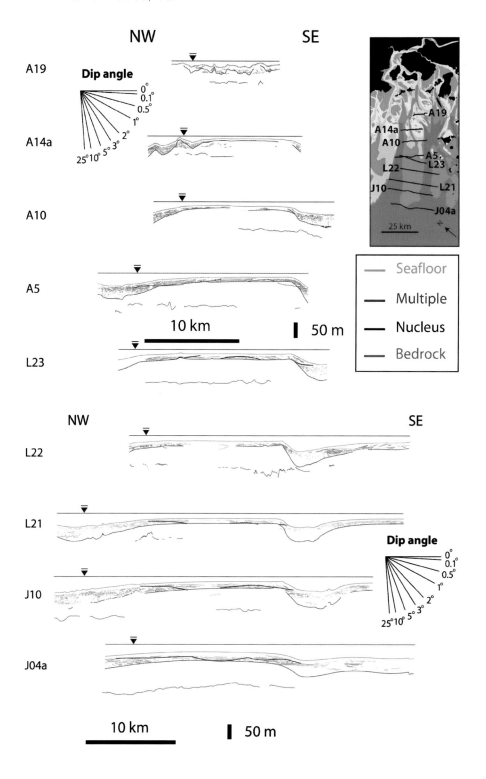

NW   SE

A19

Dip angle
0°
0.1°
0.5°
1°
2°
3°
5°
10°
25°

A14a

A10

A5

10 km   50 m

L23

NW   SE

L22

L21

Dip angle
0°
0.1°
0.5°
1°
2°
3°
5°
10°
25°

J10

J04a

10 km   50 m

Seafloor
Multiple
Nucleus
Bedrock

A19
A14a
A10
A5
L23
L22
L21
J10
J04a

25 km

(a) **Inner large tidal bar**

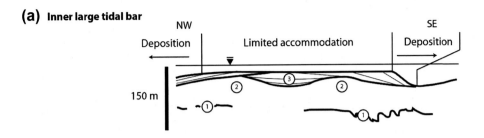

(b) **Outer large tidal bar**

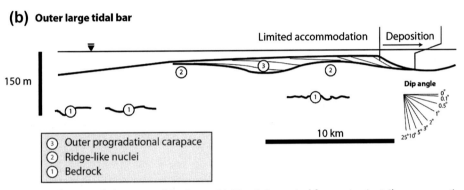

③	Outer progradational carapace
②	Ridge-like nuclei
①	Bedrock

Figure 6.2 Architectural elements of the large tidal bar interpreted from seismic strike cross-sections. (a) The inner 80 km of the large tidal bar, which is symmetric. Note lateral accretion on both sides, although more occurs on the southern flank. The elongate depression between the two ridges that form the nucleus of the modern large tidal bar has filled with concave-up draping strata that form part of the progradational carapace. (b) The outer 20 km of the large tidal bar, which is asymmetric. Note lateral accretion to the south only. The first water-bottom multiple (not shown) is typically near the top of the two ridges that form the core of the large tidal bar. As a consequence, signal-to-noise ratio between this level and bedrock is low, preventing confident interpretation of deeper stratigraphic architecture. (A) and (B) are based on seismic lines A5 and 04DC-J10, respectively (see Figure 6.1).

outbuilding consist of water-saturated mud and very fine to fine sand (Facies 3, 6, and 7; see piston cores 15, 20, 21, and 32 in Figure 5.2). Dunes are not observed on these areas.

The inner part of the large tidal bar is symmetric because it is building laterally outward on both sides. The proximal parts of the adjacent large tidal bars are also building laterally outward. This is causing the proximal parts of the intervening large tidal

Figure 6.1 Form-line traces of strike-oriented seismic transects across the large tidal bar (see Appendix 2 for original seismic lines). Note the outer carapace of lateral-accretion clinoforms, which lap onto a nucleus comprised of two elongate ridges (see also Figure 6.2). The inner part of the large tidal bar builds outward on both sides. The outer bar is erosional on the northwest (NW) side, and depositional on the southeast (SE) side. Intermediate-scale mounds (sandbars) and channels (swatchways) ornament the inner large tidal bar (transects A19 and A14). A series of irregular, high-amplitude reflections below the large tidal bar is interpreted to record bedrock.

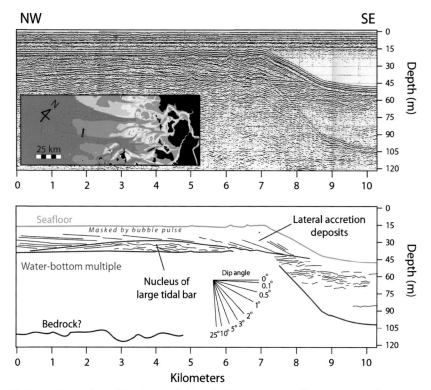

Figure 6.3 Lateral accretion of the large tidal bar. Uninterpreted and interpreted strike-oriented seismic transect from the distal south-flank of the large tidal bar (part of seismic line 04DC-J10). Note (1) that the outer carapace of the large tidal bar consists of lateral-accretion clinoforms that build southward into the adjacent large tidal channel; (2) that the outer carapace downlaps the southern one of the two ridges that core the large tidal bar (thick black line), whose top can be traced laterally to the channel thalweg; and (3) all present-day deposition is occurring on the southeast flank of the large tidal bar at this location. Top of the seismic transect is sea level.

channels to fill in and become narrower (Figure 6.5). By contrast, the outer 20 km of the large tidal bar is asymmetric and almost dune-like in profile because it is accreting exclusively on its steep (1°) southern flank, while its more gently sloping (~0.2°) northern flank is eroding (see transect J10 in Figure 6.1).

Viewed in shore-parallel seismic transects, the downlap surface that defines the base of the outer carapace can be traced out from beneath the large tidal bar, where it corresponds to the top of the ridge-shaped nuclei (see more below), into the base of the large tidal channels, where it lies at or just below the seafloor (Figure 6.3). Where dunes ornament the base of the large tidal channels, the lateral-accretion clinoforms on the bar flanks appear to downlap and bury the dunes (Figure 6.4). A distinct lithofacies change accompanies the downlap pinchout, from water-saturated mud and very fine to fine sand (mostly Facies 3 and 7; see Table 5.1) on the flank of the large tidal bar flanks to dune cross-stratified

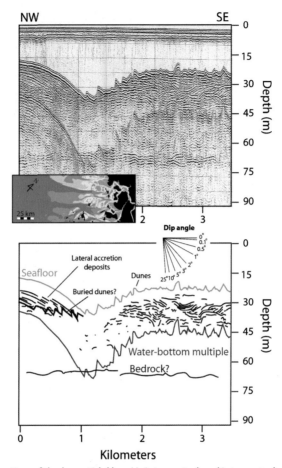

Figure 6.4 Lateral accretion of the large tidal bar. Uninterpreted and interpreted strike-oriented seismic transect from the proximal southern flank of the large tidal bar (part of line 04DC-K06). Note that lateral-accretion clinoforms appear to downlap onto very large dunes in the thalweg of the large tidal channel.

medium–coarse sand (Facies 1) in the base of the large tidal channels (compare piston cores 14 versus 15, 32 versus 31, 15 versus 14, and 21 versus 22 in Figure 5.2).

The seaward progradation of the large tidal bar is apparent in the shore–perpendicular seismic transect that extends down the axis of the bar (Figure 6.6). In this transect, continuous, low-angle (~0.1°) clinoform reflections are observed that build basinward for at least 75 km. The distal face of the large tidal bar, which together with the distal faces of the other large tidal bars makes up the deltaic clinoform/prodelta, has therefore prograded at least this distance out onto the shelf. The progradation ranged from directly basinward to slightly oblique, toward the south–southwest, given that some of the clinoforms in the progradational package dip into the large tidal channel to the south (~1°) in addition to dipping seaward (~0.1°) (Figure 6.5).

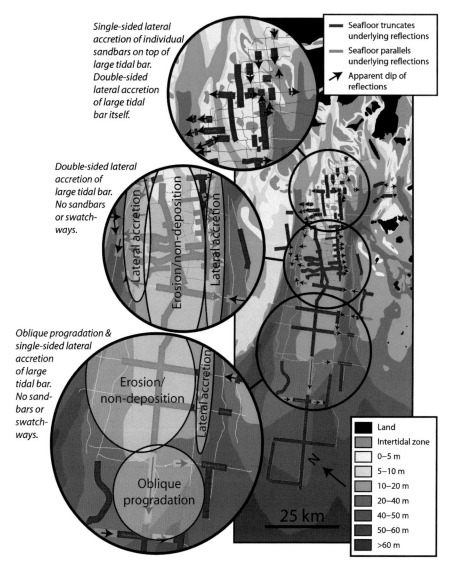

Figure 6.5 Areas of present-day deposition and erosion/non-deposition, as interpreted from seismic data. Red lines indicate areas where the seafloor truncates underlying reflections or where nondeposition and toplap occurs. The bubble pulse obscures the precise relationship. Green lines indicate areas of net deposition. Note that net deposition is confined to the distal clinoform (prodelta) and flanks of the large tidal bar, and that little (if any) sediment is accumulating on top of the large tidal bar. Portions of seismic lines where no green or red lines are drawn are zones where the seismic data were too noisy to determine whether the seafloor truncates or parallels underlying reflections.

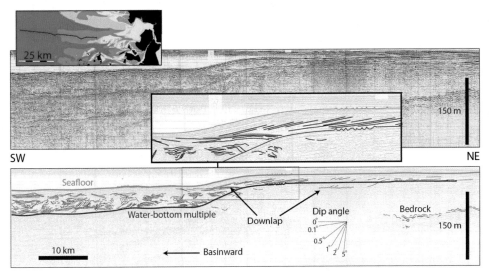

Figure 6.6 Progradation of the subaqueous delta platform. Dip-oriented seismic transect through center of the large tidal bar onto the shelf, showing downlap of low-angle (0.1°) large tidal bar clinoforms onto shelf surface (seismic line J03). Note that the basinward dip of 0.1° is only an apparent dip; in reality, the clinoforms dip much more steeply (1°) into the channel to the south. The combination of these two dips indicates that the clinoforms actually dip obliquely to the south–southwest.

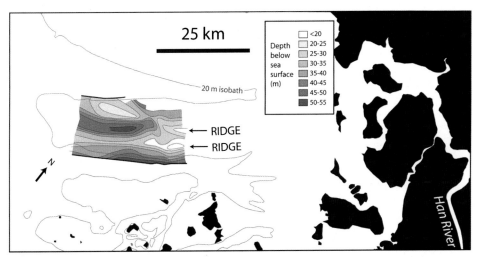

Figure 6.7 Depth below the sea surface of the top of the two ridges that form the nucleus of the large tidal bar. Drill-core data suggest that the ridges are sandy relative to the muddy outer carapace of the large tidal bar (see Chapter 8).

The two ridges that form the nucleus of the large tidal bar (Figure 6.7) extend in a shore-normal direction and trend roughly parallel to each other. They have symmetric cross-sectional profiles and relatively sharp crests, exhibit 30 m of relief from crest to the intervening swale, and are spaced 10 km apart. In plan view, they appear to form a giant horseshoe-shaped body with a large headward-terminating channel between them (Figure 6.7). A single continuous reflection defines the top of the two ridges; it is the surface onto which the outer progradational carapace downlaps. Viewed in shore-parallel seismic transects, the reflections within the channel between the two ridges build up in a trough-like fashion in the mid to inner part of the large tidal bar (e.g., line A5 in Figures 6.1 and 6.2(a)). However, in the seismic transect that extends down the axis of the large tidal bar (Figure 6.6), these reflections clearly dip basinward and record seaward progradation of the large tidal bar.

The downlap surface that defines the top of the two ridges corresponds approximately to the position of the first water-bottom multiple beneath the large tidal bar. Below this, the signal-to-noise level decreases, preventing confident interpretation of stratigraphy below this level. An irregular, high-amplitude reflection between 55 and 155 m depth is interpreted to be top of bedrock (Figure 6.8). If this interpretation is correct, between 50 and 120 m of unconsolidated sediment lies between bedrock and the seafloor. The long cores described in Chapter 8 support this estimate.

The sandbars that ornament the inner part of the large tidal bar also have a distinct seismic architecture: internally, they consist of relatively high-angle (<2°) lateral-accretion clinoforms that dip into the adjacent swathways (Figure 6.9). Because swathways cut obliquely across the large tidal bar, the migration of swathway–sandbar pairs generates lateral-accretion bedding that dips obliquely landward or obliquely basinward.

6.2 LARGE TIDAL CHANNELS

Beneath the large tidal channels, a single seismic unit is discernable between bedrock and the seafloor; this unit appears to be in lateral continuity with the sediments that lie between bedrock and the top of the dual-ridge nucleus at the core of the large tidal bar (Figures 6.1–6.4).

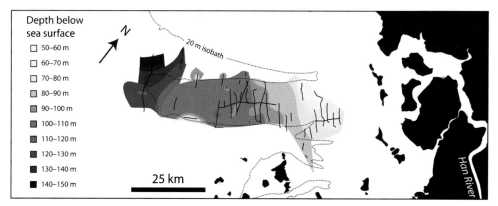

Figure 6.8 Depth to bedrock, interpreted from seismic data. Depth measured from sea surface. Solid black lines indicate where the bedrock reflection is visible in seismic data.

The unit is characterized by short, irregular, horizontal to steeply dipping (<2°) reflections. Tabular packages 0.5–5 m thick containing high-angle reflections are visible locally. Short cores from the large-tidal-channel thalwegs indicate that at least 1–2 m of dune cross-stratified medium sand occurs at the top of the unit. A discrete seismic unit associated with these deposits cannot be distinguished, either because they are too thin, or because they are seismically indistinguishable from the underlying sediments.

6.3 SHELF

The smooth, gently undulating, nearly horizontal (0.01° seaward dip) shelf seafloor is underlain by packages of high-angle (0.5°–2°) clinoform-shaped seismic reflections with intervening packages of horizontal reflections (5–20 m thick and several kilometers in

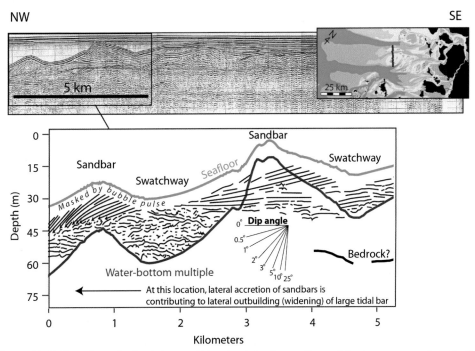

Figure 6.9 Strike-oriented seismic transect showing lateral-accretion architecture preserved within the levee-like sandbars (seismic line A14a) (Another way of thinking of the sandbars is that they are swatchway-bank deposits.). Note that these lateral-accretion packages are an order of magnitude smaller than the lateral-accretion packages produced by the large tidal bar itself (see Figure 6.3). The small (<2 m high) irregularities on the seafloor are dunes. Despite the presence of dunes on the surface, it is possible that these particular sandbars are heterolithic, given the coherent nature of the lateral-accretion reflections within them. This example is anomalous; most of the other sandbars lack such coherent internal reflections, suggesting they may be sandy throughout, a characteristic supported by the hammer cores (see Chapter 5). The faint dark grayish line at ~35 m depth in the seismic transect is not a real reflection—it is a pen mark that could not be removed from the original paper copy of the seismic transect.

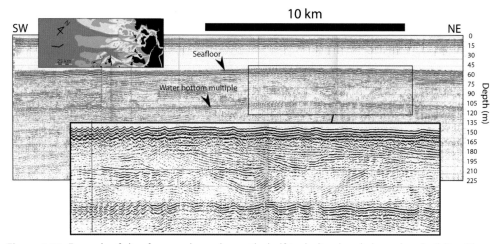

Figure 6.10 Example of clinoform packages beneath shelf, including bowl-shaped up-building. These packages are interpreted to have been generated by laterally migrating channels in a back-barrier setting during transgression. Intervening flat, horizontal reflections could potentially have been generated by vertical accretion of back-barrier tidal flats.

horizontal extent). Individual clinoform reflections are 5–15 m high (average 12 m), 0.2–2 km long (average 1 km), and slightly concave-up (Figure 6.9). Clinoform packages are bounded by relatively flat, horizontal discontinuities. Clinoform dip directions appear to be isotropic—they can dip landward, basinward, or parallel to shore. Rarely, dipping clinoforms build upward into bowl-shaped reflections (Figure 6.10). Inclined reflections appear to terminate against the seafloor.

The short core data (e.g., piston cores 11–13) and the moderate to high amplitude of clinoform reflections suggests that clinoform packages are at least somewhat heterolithic. Clinoform packages of intermediate dimension (i.e., meters to a few tens of meters thick) can form in many different ways in sedimentary systems, although the most common are progradation of delta-like bodies and lateral accretion of barforms into channels. Given their concave-up morphologies and diverse dip-directions, the clinoform packages are most reasonably interpreted as being channel deposits. This interpretation is consistent with locally observed bowl-like up-building, an architectural style commonly developed in abandoned channels, such as those imaged seismically by Rieu et al. (2005) in Holocene back-barrier deposits beneath the shelf seaward of the Wadden Sea. Intervening packages of flat, horizontal reflections could potentially have been generated by vertical accretion of back-barrier tidal flats. The presence of these reflections below the shelf supports the hypothesis that a barred, wave-dominated shoreline existed during transgression of the shelf before the shoreline translated back into Gyeonggi Bay, whereupon tides amplified and the system became tide-dominated.

CHAPTER 7

Facies Successions

Contents

Chapter Points

- The deltaic sediments at the mouth of the Han River are highly channelized, even subtidally, because of the large tidal flux into and out of the Han River.

- Seismic data suggest that these channels migrate laterally faster than they accrete vertically, provided they are unconfined by bedrock.

- As such, like other tide-dominated systems (e.g., Fly River delta, and Cobequid and Severn estuaries), outcrop-scale (i.e., >10 m thick) vertical facies successions in the Han River delta are generated primarily by lateral accretion of depositional channel banks (i.e., lateral migration of the channel-margin barforms).

- Sharp-based facies successions that commonly fine upward are produced as a result by tidal–fluvial point bars, open-coast tidal flats, large tidal bars in the subaqueous delta, and sandbars and swatchways on top of the large tidal bars.

- Progradation of the clinoform of the subaqueous delta, which involves addition of sediment to the sides and tips of the large tidal bars, produces a sharp-based succession of variably bioturbated, interlaminated mud and sand.

7.1 TIDAL–FLUVIAL POINT BARS

Twenty five years of direct observation indicates that tidal–fluvial point bars in the Han River delta (Figure 7.1) evolve by accreting laterally into distributary channels. The point bars are similar in vertical extent to channel depth, which ranges from 20 to 40 m as measured from the high tide level to the channel thalweg. Long cores have not been collected from the point bars, but a hypothetic vertical succession can be constructed based on the distribution of surficial sediment over the point bar surface (Choi et al., 2004a). In the succession, mud content and bioturbation level increase upward, whereas sand size decreases upward (Table 7.1; Figure 7.2). The base of the succession is an erosion surface that is overlain by a basal lag of gravel and coarse sand. Bowl-shaped incisions into the upper part of the point bar suggest

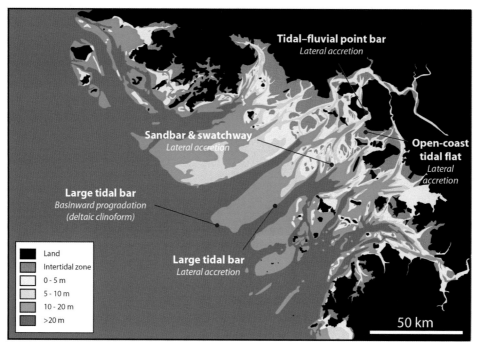

Figure 7.1 Vertical facies successions discussed in this chapter.

that slumping has occurred; the mobilized sediment may come to rest as chaotic debris at the bottom of the channel. The lower half of the succession consists primarily of dune cross-stratified medium sand (Facies 1; see Table 5.1). The dune cross-stratification is ebb-dominated in the channel thalweg but is flood-dominated throughout the point bar deposit because the flood tide becomes deflected over the point bar to the side of the main ebb outflow in the thalweg, a common situation in tidally influenced channels (e.g., Van Veen, 1950; Fitzgerald et al., 2012). The dune cross-stratified sand at the base of the point bar passes upward into interlaminated mud and sand deposited in the upper subtidal to middle intertidal zone (Facies 5 and 6). The rhythmites dip steeply (3–14°), and can be described as *inclined heterolithic stratification (IHS)* (Choi et al., 2004a), a feature common to tidally influenced fluvial point bar deposits (e.g., Thomas et al., 1987). In addition to containing evidence of *diurnal inequality*, the IHS rhythmites are packaged locally into neap–spring cycles, each consisting of <30 laminae. The neap–spring packages decrease in thickness upward, from 1 to 3 cm near the low tide level to 0.5 cm near the high tide level. The sand to mud transition within them is commonly gradational, which presumably reflects high sedimentation rates beneath the turbidity maximum. The rhythmites can also contain abundant macerated organic debris ("tea leaves"). As a whole, the IHS rhythmites become finer and more thinly laminated upward, although they are slightly coarser at the low tide level due to the associated prolonged wave action. Bioturbated mud (Facies 8) dominates the uppermost several meters of the point bar succession. Crabs are the main trace-making organism.

Table 7.1 Facies successions

Morphological element	Position relative to turbidity maximum	Vertical grain-size trend	Bed-scale cross-stratification produced by…	Lithologic character	Facies succession produced by…
Tidal–fluvial point bar	Proximal	Upward-fining	Dunes (dominant) Current ripples	Heterolithic	Lateral accretion of tidal–fluvial point bars into deltaic distributary channels
Open-coast tidal flats		Upward-fining	Current ripples (dominant) Dunes (dominant) Small wave ripples and HCS in exposed locations (minor)	Heterolithic	Accretion of open-coast tidal flats into tidal channels
Sandbars and swatchways		Upward-fining	Dunes (dominant) Current ripples	Sandy	Lateral accretion of sandbars into swatchways
Large tidal bar (lateral accretion)		No obvious vertical trend. Muddy lateral accretion deposits overlie sandy channel-thalweg dunes in places.	Current ripples	Heterolithic	Lateral accretion of large tidal bars into large tidal channels
Large tidal bar (basinward progradation of deltaic clino-form)	Distal	No obvious vertical trend. Muddy progradational deposits abruptly overlie transgressive lag.	Current ripples (dominant) Small wave ripples and HCS?	Heterolithic	Progradation of large tidal bars onto shelf

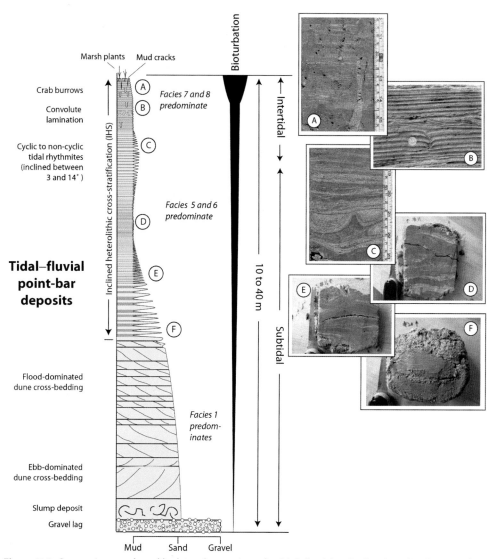

Figure 7.2 Succession produced by lateral accretion of a tidal–fluvial point bar in a distributary channel of the Han River delta. *Modified from Choi et al. (2004a).*

7.2 OPEN-COAST TIDAL FLATS

Because the shallow inner part of Gyeonggi Bay is highly channelized, most of the open-coast tidal flats pass subtidally into high-energy tidal channels, not a flat, featureless shelf (e.g., Choi et al., 2013). The tidal channels tend to be dynamic and can migrate laterally at rates of several meters per month (e.g., Choi and Jo, 2015). The tidal flats are suspected to accrete laterally into these high-energy channels in the process, producing sharp-based, upward-fining successions (Table 7.1; Figure 7.3). From a sedimentological

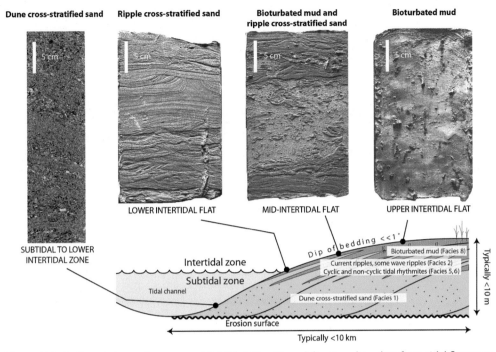

Figure 7.3 Sedimentology of open-coast tidal flat, Han River delta. Based on data from tidal flats near Ganghwa Island (Kim, 2006; Choi et al., 2011; Choi et al., 2013; Choi and Jo, 2015). Photos of sediment peels from Kim (2006) and Choi and Jo (2015). In addition to containing thin scoop-based cross-stratified beds, which may have been generated by small wave ripples (e.g., de Raaf et al., 1977), the peel from the lower intertidal flat contains undulating plane bed that is HCS-like in appearance, and may thus have been deposited under relatively large waves during a storm (e.g., Yang et al., 2005). However, such structures are not dominant; most stratification on the lower intertidal portion of the Ganghwa Island tidal flat is formed by tidally generated current-ripples. (K. Choi, unpublished data).

standpoint, therefore, open-coast tidal flats in Gyeonggi Bay are more akin to the tidal–fluvial point bars described above than to the shoreface-like open-coast tidal flats along exposed, strongly wave-influenced stretches of the Korean and Chinese coasts (e.g., Fan et al., 2004;Yang et al., 2005). Based on the distribution of facies on the open-coast tidal flats in Gyeonggi Bay (Kim, 2006; Choi et al., 2011, 2013; Choi and Jo, 2015), the successions produced as they migrate laterally should commonly pass upward from dune cross-stratified sand (subtidal; Facies 1) to current ripple cross-stratified sand (lower intertidal; Facies 2), tidal rhythmites (intertidal; Facies 5 and 6), and bioturbated mud (upper intertidal; Facies 8). Shell hash and mud pebbles may be present as a thin basal lag at the base of the successions (Choi and Jo, 2015). Compared to tidal–fluvial point bars in the Han River distributary channels, the successions produced by lateral accretion of tidal channels in the open-coast tidal flat environment will be thinner because the associated channels are shallower, typically <10 m deep. They will on average contain a greater abundance of wave-influenced sedimentary structures, higher levels of bioturbation, less-macerated organic debris, fewer gradational contacts between mud and sand laminae, and fewer cyclic tidal rhythmites. The dip of the bedding will be very subtle (<1°) and thus difficult to detect in either core or outcrop. See Choi (2014) for a comparison of tidal flats elsewhere along the Korean coast.

7.3 SANDBARS AND SWATCHWAYS

Lateral accretion of sandbars (swatchway banks) into the adjacent swatchways is expected to produce sharp-based, relatively mud-free upward-fining successions that are 5–20m thick and composed of coarse to fine, dune cross-stratified sand (Facies 1; Figure 7.4). Based on dune heights interpreted from seismic data (Figure 4.10), bed thickness may commonly

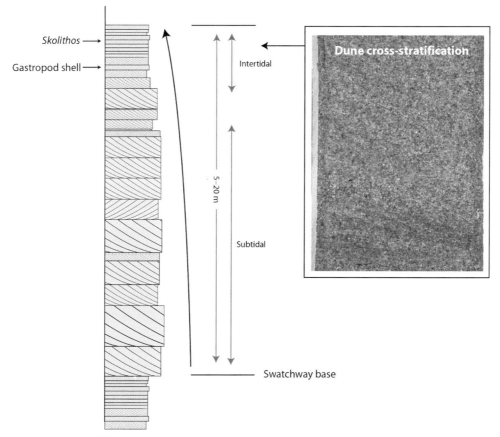

Figure 7.4 Upward-fining succession produced by lateral migration of a sandbar (swatchway channel bank) into the adjacent swatchway channel. The succession is conceptual, and is based on several different types of data, including side-scan sonar data (see Figure 4.11; Kim, 2009), seismic data (e.g., Figure 6.9), core data from the intertidal zone (hammer cores in Figure 5.2), and grab samples (Figure 4.14). Core photo is from 05 JB to 03 hammer core (see Appendix 1C, http://booksite.elsevier.com/9780128007686/).

decrease irregularly upward in an upward-fining sandbar–swatchway succession, from decimeters to meters thick near the base to centimeters to decimeters thick near the top. The lateral accretion bedding, as defined by the bases of sets of cross-stratification, would dip very gently, less than 1° (Figure 6.9). Sandbar–swatchway successions could potentially be difficult to differentiate from fluvial channel-fill successions, except that they would lack overbank muds and paleosols, contain rare *Skolithos* burrows and gastropod shells, and might possibly display a larger spread in the dip direction of dune cross-strata given the back-and-forth, slightly rotary motion of tidal currents in Gyeonggi Bay (see Figure 2.19).

7.4 LARGE TIDAL BAR

As revealed by seismic data (Chapter 6), the shore-attached large tidal bar is growing outward in all directions (except landward): it is accreting laterally into the adjacent large tidal channels and it is prograding basinward onto the continental shelf (Figure 7.5). The YSDP-107 and NYSDP-101 drill cores sample successions produced by the former and latter process, respectively (Figure 7.5). Short cores (Figure 5.2) supplement these data.

7.4.1 Large Tidal Bar: Lateral Accretion

Lateral accretion of the large tidal bar produces a 15–30 m thick succession of variably bioturbated interlaminated mud and fine sand (Figure 7.5; see also Figure 5.2). Sand laminae are commonly current ripple cross-stratified. Wave ripple cross-stratification is apparently absent. Also absent are gradationally based rhythmites, cyclic rhythmites, and abundant macerated woody organic debris—features that seem to be more commonly associated with the turbidity maximum closer to the river mouth. Dune cross-stratification was not observed. This appears consistent with the absence of dunes on the depositional flanks of the large tidal bar (Figure 4.10). The lack of obvious vertical trends in the YSDP-107 core is striking. Moving up section, the sand remains very fine to fine grained, mud content increases only slightly, and bioturbation decreases slightly. These trends are not obvious, however, in piston cores from the modern depositional bar flanks. The lower boundary, which corresponds to a downlap surface in seismic data (Figures 6.1–6.4), is expressed as a marked upward increase in mud content in the YSDP-107 core at 18.5 m depth (Figure 7.5). A similarly rapid facies change is observed in piston cores at the downlap pinchout between muddy heterolithics on the depositional flanks of large tidal bar and dune cross-stratified sand in the adjacent large tidal channel thalweg (Figure 5.2).

7.4.2 Large Tidal Bar: Basinward Progradation

Progradation of the large tidal bar produces a ~27 m-thick muddy heterolithic succession that is very similar to that produced by lateral accretion (Figure 7.5). The succession, which is sampled by the NYSDP-101 drill core, consists primarily of variably bioturbated interlaminated mud and very fine to fine sand (Facies 6 and 7 are dominant; see Table 5.1). Vertical

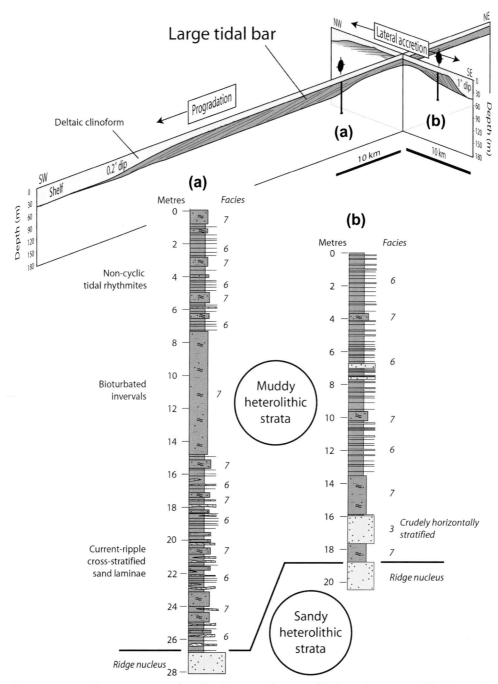

Figure 7.5 Vertical successions produced by (a) progradation and (b) lateral accretion of the large tidal bar. The successions in (a) and (b) are based on the NYSDP-101 and YSDP-107 drill cores, respectively. See Appendices 1A and 1B for details. Seismic architecture traced from seismic lines J03 (dip section) and L23 (strike section), respectively (see Appendix 2). Facies 6 and 7 refer to interlaminated mud and sand and a bioturbated version thereof, respectively (see Table 5.1 for facies details).

trends in sand grain size, mud content, and bioturbation are similarly difficult to detect. Sharp-based, noncyclic tidal rhythmites abound, and gradationally based rhythmites and cyclic rhythmites were not observed. Most sand laminae are current ripple cross-stratified, although some wave-generated and combined-flow ripples are observed. Woody organic detritus is rare.

The short cores from the prograding deltaic clinoform are decidedly different from the NYSDP-101 drill core that samples older versions thereof. The short cores consist predominantly of fine sand (e.g., piston cores 39 and 40 and vibrocore 6 in Figure 5.2) that is similar in appearance to the transgressive lag on the shelf. Clearly, a significant change in sedimentation regime has occurred. Mud is no longer accumulating on the deltaic clinoform in significant quantities. The sand is suspected to have been reworked landward from the shelf by flood-dominated tidal currents, a situation common to the outer portions of tide-dominated estuaries, and possibly some tide-dominated deltas, due to deformation of the tidal wave (Dalrymple and Choi, 2007). The reason for the switch in sedimentation regime is unclear. One potential cause is the apparent increase in aridity that has affected the greater region since the mid-Holocene (see Figure 2.32 and Box 2.2). Extensive damming of the Han River since the early 1900s (Kim et al., 2012) might also have contributed to an overall decrease in the discharge of water and suspended sediment, leaving the deltaic clinoform starved of mud. We discuss this switch in sedimentation further in subsequent chapters.

CHAPTER 8

Sequence Stratigraphy

Contents

Chapter Points

- A thick (>100 m) package of sediment overlies bedrock at the mouth of the Han River in Gyeonggi Bay.
- It consists of at least two sequences, one Late Pleistocene in age, one postglacial/Holocene. A prominent high-relief erosion surface that is locally pedogenically altered (the sequence boundary) separates the two.
- The two sequences contain a similar succession of units. Each is interpreted to record (1) interfluve pedogenesis and fluvial incision during lowered sea level; then (2) onlapping of transgressive systems tract (TST) deposits, which consist of fluvial gravel, freshwater mud, and backbarrier estuarine strata; and finally as sea-level rise slowed, the shoreline retreated into Gyeonggi Bay and the system switched to an unbarred, macrotidal state, (3) aggradation and progradation of highstand systems tract (HST) deposits, which consist of muddy tidal flat and subaqueous delta strata.

Like all models that attempt to convey the behavior of a complex, natural system in a simple fashion, the classic sequence stratigraphic models (e.g., Van Wagoner et al., 1990) make several simplifying assumptions. One such assumption is that tides are only important during transgression, and only then within incised valleys. This clearly does not hold true in all cases; such an assumption is clearly not appropriate for tide-dominated settings such as the Han River delta. As mentioned previously, most of the world's large deltas are tide-dominated or strongly tide-influenced (Middleton, 1991) and tidal dominance can switch on at any time during a relative sea-level cycle—regression or transgression—if the configuration of the basin, which is ever-changing as sea level rises and falls, enters into resonance with the tides or creates a morphology that accentuates the tidal currents without necessarily becoming resonant (Yoshida et al., 2007; Dalrymple, 2010). Korea's west coast can provide insight into this issue because the present coastline is for the most part tide-dominated (but see Yang et al., 2005) and, in some areas, including Gyeonggi

The Tide-Dominated Han River Delta, Korea
http://dx.doi.org/10.1016/B978-0-12-800768-6.00008-0

Bay, it appears to have remained tide-dominated during the last two sea-level highstands (Oh and Lee, 1998; Uehara and Saito, 2003; Choi and Dalrymple, 2004).

The stratigraphic organization of the succession within Gyeonggi Bay is approached here in three steps. First, the stratigraphic succession observed in long cores from tidal flats in inner Gyeonggi Bay is described (e.g., Choi and Dalrymple, 2004). Second, the stratigraphic succession observed in long drill cores from outer Gyeonggi Bay (e.g., Jin, 2001) is described, and is correlated with the succession in the inner bay. Third, a sequence-stratigraphic model is proposed to explain the observations. Cores and seismic transects used in this analysis are shown in Figure 8.1.

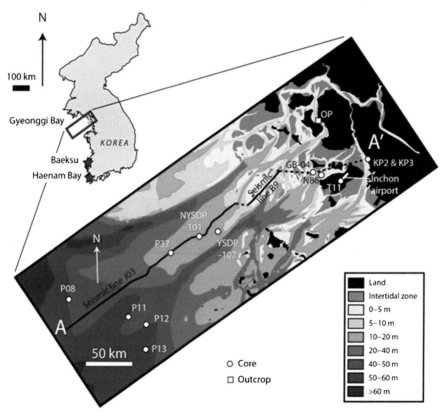

Figure 8.1 Location of the cores and seismic transects discussed in this chapter. KP2 is from Choi and Kim (2006b). KP3, T11, N86, and OP are from Choi and Dalrymple (2004). GB-04 is from Kwon (2012). YSDP-107 is from Jin (2001). Piston cores (P08 to 37), seismic lines J03 and B9, and the NYSDP-101 core are from the current study. Cores from tidal flats in Haenam Bay and Baeksu, which are also discussed in the text, are from Lim and Park (2003) and Chang et al. (2014), respectively. Dashed lines indicate parts of the A–A′ cross section (Figure 8.9) that lack seismic coverage.

8.1 STRATIGRAPHY OF INNER GYEONGGI BAY

Long cores collected by previous workers from tidal flats near the mouth of the Han River reveal a mud-rich succession over bedrock (Choi and Park, 2000; Park and Choi, 2002; Choi and Dalrymple, 2004; Lim et al., 2003; Choi and Kim, 2006a,b; Kwon, 2012; Choi et al., 2012; Choi, 2014). The succession thickens irregularly basinward as a function of the topographically irregular bedrock surface, from near zero at the landward limit of the tidal flats (active and/or reclaimed) to a maximum of 40 m, sometimes within only a few kilometers of the high tide shoreline (Choi, 2014). It has been most intensively studied at the reclaimed tidal flat between Youngjong and Yongyou islands, where kilometer-long trenches and thousands of geotechnical probes and boreholes were completed during construction of the Incheon airport in the 1990s (KACA, 1996). Five units are observed at this location, as described below and depicted in Figures 8.2–8.5. See Choi (2001), Choi et al. (2001), Park and Choi (2002), Choi and Dalrymple (2004), Choi (2005), Choi and Kim (2006a), and Choi (2014) for details.

At the base of the succession, pebbly fine to medium sand (Unit 5), likely of fluvial origin, unconformably overlies weathered bedrock (Park and Choi, 2002; Choi and Kim, 2006a). The unit is 2–7 m thick, laterally continuous, and present over a considerable elevation range (Figure 8.2). Gravel content varies from 8–45%. Marine fossils and tidal signatures are absent. In some locations, the gravel is oxidized (Kwon, 2012; cf. Park and Choi, 2002).

Overlying the pebbly sand is stiff dark-gray mud (Unit 4) interpreted to be a freshwater bog deposit (Park and Choi, 2002; Choi and Kim, 2006a). The unit is <3 m thick and is

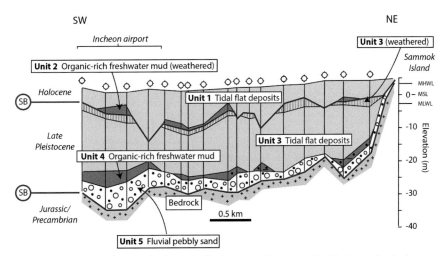

Figure 8.2 Cross section of boreholes from the Youngjong–Yongyou tidal flat near the Incheon airport. SB, sequence boundary. See text for description of the units and Figure 8.1 for location of the Incheon airport. MSL, mean sea level; MHWL, mean high water level; MLWL, mean low water level. *Modified from Choi and Dalrymple (2004).*

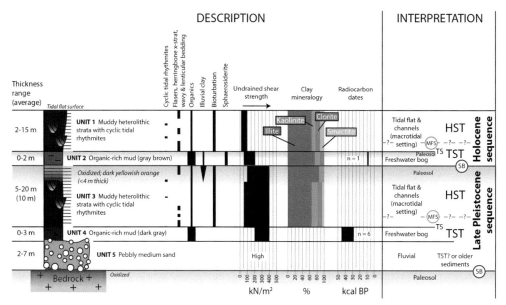

Figure 8.3 Description and interpretation of the stratigraphic succession over bedrock at the Youngjong–Yongyou tidal flat near the Incheon airport, inner Gyeonggi Bay. Most of the information was derived from Choi (2001), Park and Choi (2002), Choi and Dalrymple (2004), Choi (2005), and Choi and Kim (2006a). Black rectangles in the "radiocarbon dates" column depict the range of radiocarbon ages obtained for that particular unit (see age scale at bottom of column). The interpreted positions of the systems tracts and maximum flooding surfaces (MFSs) have been added. TST, transgressive systems tract; HST, highstand systems tract; SB, sequence boundary; TS, transgressive surface (*sensu* Posamentier and Allen, 1999). See Figure 8.1 for location of the Incheon airport.

discontinuously distributed, occurring only within topographic lows on the underlying surface. The unit lacks marine fossils and tidal signatures, and it is rich in woody organics (plant remains, rootlets, and peat fragments), which have yielded radiocarbon dates of 30–45 kcal BP. Pollen from the unit indicates a warm, humid climate (Park and Choi, 2002).

Sharply overlying Unit 4 is a muddy heterolithic unit (Unit 3). The unit is thick (5–20 m), stiff to moderately stiff, and laterally continuous, except where Holocene tidal channels scour into it (Figure 8.4). It consists of (1) variably bioturbated, noncyclic tidal rhythmites interpreted to be tidal-flat deposits, and (2) sharp-based upward-fining successions (<2 m thick) interpreted to be intertidal creek fills (Choi et al., 2001; Park and Choi, 2002; Choi and Dalrymple, 2004; Choi and Kim, 2006a). The latter tends to be lined with a thin basal lag (gravel, mud pebbles, and/or shell hash) that passes up into muddy fine sand that is flaser- and wavy-bedded, ripple cross-stratified, and/or herringbone cross-stratified, which in turn passes up into muddy cyclic and/or noncyclic tidal rhythmites. Given that cyclic rhythmites on modern tidal flats in Gyeonggi Bay are only known to accumulate at the tops of tidal-creek point bars between mean neap high water level and mean high water level, Unit 3 is interpreted to have formed in a similar,

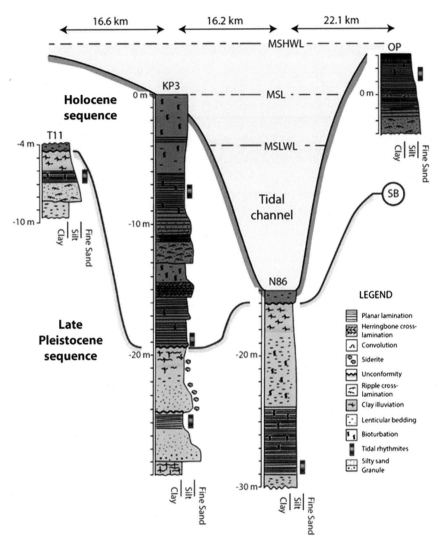

Figure 8.4 Cores from tidal flats in inner Gyeonggi Bay showing irregularity of the sequence boundary (SB) separating the Late Pleistocene and Holocene sequences. The irregularity is likely the product of fluvial incision during the lowstand coupled with later scour by tidal channels, such as that pictured, during the postglacial transgression and highstand. See Figure 8.1 for location of cores. MSL, mean sea level; MSHWL, mean spring high water level; MSLWL, mean spring low water level. *Modified from Choi and Dalrymple (2004).*

mid to upper intertidal setting. Furthermore, because cyclic tidal rhythmites tend to occur in macrotidal settings (Coughenour et al., 2009), the setting was likely macrotidal, similar to today.

Strata at the top of Unit 3 are locally oxidized to a yellow-brown color. The oxidized zone is <4 m thick and its color fades out downward. It is stiffer than the strata above and below, and it contains illuvial clay coatings, rare rootlets, a paucity of mobile elements and clay minerals that are unstable in subaerial conditions (e.g., smectite and chlorite; see Boles and Franks, 1979), and an enrichment in ferric iron (Park and Choi, 2002; Choi, 2005). It is interpreted to be a paleosol generated in a well-drained upland (Figure 8.5).

Overlying Unit 3 is a thin (<3 m) dark yellowish brown mud unit (Unit 2) (Choi, 2005). Like Unit 4 lower in the succession, it is discontinuously distributed in topographic lows on the underlying surface. Also like Unit 4, it is organic-rich, with abundant peaty fragments, one of which yielded a radiocarbon date of 9037 ± 276 cal BP (Choi, 2005). Unit 2 is softer and darker than Unit 3, and it contains less ferric oxide and fewer illuvial clay coatings. Some of the sphaerosiderite within it is coated with illuvial clay, suggesting that conditions switched from reducing to oxidizing at some point following deposition (Choi, 2005). The unit is interpreted to be a freshwater

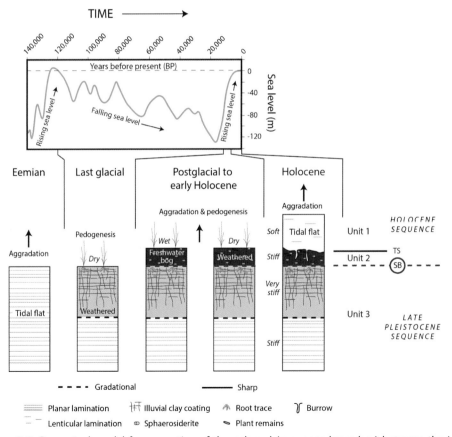

Figure 8.5 Conceptual model for generation of the paleosol (sequence boundary) between the Late Pleistocene and Holocene sequences in inner Gyeonggi Bay. SB, sequence boundary; TS, transgressive surface (*sensu* Posamentier and Allen, 1999). See text for description of Units 1–3. *Modified from Choi (2005).*

swamp deposit that became pedogenically altered, perhaps during episodic periods when the water table was low (Figure 8.5).

Sharply overlying Unit 2 is a soft, thick (5–15 m), muddy heterolithic deposit (Unit 1). Its internal laminations are almost identical to those of Unit 3, even down to the detailed characteristics of the cyclic tidal rhythmites (Choi and Dalrymple, 2004), suggesting that it likewise formed in a mid-upper intertidal flat setting ornamented by tidal creeks. The bulk grain size of the unit coarsens up slightly, from sandy mud to muddy sand.

The succession of units and surfaces described above is best explained as the product of at least two major shelf-crossing relative sea-level cycles, each of which deposited a single unconformity-bound sequence (Figure 8.6). The lower sequence consists of Units 3, 4, and perhaps part of 5, whereas the upper sequence consists of Units 1 and 2. The paleosol that separates the two sequences (i.e., at the top of Unit 3) is a regionally extensive surface observed in cores from other tidal flats along the west Korean coast, including Haenam Bay and Baeksu (Lim et al., 2003; Chang et al., 2014). Given the radiocarbon and optically stimulated luminescence (OSL) dates from these cores (Table 8.1; Figure 8.7), the paleosol likely formed when sea level was low during the last glaciation, and the upper and lower sequences are likely Holocene and Late Pleistocene in age, respectively (e.g., Park and Choi, 2002; Choi and Dalrymple, 2004; Chang et al., 2014). Because global sea level is only thought to have reached its present level once during the Late Pleistocene, namely during the Eemian interglacial, most workers interpret the lower sequence as being Eemian in age (e.g., Lim and Park, 2003; Choi and Dalrymple, 2004). This is supported by OSL dates from Baeksu (Chang et al., 2014). Taken at face value, finite radiocarbon dates from below the oxidized surface, which range from 26,358 to 50,900 cal BP (Table 8.1), suggest the lower sequence is much younger than this. However, these dates have always been held in suspicion by previous workers (e.g., Lim and Park, 2003; Choi and Dalrymple, 2004) because (1) they do not fall on the global sea-level curve (Figure 8.7); (2) the substantial vertical tectonic motion needed to place them on the global curve is unrealistic, undocumented, and unlikely; and (3) similar pre-Holocene radiocarbon dates from marine deposits in neighboring regions have been put into question by alternative dating methods (Yim et al., 1990). Therefore, we follow previous workers and disregard these ^{14}C dates.

The Late Pleistocene and Holocene sequences in inner Gyeonggi Bay have similar internal organizations. Each overlies a pedogenically altered, fluvially incised sequence boundary generated during the preceding lowstand. Nonmarine deposits accumulated over each sequence boundary during the postglacial transgressions as accommodation was generated landward of the shoreline by the rise and landward migration of the *backwater zone* (i.e., the terminal part of the river where flow is modulated by tides and other basinal processes). The freshwater mud units (Units 4 and 2), and likely at least part of the fluvial pebbly sand unit (Unit 5), are interpreted to comprise early TST strata at the base of each

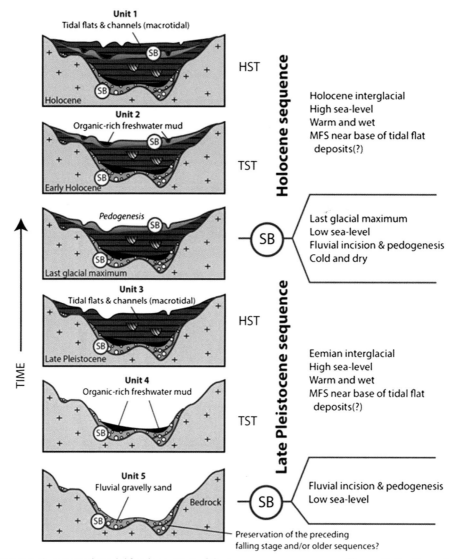

Figure 8.6 Conceptual model for deposition of the two sequences preserved beneath the Youngjong–Yongyou tidal flats, inner Gyeonggi Bay. SB, sequence boundary; MFS, maximum flooding surface; TST, transgressive systems tract; HST, highstand systems tract. The cartoons show shore-parallel cross-sections. *They are adapted from Park and Choi (2002), to which we have added sequence stratigraphic interpretations.*

sequence (Figure 8.6), deposited during a rise in sea level and water table as the shoreline retreated into Gyeonggi Bay. As base level continued to rise, the shoreline translated landward to the inner bay. Tidal-flat strata (Units 3 and 1) started to be deposited under macrotidal conditions, and tidal channels scoured locally into the substrate, in places cutting through the entire sequence and enhancing relief on the underlying sequence boundary.

Table 8.1 Radiocarbon dates from tidal-flat cores, west Korea

Location	Core	Radiocarbon age (years BP)	Calibrated ages (years BP)	Elevation relative to present-day mean sea level	Position relative to top of oxidized layer	References
Gyeonggi Bay (Jangbong Island)	GB-B04	536±25	573	0 m	Above	Kwon (2012)
Haenam Bay	G7	1362±48	1284	−1.5 m	Above	Lim and Park (2003)
Gyeonggi Bay (Incheon Airport)	T1	8130±200	9037	−2 m	Above	Choi (2005)
Gyeonggi Bay (Kimpo tidal flat)	KP2	5230±40	6018	−5.55 m	Above	Choi and Kim (2006b)
Gyeonggi Bay (Jangbong Island)	GB-B04	2443±25	2532	−7 m	Above	Kwon (2012)
Gyeonggi Bay (Kimpo tidal flat)	KP2	5850±80	6661	−7.2 m	Above	Choi and Kim (2006b)
Gyeonggi Bay (Kimpo tidal flat)	KP1	7310±170	8143	−7.95 m	Above	Choi and Kim (2006b)
Gyeonggi Bay (Kimpo tidal flat)	KP2	7890±40	8723	−13.02 m	Above	Choi and Kim (2006b)
Gyeonggi Bay (Incheon Harbor)	BH26	7220	8060	−14 m	Above	Lim et al. (2003)
Haenam Bay	BH1	8524±69	9513	−17 m	Above	Lim and Park (2003)
Haenam Bay	BH2	32,200±770	36,698	−10.2 m	Below	Lim and Park (2003)
Haenam Bay	BH2	21,940±800	26,358	−11 m	Below	Lim and Park (2003)
Haenam Bay	BH2	>44,500	N/A	−15 m	Below	Lim and Park (2003)
Gyeonggi Bay (Incheon Harbor)	BH26	36,780±450	41,749	−17 m	Below	Lim et al. (2003)
Gyeonggi Bay (Kimpo tidal flat)	KP2	41,100±1100	44,640	−17.65 m	Below	Choi and Kim (2006b)
Gyeonggi Bay (Jangbong Island)	GB-B04	44,700±1600	48,229	−19.5 m	Below	Kwon (2012)
Gyeonggi Bay (Kimpo tidal flat)	KP3	32,370±430	36,897	−20.3 m	Below	Choi and Kim (2006b)
Gyeonggi Bay (Jangbong Island)	GB-B04	>47,000	N/A	−23.5 m	Below	Kwon (2012)
Gyeonggi Bay (Kimpo tidal flat)	KP3	50,900±3500	N/A	−25.65 m	Below	Choi and Kim (2006b)

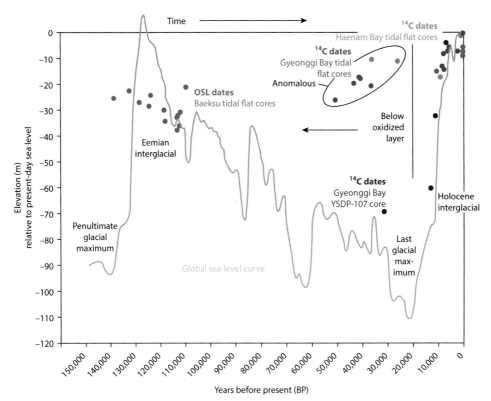

Figure 8.7 Radiocarbon dates and optically stimulated luminescence (OSL) dates from tidal-flat cores, west coast of Korea, plotted against the global sea-level curve derived from oxygen isotope data (Shackleton, 1987). Radiocarbon dates for YSDP-107 drill core from the Han River subaqueous delta (Jin, 2001) are also shown. Red circles = calibrated radiocarbon dates from Gyeonggi Bay tidal-flat deposits (Lim et al., 2003; Choi and Kim, 2006b; Kwon, 2012). Green circles = calibrated radiocarbon dates from Haenam Bay tidal-flat deposits (Lim and Park, 2003). Blue circles = OSL dates from Baeksu tidal-flat deposits (Chang et al., 2014; cf. Kim et al., 2014). Dark purple circles = radiocarbon dates from the YSDP-107 core, outer Gyeonggi Bay (Jin, 2001).

Although the base of the tidal-flat units in inner Gyeonggi Bay are likely TST deposits, the upper portions of these units are thought to be progradational and to form the HST at the top of each sequence. In the upper sequence, tidal-flat progradation is suspected to have started during the second half of the Holocene, after sea level reached its approximate current level (Chough et al., 2004) and accommodation became limited accordingly. Several lines of evidence suggest the tidal flats are progradational. For one, extensive tidal-flat reclamation, which started near the Han River mouth in the 1200s and has displaced the shoreline by up to 50 km basinward locally (Koh, 2001; Koh and Khim, 2014), would likely not have been possible if the system was still in a transgressive state. Continued reclamation is suspected to have exploited the natural tendency for progradation. Also, there is evidence that some of the funnel-shaped channels are translating basinward, as would be expected if the tidal flats were prograding. When tide-dominated channels prograde, the entire funnel tends to be displaced basinward, causing a decrease in cross-sectional area over time at any given location (e.g., Dalrymple and Zaitlin, 1994). Seismic and core data indicate that the Sukmo Channel, the main deltaic distributary, has been infilling since the early to mid Holocene (Choi, unpublished data), which seems to support this. The upward-coarsening trend documented at the Youngjong–Yongyou tidal flat (Figure 8.3), which is atypical of progradational tidal-flat successions (e.g., Van Straaten, 1961; Dalrymple, 2010b), may have been generated by migration of the adjacent tidal channel toward the tidal flat, leading to an increase in tidal and wave energy (and thus sand content) over time. Such a process has been observed to generate upward-coarsening successions in other tide-dominated systems (e.g., Bay of Fundy; Dalrymple et al., 1991).

Oxygen isotope data (e.g., Shackleton, 1987) suggest that multiple Quaternary high-stands have inundated the west coast of Korea to approximately the same level, raising the possibility that fragments of older sequences may be preserved locally in inner Gyeonggi Bay. Although parts of the fluvial unit that overlies bedrock (Unit 5) potentially represent early TST deposition in the Late Pleistocene sequence as discussed above, parts of this unit are oxidized and occur at relatively high elevations (e.g., Figure 8.2), implying that they predate the sea-level fall that preceded the deposition of the terrestrial deposits of Unit 4. If this is the case, then at least some parts of Unit 5 represent remnants of one or more sequences that predate the deposits associated with the Eemian transgression and highstand, as depicted in Figures 8.3 and 8.6.

8.2 STRATIGRAPHY OF OUTER GYEONGGI BAY

Two drill cores, YSDP-101 (60.5 m long) and NYSDP-101 (74.3 m long), penetrate the centermost large tidal bar in the Han River delta (Figure 8.8). They provide insight into the stratigraphic succession in outer Gyeonggi Bay, as do the seismic transects described in Chapter 6. Unlike the long cores from inner Gyeonggi Bay described above, neither YSDP-107 nor NYSDP-101 penetrates the entire sediment column. Rather, seismic data suggest an additional 15–30 m of sediment lies between their bases and bedrock (Figure 8.9). The two cores contain a similar succession of stratigraphic units, each of which bears evidence of tidal influence. These units are described below, from bottom to top.

A sandy heterolithic unit composed of orange sand with mud drapes is present at the base of the cores (Unit D in Figure 8.8). The unit is at least 18 m thick, and its sand-to-mud ratio is approximately 90:10. Core recovery is 70%. Drilling-mud injection and drilling-induced deformation are common in NYSDP-101. The sand is generally fine to medium grained and is well sorted. Gravel is absent, except for rare, isolated granules. Bioturbation is absent. Where stratification is visible, beds are centimeters to several

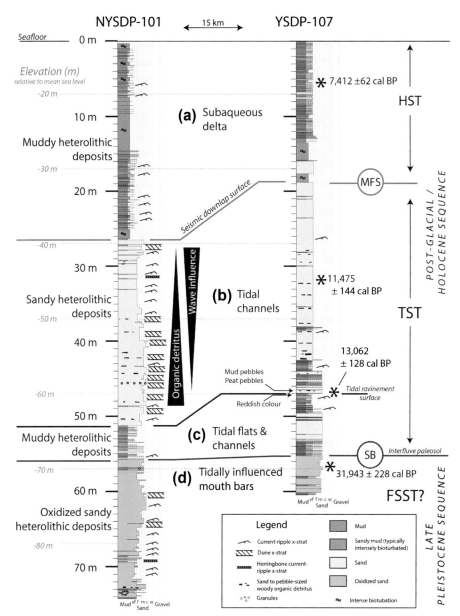

Figure 8.8 Description and interpretation of stratigraphic units in the YSDP-107 and NYSDP-101 drill cores from the large tidal bar, Han River subaqueous delta, outer Gyeonggi Bay. See Figure 8.1 for core locations. Units a–d are described in the text. Missing intervals (core recovery varied from 70% to 80%) are not shown; see Appendices 1A and 1B for details. The elevations relative to mean sea level should be considered accurate to within several meters because water depths at the core locations were estimated from bathymetric maps. MFS, maximum flooding surface; TST, transgressive systems tract; HST, highstand systems tract; SB, sequence boundary; FSST, falling-stage systems tract.

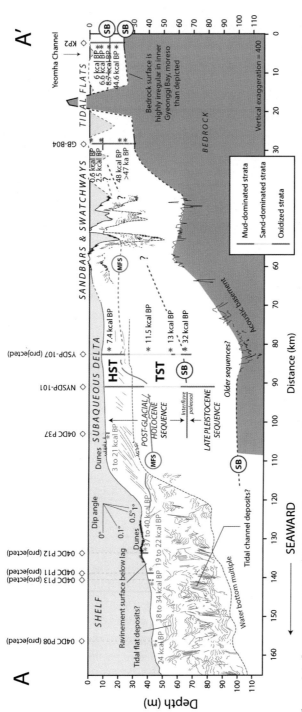

Figure 8.9 Sequence stratigraphic interpretation of the sedimentary succession above bedrock in Gyeonggi Bay. The seafloor and the subsurface reflections are traced from seismic lines J03 and B9. Dashed lines indicate interpolated surfaces between areas of seismic and/or core control. Bathymetry from the inner bay (dashed lines) is approximated from bathymetric charts. For location of seismic lines and cores used to construct this figure, see Figure 8.1. Radiocarbon dates from piston cores from the shelf and subaqueous delta, which are plotted in light gray, are not considered in this analysis (cf. Kwon, 2012) because they, unlike the radiocarbon dates from the YSDP-107 core, fall tens of meters above the global sea-level curve (curve shown in Figure 8.7) and do not decrease systematically upward in individual cores. The material that yielded these dates is suspected to have been contaminated by old organic carbon reworked landward during the postglacial transgression by flood-dominated tidal currents. MFS, maximum flooding surface; TST, transgressive systems tract; HST, highstand systems tract; SB, sequence boundary.

tens of centimeters thick and are commonly dune or current-ripple cross-stratified (Figure 8.10(a) and (b)). Dune cross-sets are thin, generally 5–10 cm thick. Rare herringbone current-ripple cross-stratification is observed. Mud occurs primarily within intervals of interlaminated mud and sand. These intervals are thin, generally only 1 or 2 cm thick (maximum 7 cm). The mud and sand laminae within them are sharp-based and a few millimeters thick on average, and they extend horizontally across the core, although some of the intervals dip up to 20°, suggesting they drape dune foresets. The mud–sand interlamination in these intervals is noncyclic, meaning that thickening and thinning trends suggestive of longer-term cyclic processes (e.g., neap–spring cycles) are absent. Sand-sized woody organic detritus is locally present in low to moderate amounts (Figure 8.10(c)). Rare pebble-sized woody organic fragments are also present. Woody organics from the YSDP-107 core yielded a radiocarbon date of 31,943 ± 228 cal BP. On the scale of the unit itself, obvious upward-fining or upward-coarsening trends are absent (Figure 8.8), except for a slight increase of mud in the upper several meters below the top contact in the NYSDP-101 core. Several stacked, sharp-based, meter-scale upward-fining units are present in YSPD-107. They pass from medium-sand up into horizontally interlaminated ("banded") fine sand and muddy fine sand. This type of succession, coupled with the evidence of strong tidal currents, suggests that this unit consists of channel-fill deposits. The thin intervals of interlaminated mud and sand are likely dune-trough mud deposits. Lack of bioturbation may reflect higher energy conditions and shifting substrates common to such settings (Dalrymple et al., 2012), as well as reduced salinity values. The orange color of the unit, which is in sharp contrast to the brown sand above (see below, and Figure 8.10), is suggestive of prolonged oxidation following deposition, and the upper contact is therefore suspected to be a paleosol. Radiocarbon dates suggest that the unit is Late Pleistocene in age and that the paleosol is a sequence boundary related to the last-glacial-maximum lowstand.

The exact timing of Unit D deposition should perhaps be taken as tentative pending additional dating, given the uncertainty surrounding radiocarbon dates from below the last-glacial-maximum paleosol in cores collected along the west coast of Korea (e.g., Lim and Park, 2003; Choi and Dalrymple, 2004; Chang et al., 2014). However, if the single radiocarbon date from the YSDP-107 core is taken as valid, which is possible because its elevation falls close to the global sea-level curve (Figure 8.7), the unit may represent Late Pleistocene falling-stage systems tract (FSST) deposit emplaced in a deltaic mouth-bar setting analogous to the sandbar and swatchway area of the modern delta (cf. the facies descriptions of this area in Chapter 5). Under this interpretation, the Late Pleistocene FSST deposits in the outer bay (Unit D) would link updip to the Late Pleistocene HST tidal-flat deposits in the inner bay (Unit 3), together forming a single regressive package.

A 5- to 8-m-thick muddy heterolithic unit (Unit C in Figure 8.8) overlies the orange-yellow heterolithic unit. Unit C extends from 60 to 68 m and 64–69 m below sea level in YSDP-107 and NYSDP-101, respectively. Core recovery is 80%. In YSDP-107,

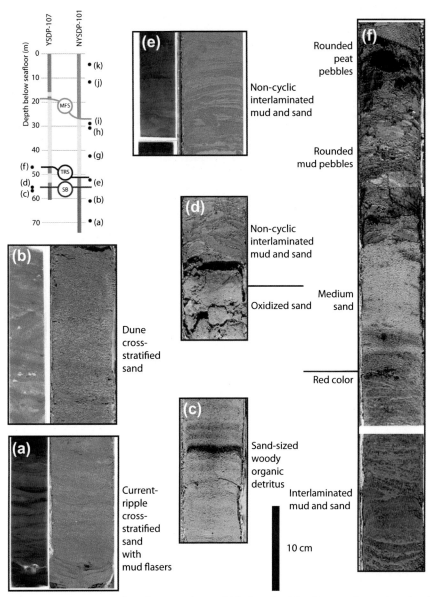

Figure 8.10 Representative color photographs and black-and-white X-ray radiographs of sediment from different stratigraphic units, YSDP-107 and NYSDP-101 drill cores, Han River subaqueous delta, outer Gyeonggi Bay. See Figure 8.1 for core locations. (a) Current-ripple cross-stratified very fine sand with mud flasers, 69.2 m depth, NYSDP-101 core. (b) Dune cross-stratified medium sand, 60.8 m depth, NYSDP-101 core. (c) Interlaminated ("banded") fine and medium sand, 56.4 m depth, YSDP-107 core. The black consists of a mix of sand-sized woody organic particles and siliciclastic grains. (d) Sharp contact between orange sandy heterolithic strata and overlying muddy heterolithic strata, 55.35 m depth, YSDP-107 core. Note reddish color immediately below contact. (e) Interlaminated mud and fine sand, 51.6 m depth, NYSDP-101 core. (f) Sharp contact between muddy heterolithic unit and overlying sandy heterolithic unit, 47 m depth, YSDP-107 core.

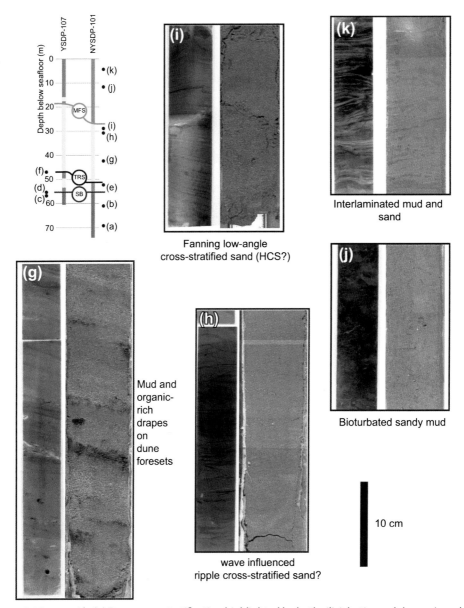

Figure 8.10—cont'd. (g) Dune cross-stratification highlighted by both siliciclastic mud drapes (gray layers) and drapes rich in sand-sized woody organic detritus (black layers), 42.1 m depth, NYSDP-101 core. Note sand-on-sand reactivation surface approximately 12 cm below top of photo. (h) Scoop-based ripple cross-sets, possibly wave ripples or wave-influenced current ripples, 30.8 m depth, NYSDP-101 core. (i) Fanning low-angle cross-stratification, possibly hummocky cross-stratification (HCS), 29.4 m depth, NYSDP-101 core. (j) Bioturbated sandy mud, 11.65 m depth, NYSDP-101 core. White bits are shell fragments (mostly bivalve). (k) Interlaminated mud and sand, 4.5 m depth, NYSDP-101 core. As shown in the inset figure showing the YSDP-107 and NYSDP-101 drill cores (see also Figure 8.8), Photos (a–c), and the lower half of Photo (d) are from Unit D; the upper half of Photos (d), (e), and the lower portion of Photo (f) are from Unit C; the upper portion of Photo (f) and Photos (g–i) are from Unit B; and Photos (j) and (k) are from Unit A. SB, sequence boundary; MFS, maximum flooding surface; TRS, tidal ravinement surface.

the lower contact is sharp, marked by an abrupt increase in mud content and an abrupt change in sand color from orange to brown (Figure 8.10(d)). The unit consists primarily of noncyclic interlaminated mud and sand (Figure 8.10(e)). Mud and sand laminae are sharp-based and are generally one to several millimeters thick. They extend horizontally across the core. Thin sand layers (5–10 cm thick) are also present. These are commonly current-ripple cross-stratified. Rare herringbone current-ripple cross-stratification is observed. Mud flasers (<1 mm thick) are common on ripple foresets. Bioturbation ranges from low to moderate (Table 5.2), with rare, thin (<10 cm thick), intensely bioturbated intervals. Sand-sized woody organic detritus is present in low concentrations. In YSDP-107, a sharp-based, ~3-m thick sandy interval is present in the middle of the unit (Figure 8.8). On a whole, the muddy heterolithic unit fines up slightly from bottom to top, the sand-to-mud ratio decreases from 70:30 to 50:50 and the average sand grain size decreases from medium to very fine grained. The upper 10 cm of the unit in YSDP-107 is slightly mottled in appearance and the upper 2 cm of this zone is reddish (Figure 8.10(f)). Seismic data at the elevation of this unit have a low signal-to-noise ratio in the vicinity of the YSDP-107 and NYSDP-101 cores, which precludes further insight into its stratigraphic architecture. However, below the shelf at this elevation, packages of flat-lying reflections that are believed to be associated with this unit and other units like it are intercalated with the lateral-accretion clinoform packages interpreted to be channel fills (Figure 8.9).

The muddy heterolithic unit is interpreted to be an estuarine tidal-flat deposit. The sharp-based sandy interval within it is likely a tidal-channel deposit. The tidal-flat unit above the paleosol in YSDP-107 and NYSDP-101 is interpreted to record the initial onlap of strata onto this unconformity during the postglacial transgression, and therefore represents the basal part of the postglacial TST (Figure 8.8).

A 24- to 28-m thick sandy heterolithic unit composed of brownish sand with mud drapes (Unit B in Figure 8.8) overlies the muddy heterolithic unit. Its sand-to-mud ratio ranges between 75:25 (YSDP-107) and 95:5 (NYSDP-101). The unit extends from 40 to 64 m and 32–60 m below sea level in NYSDP-101 and YSDP-107, respectively. Core recovery is 70%. The lower contact of the unit is sharp in YSDP-107 (Figure 8.10(f)); it was not recovered in NYSDP-101. The sand is medium to very fine grained. Siliciclastic gravel is absent, except for rare isolated granules. Where stratification is visible, sand beds are 5–40 cm thick (average 10 cm) and are dune or current-ripple cross-stratified. Dune cross-sets range in thickness from 5 to 40 cm (average 10 cm). Rare siliciclastic mud and organic-rich drapes are present on dune foresets (Figure 8.10(g)). Rare herringbone current-ripple and dune cross-stratification are observed. In places in NYSDP-101, some current ripples are scoop-based, suggesting they are wave influenced (Figure 8.10(h)), and rare intervals (<15 cm thick) of fanning, low-angle (<10° dip), undulating stratification interpreted to be hummocky cross-stratification are present (Figure 8.10(i)). Sand-sized woody organic detritus is commonly mixed into the siliciclastic sand, forming layers <5 cm thick that can either extend horizontally across the core or highlight dune foresets. Rare angular pebble-sized woody organic fragments are also present. Woody organic detritus from this unit has yielded radiocarbon dates of $13,062 \pm 128$ and $11,475 \pm 144$ cal BP. Mud is present primarily in 5–40 cm thick intervals of noncyclic interlaminated mud and sand. The mud and sand laminae within them are sharp-based, a few millimeters thick on average, and extend horizontally across the core. Muddy intervals are common in YSDP-107 and rare in NYSDP-101. As a whole, the unit fines upward from medium to very fine sand (Figure 8.8). The abundance of organic detritus decreases upward, and evidence of wave influence increases upward. Rounded pebble-sized mud balls and peat balls are present in the lowermost 3 m of the unit in YSDP-107 (Figure 8.10(f)), although the abundance of mud laminae does not obviously increase or decrease upward.

The sandy heterolithic unit (Unit B in Figure 8.8) is interpreted to have been deposited in laterally migrating estuarine tidal channels. Its base is interpreted to be a *tidal ravinement surface* formed as the tidal channels scoured across underlying inner-estuary tidal flats (Unit C in Figure 8.8) during transgression. The appearance of wave-influenced stratification near the top of the unit may actually be related to a gradual switch from wave- to tide-dominance associated with retreat of the shoreline into Gyeonggi Bay. Progressive flooding of the low-lying bay would have brought about an increase in the volume of water moving in and out of the coastal zone (i.e., an increase in *tidal prism*), which could have instigated barrier island destruction and caused the switch to the tide-dominated state that persists today. Numerical modeling results support this; they suggest that nearshore tidal bottom-stress likely increased significantly as the shoreline retreated into Gyeonggi Bay (Uehara and Saito, 2003).

The red color of the top of the tidal-flat unit (Unit C) suggests that the tidal-flat deposits were pedogenically altered before the tidal channels scoured into them. An alternative interpretation of this surface, therefore, is that it is in fact the last-glacial-maximum sequence boundary, an option left open in part because of lack of dates from the tidal-flat unit. However, the lack of orange, oxidized sand in the tidal-flat unit, and its abundance in the sandy heterolithic unit below (Figures 8.10), suggests that the tidal-flat–capping paleosol may be better interpreted as having formed briefly during the postglacial transgression. Similarly, poorly developed paleosols in the tidal-flat deposits in inner Gyeonggi Bay (e.g., Figure 8.5) are interpreted as having formed during transgression. In concert, the tidal-flat unit (Unit C in Figure 8.8) and the overlying tidal-channel unit (Unit B in Figure 8.8) are, therefore, interpreted to make up the TST at the base of the postglacial/Holocene sequence in the outer part of Gyeonggi Bay.

A 19–27 m thick muddy heterolithic unit (Unit A in Figure 8.8) is present at the top of the succession. The surficial portion of this unit has been described in previous chapters (e.g., Figure 7.5); we summarize its salient features here. It extends from 13–39 m and 13–32 m below mean sea level in NYSDP-101 and YSDP-107, respectively. Core recovery is 95%. Its lower contact is bioturbated in YSDP-107 and was not recovered in NYSDP-101. The unit consists of intercalated intervals of non-cyclic interlaminated mud and sand (each package of laminae 5–60 cm thick; average 10 cm) and intervals of bioturbated sandy mud (each 5–700 cm thick; average 10 cm) (Figures 8.10(j) and (k)). The latter are suspected to be an intensely bioturbated version of the former. The interlaminated mud and sand is variably bioturbated. Mud and sand laminae are one to several millimeters thick and extend horizontally across the core. The sand is brown and is very fine to fine grained. Thicker sand layers (1–10 cm) are common. Where stratification is visible, sand layers are current-ripple cross-stratified. Mud flasers commonly highlight ripple foresets. Bioturbated layers commonly contain isolated shell fragments (primarily bivalve). Woody organic detritus and wave-influenced sedimentary structures were not observed. A shell from YSDP-107 yielded a radiocarbon date of 7412 ± 62 cal BP.

This muddy heterolithic unit (Unit A) lacks obvious vertical trends in grain size, mud content, bioturbation, and other visible physical properties. Unlike the units below, the architecture of this unit is clearly visible in seismic transects because it lies largely above the first water-bottom multiple (Figures 6.1, 6.2, and 6.6). The unit consists of low-angle ($<1°$ dip) clinothems that record both lateral widening and basinward progradation of the large tidal bar (Figure 8.9), as discussed in previous chapters (e.g., Figure 7.5). Its lower contact is a prominent seismic downlap surface interpreted to be the maximum flooding surface in the middle of the postglacial/Holocene sequence, and the unit itself is interpreted to consist of Holocene-aged, HST sediment deposited by progradation of the Han River subaqueous delta.

8.3 SEQUENCE STRATIGRAPHIC EVOLUTION

The following sequence of events is envisioned to have generated the observed stratigraphic succession in Gyeonggi Bay.

8.3.1 Older Deposits

The bedrock surface both on land and in the shallow, inner part of Gyeonggi Bay shows evidence of weathering and pedogenic alteration that presumably reflects exposure during several Quaternary lowstands. As such, it represents a composite sequence boundary. The fluvial deposits that overlie this surface in many places are of uncertain age, but the presence of oxidation indicates that they too were exposed and weathered during one or more lowstands. These older fluvial deposits are thought to represent remnants of falling-stage and/or transgressive deposits that have escaped erosion during subsequent lowstands and transgressions. Such remnants are most likely to occur on fluvial and tidal interfluves, where channelization of flow is less likely to have occurred.

8.3.2 Late Pleistocene Sequence

During the transgression that preceded the Eemian interglacial highstand, a wave-dominated, barred shoreline is believed to have retreated to a position just outside the mouth of Gyeonggi Bay. Fluvial gravel and freshwater mud in the inner bay (Units 5 and 4 in Figures 8.2 and 8.3) accumulated either during and/or prior to this, when accommodation started to be generated in the lagoon and farther landward. As transgression proceeded and the shoreline moved into the bay, the tidal prism increased, causing the barrier to be destroyed, and the bay transitioned to a tide-dominated state. Macrotidal flats aggraded along the coast, supplied by muddy sediment from the Han River and from erosion of the shelf surface by flood-dominated tidal currents (Figures 2.34 and 2.37). Once sea level neared its maximum, accommodation became limited and the tidal flats started to prograde basinward, generating a prograding deltaic/tidal-flat HST deposit in the inner part of Gyeonggi Bay. Although global sea level during the Eemian was several meters higher than today (Figure 8.7), no older raised deposits have been identified along the west Korean coast, suggesting that minor subsidence may have occurred over the course of the Quaternary (Choi and Dalrymple, 2004).

With the start of the last glaciation, relative sea-level fell, causing a forced regression of the shoreline. Continued progradation of the highstand delta at increasingly lower elevations generated a falling-stage deltaic deposit in the outer part of Gyeonggi Bay (Unit D in Figure 8.8). Ultimately, the floor of Gyeonggi Bay became exposed, generating a widespread paleosol. Incised valleys associated with this period of lowered sea level likely formed; these have not yet been intersected by long cores. The limited amount of deeply penetrating, good-quality seismic data suggests that such valleys most likely lay close to the present-day positions of the large tidal channels.

8.3.3 Postglacial/Holocene Sequence

Accommodation was generated again in Gyeonggi Bay during the postglacial transgression leading up to the current (Holocene) interglacial. This triggered the accumulation of freshwater mud over the sequence boundary in interfluve locations (Unit 3 in Figures 8.2 and 8.3). Aggradation of fluvial deposits presumably also occurred within the incised valleys. Again, a wave-dominated, barred shoreline retreated to the mouth of the bay, and associated estuarine tidal-flat mud and tidal-channel sand was deposited behind the barrier (Units C and D in Figure 8.8; see also Figure 6.10). As the shoreline moved into the bay, the barrier was destroyed and the coast became tide-dominated. Tides reworked the substrate, forming long, narrow sand ridges (Figure 6.7). Progradational pulses during this period blanketed at least one of the sand ridges in the south end of Gyeonggi Bay with mud, forming a broader, mud-covered "pseudo-tidal sand ridge" (Jung et al., 1998)—essentially a parasequence—that became detached from the shoreline as transgression proceeded. Eventually, muddy macrotidal flats started to aggrade along the coast. Once sea level neared its maximum, the flats started to prograde outward (Unit 1 in Figures 8.2 and 8.3) and a subaqueous delta composed of muddy, broad (>10 km wide) large tidal bars prograded out from the Han River mouth in the western bay, blanketing several narrow tidal sand ridges beneath (Unit A in Figure 8.8; see also Figure 6.10). Together, the prograding tidal flats and subaqueous delta form the HST of the Holocene sequence. The floor of the eastern bay has yet to receive abundant muddy HST sediment. It remains sandy, largely covered by dunes and narrow (typically 1–2 km wide) tidal ridges (e.g., Kim and Lim, 2009; Kum and Shin, 2013), and is fundamentally estuarine in nature.

CHAPTER 9

Discussion and Concluding Remarks

Contents

Chapter Points

- **Dispersal of sediment** from the river to the subaqueous delta likely occurs in a series of steps. (1) Most sediment is delivered to Gyeonggi Bay when fluvial discharge is high, in the summer. This sediment accumulates initially in storage sites in the proximal delta, such as the sandbar–swatchway zone, which appears to be functioning as a mouth bar zone, in addition to the nearby muddy tidal flats and the inner parts of the large tidal channels. Most of the bedload (medium sand and coarser) remains trapped in the proximal delta. A fraction of the mud is likely dispersed more broadly in the surface plume. (2) Subsequently and especially during fair-weather periods, tides sweep much of the mud and finer sand from the deep parts of the large tidal channels, transferring it to the adjacent large tidal bars and intertidal flats, which develop in areas of smaller tidal prism and lower tidal current speeds. (3) Winter waves, in combination with tides, reverse this trend: they limit accommodation on top of the large tidal bars, causing sediment above a certain depth threshold to move to the flanks and distal tips of the large tidal bars, the main, long-term subtidal sink for fine-grained sediment in the system. The large tidal bars widen and grow basinward accordingly; progradation of the subaqueous delta involves both processes.

- The Han, like all tide-dominated and strongly tide-influenced deltas studied to date, exhibits the following **key traits**: a large subaqueous delta whose clinoform is the main subtidal locus of mud deposition in the system; gently inclined (<1°) heterolithic clinoform deposits that record deltaic progradation; a high degree of channelization in the proximal, shallow subaqueous delta topset, an area that functions as the mouth bar zone, where most fluvial bedload is trapped and where channels commonly migrate laterally, generating sharp-based, upward-fining, channel–bar successions.

- As in the Fly River delta, the mouth bar zone (i.e., sandbar–swatchway zone) of the Han River delta appears to be a zone of **bedload convergence**: sand moves basinward to this area through ebb-dominated swatchways that are connected to the main distributary, and sand also moves landward to this area through flood-dominated swatchways that are connected to a large tidal channel.

- As it continues to prograde, the Han River delta will likely generate a **two-part vertical facies succession** composed of (1) an upward coarsening to ungraded unit of interlaminated mud and

The Tide-Dominated Han River Delta, Korea
http://dx.doi.org/10.1016/B978-0-12-800768-6.00009-2

sand (20–25 m thick) generated by progradation of the tips and flanks of the large tidal bars that make up the subaqueous delta, and (2) a sharp-based, upward-fining top portion (typically 5–20 m thick) generated by lateral accretion of channels and bars. Similar successions will likely be produced by other tide-dominated and strongly tide-influenced deltas.

- The Han is different than other strongly tide-influenced deltas in that its subaqueous delta is highly three-dimensional—it consists of a series of **large tidal bars** separated by **large tidal channels.** These features likely result from along-coast variations in tidal prism that are a result of localized coastal embayments that have not been filled by deltaic progradation. The anomalously large size of the large tidal channels, and especially their extension through the entire prodeltaic region, might be due to the large tidal prism generated by the extreme tidal range. Because the subaqueous delta is complex, progradation generates an architecturally complex package of deltaic clinoform deposits. Also, the system likely receives proportionally more sediment from flood-dominated tidal currents than do other strongly tide-influenced deltas, again possibly because of the extreme macrotidal setting coupled with the moderate size of the Han River.

The extensive work performed on the inner part of the Han River delta by previous workers (e.g., Park and Choi, 2002; Choi et al., 2004a; Choi and Dalrymple, 2004; Choi, 2014; Choi and Jo, 2015), coupled with the work performed on the outer part of the delta in this study, makes the Han one of the most comprehensively studied tide-dominated deltas to date. The data set reveals that the Han is similar to previously studied tide-dominated deltas in several important ways.

- Most of the delta is subaqueous: a wide, mud-rich subaqueous delta has prograded into Gyeonggi Bay, burying a thick, sand-rich transgressive estuarine succession.
- Elongate tidal bars (sandbars) characterize the mouth bar zone (i.e., the sandbar–swatchway zone), the terminal sink for most of the fluvial bedload and an area where bedload convergence is occurring.
- Open-coast tidal flats are present in tidal interfluve areas and along the coastline away from the delta proper.
- Inclined heterolithic stratification (IHS) is forming by lateral accretion of point bars in the tidal–fluvial transition.
- Fluid mud appears to be present locally in channel bases beneath the turbidity maximum.
- Interlaminated mud and sand is accumulating in several different subenvironments, in both the intertidal and subtidal zones.
- Cyclic tidal rhythmites are rare, and are only accumulating on point bars (sheltered zones of rapid sedimentation), primarily in the intertidal zone.
    Despite these similarities, the depositional setting is, nonetheless, unique (Figure 2.1), in particular because of the large ratio of tidal to fluvial energy in Gyeonggi Bay (Figures 1.1 and 1.3). This imparts the Han River delta with several distinguishing traits.
- Instead of being a morphologically simple wedge of sediment, the Han's subaqueous delta is highly three-dimensional: it consists of a series of flat-topped large tidal bars separated by large tidal channels that dissect the prodelta region.

- The Han's subaqueous delta is smaller than those of previously studied tide-dominated and strongly tide-influenced deltas (see below). It is similar in height (~25 m) and off-shore extent (~100 km) (Figure 1.1), but appears to be it is less extensive in an along-shore direction: it extends approximately 100 km parallel to shore, whereas those of previously studied modern tide-dominated and strongly tide-influenced deltas tend to be hundreds of kilometers to over 1000 km in extent (Nittrouer et al., 1986; Wright and Nittrouer, 1995; Walsh et al., 2004). (This is only the "known" extent. There might be mud to the north along the North Korean coast.) The aforementioned presence of large tidal channels further reduces the volume of sediment in the Han's subaqueous delta.
- The mouth bar zone in the Han River delta, which is synonymous with the sandbar–swatchway zone (see below), occurs in a region of shoreline-oblique flow, not shoreline-perpendicular flow (cf. Wright, 1977; Dalrymple et al., 2003).
- Sand in the Han River delta can at times be reworked landward up the deltaic clino-form by flood-dominated tidal currents.
- Extensive prodelta bottomsets are absent.

The substantial data set from the Han River delta allows us to explore a series of fundamental questions relating to the morphology, sediment dispersal, facies, and strati-graphic architecture of the system, and of tide-dominated deltas in general, as follows.

9.1 DYNAMICS OF SEDIMENT DISPERSAL

Sandy *bedload* in tide-dominated estuaries and deltas tends to disperse differently than muddy *suspended load* (Van Straaten and Kuenen, 1958; Postma, 1967; Wang, 2012). Bedload transport rates are highly sensitive to flow velocity—they scale to the third power of current speed (Meyer-Peter and Mueller, 1948; Bagnold, 1966). Because of this, sand in tide-dominated estuaries and deltas tends to move over time in the direction of the fastest current (the "dominant" current). In the proximal portions of the system, this direction is generally seaward due to the combined influence of river outflow and ebb-dominated tidal currents, supplemented by hypsometric effects (Friedrichs and Aubrey, 1988; Dalrymple et al., 2012), whereas in more distal portions of the system, the direction of sand transport is commonly landward because of reduced river influence coupled with landward-directed shear from the short, fast flood tide (Figure 2.21). A zone of *bedload convergence* therefore occurs between these two *bedload-transport paths*, which generally coincides with the tidal–fluvial transition in tide-dominated estuaries (Dalrymple et al., 2012) and (possibly) the mouth bar zone of tide-dominated deltas (Dalrymple et al., 2003; Dalrymple, 2010b). As argued below, the morphology of the sandbar–swatchway zone in the Han River delta suggests that this area is functioning as the mouth bar zone and that bedload convergence is occurring there.

Unlike sand, mud generally travels as suspended load in tide-dominated estuaries and deltas because of its lower *settling velocity*. It can therefore be advected back and forth with the tides and, over time, travels slowly in the direction of any residual current that may exist

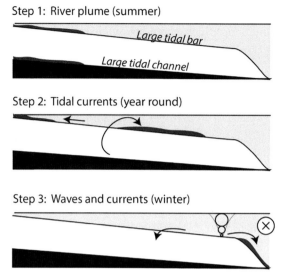

Figure 9.1 Depositional model for the Han River delta. Step 1: During periods of high fluvial discharge in summer, sediment from the river is deposited in the inner portion of the subaqueous delta, including the mouth bar zone (sandbar–swatchway zone), the inner parts of the large tidal channels, and tidal flats adjacent to the river mouth. Most bedload is trapped here, primarily in the sandbar–swatchway zone. Step 2: Sediment is remobilized year-round by tidal currents. This moves the finer fractions (mud, fine sand) from the large tidal channels (high tidal flux) to the large tidal bars (low tidal flux). Flood-dominated tidal currents also deliver mud to tidal flats, and bedload to the bedload convergence zone in the mouth bar (sandbar–swatchway) zone. Step 3: Wave–current base deepens in the winter, limiting accommodation on top of the large tidal bars. Any sediment that had accumulated here is reworked to the flanks (not shown because of 2-D nature of cartoon) and distal tips of large tidal bars, where it is deposited because bed shear stress from combined wave–current action is lower. The South Korean Coastal Current also strengthens during winter, transporting sediment in the outer portions of the large tidal bars southward, eroding sediment from their upflow flanks and depositing on their downflow flanks. The X indicates the coastal current moving into the page (i.e., southward along the coast outside of Gyeonggi Bay).

(Wang, 2012), either as a result of any time–velocity asymmetry of the tidal currents or of any superimposed current caused by the Coriolis effect or oceanic circulation. Flocculated mud can settle out relatively rapidly below the turbidity maximum and form mobile fluid mud pools in high-energy channel base settings (Box 5.1). Fluid mud can also travel downslope as wave–current-supported sediment gravity flows to feed offshore or channel margin depocenters (e.g., Kineke et al., 1996). The net result is that mud and sand can travel in different directions in the same system. Below, a conceptual model is presented that describes how mud is envisioned to disperse from the Han River to the delta (Figure 9.1).

Step 1: Deposition of mud from the river plume (summer)
Mud from the Han River likely enters Gyeonggi Bay primarily during summer floods. Its delivery to the basin is likely quite efficient because floodplains are limited in extent, the

catchment is relatively steep and precipitation events are intense (see Chapter 2). Most of the mud is suspected to accumulate initially in storage sites on the inner delta given the moderate size of the river plumes (Figure 5.4) coupled with the various processes that tend to limit the offshore dispersal of mud in such plumes (e.g., Coriolis effect, flocculation, estuarine circulation, see Geyer et al., 2004). Storage sites on the inner delta likely include tidal flats near the river mouth; the proximal parts of the large tidal channels, which are infilling on both sides (Figure 6.5); pools of mobile fluid mud, which are suspected to exist in channel base locations beneath the turbidity maximum (e.g., Piston core 19 in Figure 5.3; see also Choi and Jo, 2015); and the distal portion of the sandbar–swatchway zone.

Step 2: Remobilization of mud by tidal currents (year-round)
Following initial deposition from the river plume, much of the mud is suspected to be remobilized by tidal currents. A portion of the remobilized mud is likely transported land-ward, where it is deposited on tidal flats. Several processes likely contribute to this, includ-ing flood dominance of tidal currents and the associated *tidal pumping* (Yu et al., 2014), and settling and scour lags (Van Straaten and Kuenen, 1958; Postma, 1967). A substantial por-tion of the remobilized mud is also likely moved from the large tidal channels, where tidal currents are faster, to the large tidal bars, where tidal currents are slower.

Step 3: Remobilization by waves and coastal current (winter)
Although the large tidal bars function as the main subtidal sink for fine-grained sediment in the delta, something prevents them from aggrading to sea level; indeed, the seismic data (Figures 6.1 and 6.2) suggest that they have reached a dynamic equilibrium and have not accreted significantly for sometime. Their flanks and distal tips are, however, actively out-building, but little if any sediment is accumulating on their top surfaces (Figures 6.5 and 7.5)—a feature common to tidal bars in shallow water that are growing in size (Harris, 1988). Winter storm waves acting in conjunction with tidal currents are believed to be responsible. Based on the equations in Box 2.1, the depth to which winter waves can stir fine sand is estimated to be approximately equal to the depth of the bar tops. Tides would only accentuate this because their interaction with the wave boundary layer is nonlinear and generates bed shear stresses greater than the sum of the two taken individually (Grant and Madsen, 1979, 1986). The drop in winter water temperature (and associated increase in water viscosity and density) may also be significant. For example, in lab experiments, a temperature drop of 30 °C, which is similar to the magnitude of the seasonal drop in water temperature in inner Gyeonggi Bay (~25 °C in summers, ~3 °C in winters; see Yang et al., 2008), can increase the transport rate of fine sand by up to an order of magnitude (Hong et al., 1984). Suspended by waves and advected back and forth by the tides, any fine sedi-ment that had accumulated on top of the bars in summer is therefore envisioned to slowly zigzag downslope under the force of gravity in winter and accumulate on the periphery (flanks and distal tip) of the large tidal bars. The downslope movement of wave-assisted,

muddy density currents is also possible (cf. Traykovski et al., 2000). Such rapid delivery of mud to the bar periphery may facilitate the burial of large sandy dunes by these muddy deposits (Figure 6.4), a trait observed in several other deltaic systems (cf. Hübscher et al., 2002; Shinn et al., 2004). Delivery of sediment to the tips and flanks of the large tidal bars causes the bars to widen and extend basinward over time. Both processes are involved in progradation of the subaqueous delta.

The southward-flowing South Korean Coastal Current (Figure 2.28) is also suspected to contribute to sediment transport. In particular, it likely causes along-coast advection of sediment in the outer portion of the delta, especially during winter when strong winds remobilize seafloor sediment and cause the South Korean Coastal Current to intensify up to speeds of 10 cm/s at the sea surface (Wells, 1988). Although weak, the current offers the most reasonable explanation for the asymmetric dune-like profile of the outer tips of the large tidal bars (Figures 4.1 and 6.2(b)). Aided by the combined bed shear of winter waves and tides, which would be needed to put sediment into motion, the current is envisioned to slowly transport mud and fine sand up the gently dipping (0.1°) upflow (northern) flanks of the large tidal bars and deposit it on the more steeply dipping (1°) downflow (southern) flanks, where the combined bed shear stress from waves and currents decreases. (It is of note that the piston cores from this flank of the large tidal bar were the most water saturated of the cores collected and, as such, experienced the highest levels of coring-induced deformation.) Unlike the outer parts of the large tidal bars, the inner parts of the large tidal bars are symmetric and outbuilding on both sides, likely because they are not affected by the South Korean Coastal Current and the large waves incoming from the fetch-unlimited Yellow Sea (Figure 6.2(a)).

9.2 MORPHOLOGY

The Han River delta is similar to previously studied tide-dominated deltas in that it has developed a large subaqueous delta, within which most of its sediment is stored (cf. Nittrouer et al., 1996; Hori et al., 2001; Walsh et al., 2004; Kuehl et al., 2005; Walsh and Nittrouer, 2009). Subaqueous deltas, such as those in Figure 1.1, appear to form in response to "energetic" conditions in the receiving basin: the combined action of waves and tidal currents limits deposition of finer grained sediment above a certain depth threshold (i.e., the wave–current base), which causes the clinoform rollover to become submerged and displaced offshore (Pirmez et al., 1998; Friedrichs and Wright, 2004; Swenson et al., 2005; Walsh and Nittrouer, 2009). The Han's subaqueous delta is unique, however, because it is highly three-dimensional—it consists of a series of large tidal bars separated by large tidal channels that continue through the prodelta region and merge with the shelf surface. By contrast, the subaqueous deltas of the Fly, Amazon, Changjiang, and Ganges–Brahmaputra rivers are morphologically simple wedges of sediment: like two-dimensional dunes, their shape can be captured relatively accurately by drawing an arbitrary two-dimensional profile across the delta in an onshore–offshore direction (e.g., Figure 1.1); in such situations, the shore-normal

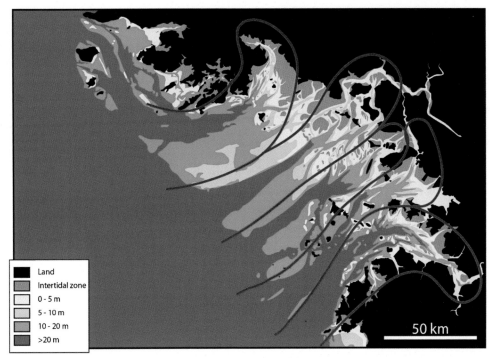

Figure 9.2 Tidal drainage basins (outlined in red)—areas of elevated tidal prism associated with invaginations of the rocky coastline. Note that each tidal drainage basin links to a large tidal channel. The red lines approximate the location of "tidal interfluves," areas of reduced tidal prism where the large tidal bars have developed.

channels do not continue through the entire subaqueous delta, but rather peter out on its top surface (topset). Why is the Han's subaqueous delta so highly channelized in comparison? The answer likely relates to the large tidal prism and its significant variation in an along-coast direction. The Han has the largest tidal range, by far, of any tide-dominated delta studied to date (Figure 1.3). This causes a large volume of water to move across the low-gradient subaqueous delta during each tide. This tidal flux is not uniform; rather, it is greater through the large tidal channels because they link landward to either a bedrock-controlled coastal invagination or a bedrock-controlled distributary channel. These define "tidal drainage basins"—areas where tidal prism and tidal current speeds are elevated (Figure 9.2). These drainage basins are believed to generate the large tidal channels and control their spacing. The large tidal bars occupy "tidal interfluves," the shadow zones between the tidal drainage basins where tidal prism and tidal current speeds are reduced.

Another morphological difference between the Han and previously studied tide-dominated deltas relates to its mouth bar region. As a general rule, "classic" mouth bars—progradational lobe-shaped sediment bodies formed by decelerating flows at distributary mouths (Bates, 1953; Hoyal et al., 2003; Bhattacharya, 2010)—do not form in tide-dominated deltas (cf. Wright, 1977). Such features would impede the passage of tides in and out of distributary channels and are consequently reworked by tidal currents, generating elongate tidal bars (e.g., Dalrymple et al., 2003). In the Fly, Ganges–Brahmaputra and Changjiang deltas, the mouth bar region is characterized by a divergent array of these elongate tidal bars that forms downflow of the first fluvial channel bifurcation (i.e., downflow of the delta apex). The Han is somewhat different. In the Han, the river bifurcates around bedrock islands, forming bedrock-controlled distributaries. Each distributary then passes beyond the zone of bedrock control and into Gyeonggi Bay where it merges with a large tidal channel, each defining the axis of a tidal drainage basin (Figure 9.2). Elongate mouth bars are lacking at these locations. However, the main distributary veers up into ebb-dominated swatchways and feeds into the sandbar–swatchway zone (Figure 4.14). This zone, which is in the main location of sand accumulation, is suspected to function as the main mouth bar zone in the delta. Additional fluvial bedload may be accumulating in the proximal parts of the large tidal channels.

In addition to functioning as a mouth bar zone, the sandbar–swatchway area in the Han River delta appears to be functioning as a zone of *bedload convergence*: ebb-dominated swatchways connected to the ebb-dominated main distributary feed into it from onshore, and flood-dominated swatchways feed into it from offshore (Figure 4.14). The swatchways facilitate across-bar flow, which is likely integral to the maintenance of the large tidal bar (Huthnance, 1982a,b; Dalrymple and Rhodes, 1995). The mouth bar zone in the Han River delta is therefore similar to that of the Fly River delta in that both act as a zone of bedload convergence (Dalrymple et al., 2003). This might be a common theme for the mouth bar region of tide-dominated deltas, just as it is for the tidal–fluvial transition of tide-dominated estuaries (Dalrymple and Choi, 2007; Dalrymple, 2010b; Dalrymple et al., 2012). However, because the sandbars (mouth bars) in the Han form in a zone of across-bar flow, an apparent consequence of the extreme tidal prism, they are oriented oblique to the main channels, whereas mouth bars in the Fly are oriented parallel to the main channels.

Up to this point, we have placed emphasis on the Han River's large subaqueous delta. Such features, which are also known as subaqueous clinoforms or subaqueous delta platforms, are common to all large tide-dominated and strongly tide-influenced deltas (e.g., Figure 1.1; Walsh and Nittrouer, 2009). However, a smaller, shoreline-attached clinoform also tends to be present in these systems (Nittrouer et al., 1996; Allison, 1998). It is commonly referred to as the subaerial clinoform (even though only its topset is subaerial) in order to differentiate it from the much larger subaqueous clinoform. Together, the two form what is known as a compound clinoform (Helland Hansen and Hampson, 2009). Subaerial clinoforms in tide-dominated deltas tend to be synonymous with open-coast tidal flats or muddy shorefaces developed adjacent to the fluvial axis of the system. A large supply of mud from the river, currents that advect this mud downcoast, and reduced tidal prism in off-axis areas appear to promote their development. Waves generally take on a greater role off-axis relative to tides (e.g., Dalrymple et al., 2012), and the associated clinoforms can be significantly wave-influenced or even wave-dominated as a consequence, although tides can still influence sedimentation, sometimes significantly (Fan, 2012). The Amazon, Changjiang, and Han River deltas illustrate a spectrum of subaerial clinoforms that can develop in these systems. The subaerial clinoform downcoast of the Amazon's river mouth is 5 m high, dips <<1° basinward, and actively progrades basinward overtop of the unchannelized, sediment-starved topset of the subaqueous delta (Allison et al., 1995; Nittrouer et al., 1996). Despite being composed almost entirely of mud, it is wave-dominated, both in terms of its geomorphology (it lacks tidal channels) and sedimentology (it contains sedimentary structures generated primarily by waves and wave-generated currents), and is therefore commonly referred to as a muddy shoreface (e.g., Rine and Ginsburg, 1985). The subaerial clinoform downcoast of the Changjiang River mouth is also approximately 5 m high and dips basinward at <<0.1°, but it is heterolithic and is influenced by both tides and waves. Like a sandy shoreface, it lacks well-developed tidal channels, contains sandy tempestites, and faces a subaqueous delta topset that is largely unchannelized by tidal currents. However, like a "classic" (back-barrier) tidal flat, its surficial sediment fines landward from sand to mud, and it contains intervals of tide-generated interlaminated mud and sand. It is commonly referred to as an open-coast tidal flat (Fan et al., 2004), and is similar to, but muddier than, open-coast tidal flats along open stretches of the Korean coast (e.g., Yang et al., 2005). The mouth of the Han is also bordered by open-coast tidal flats (cf. Choi, 2014) that are 5–10 m high, dip basinward at ~0.1°, and fine landward from sand to mud. These open-coast tidal flats are also influenced by both waves and tides, but because of the macrotidal, fetch-limited setting in Gyeonggi Bay, they are much more tidally influenced than previously described Chinese (Fan et al., 2004) or Korean (Yang et al., 2005) open-coast tidal flats. In particular, they are highly channelized: tidal channels are ubiquitous, both intertidally and subtidally (Figure 4.5). Fewer wave-generated strata (e.g., tempestite beds) are present compared to true open-coast tidal flats discussed above, and current-ripple and dune cross-stratified beds are common, as are intervals of tidally generated interlaminated mud and sand. As with all open-coast tidal flats studied to date (e.g., Fan, 2012), cyclic tidal rhythmites (e.g., with neap–spring thickening and thinning trends) are rare, but they do form locally on tidal creek point bars.

9.3 MORPHOLOGICAL AND STRATIGRAPHIC EVOLUTION

Because the large tidal bars that make up the Han's subaqueous delta clinoform are highly three-dimensional, and because their growth is a highly three-dimensional process, continued progradation of the subaqueous delta should generate a thick (20–25 m, precompaction), highly complex package of low-angle (0.1°–1°) mud-rich heterolithic clinoform deposits (Figure 9.3) formed by accumulation of mud-rich sediment on the margins of the large tidal channels and on the tips of the large tidal bars. The architecture of these deposits will likely be more complex than those of other tide-dominated deltas, which have comparatively simple 2-D subaqueous delta clinoforms and comparatively simple, basinward-accreting patterns. As progradation continues, the open-coast tidal flats and mouth bars of the subaerial clinoform will become superimposed on the subaqueous clinoform deposits (Figure 9.3), generating a 5–10 m thick upward-fining succession that will be predominantly sandy in its lower part. This superposition is likely to be an erosive one, given the extensive channels both on- and off-axis (Figure 9.3). Because these channels are highly dynamic and will exhibit some lateral migration during the progradation (cf. Song et al., 2015), the subaerial "clinoform" in this system may therefore consist largely of lateral-accretion deposits as opposed to basinward-accretion deposits, a potentially common theme in the mouth bar zone of tide-dominated deltas (e.g., Dalrymple et al., 2003; Dalrymple, 2010b). Partial to complete cannibalization of the on-axis deposits by the fluvial system as progradation continues is a distinct possibility.

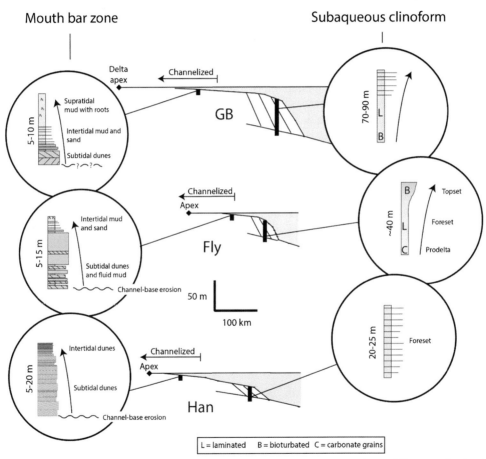

Figure 9.3 Comparison of stratigraphic successions produced on-axis in three different tide-dominated deltas, the Ganges–Brahmaputra (GB; Kuehl et al., 2005), the Fly (Dalrymple et al., 2003; Walsh et al., 2004), and the Han (this study). Fluvial and tidal–fluvial deposits (e.g., Choi et al., 2004a) that would be produced farther landward are not shown. Continued regression would erosively superimpose the mouth bar and/or fluvial deposits on top of the subaqueous clinoform deposits, generating a two-part stratigraphic succession separated by an erosion surface. Off the fluvial axis of these systems, open-coast tidal flat or muddy shoreface deposits (e.g., Fan, 2012) would commonly replace mouth bar and fluvial deposits at the top of the succession (see text for details).

9.4 PROXIMAL–DISTAL TRENDS

A major goal of the study was to identify proximal–distal trends in the facies of the Han River delta: if one cannot differentiate between facies deposited in proximal versus distal settings, one cannot interpret whether a given succession was deposited during transgression or regression, which renders sequence stratigraphic analysis impossible (Dalrymple and Choi, 2007). Based on the data collected, the best proximal–distal indicator in the Han River delta appears to be the nature of the interlaminated mud and sand layers. Interlaminated mud and sand is actively accumulating in many parts of the depositional system: on tidal–fluvial point bars in distributary channels in the turbidity maximum zone, at the base of the proximal parts of large tidal channels, on open-coast tidal flats, and on the flanks of the large tidal bars (Figure 9.4; Table 9.1). Because such heterolithic deposits are abundant, forming as they do in a wide range of settings throughout the entire delta, the ability to use them to determine proximal–distal position is especially powerful. In a proximal–distal transect, the following trends are noted.

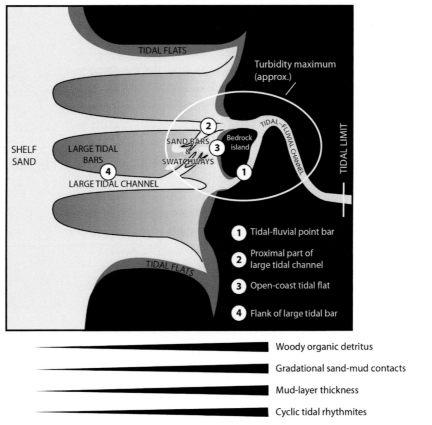

Figure 9.4 Proximal–distal trends in interlaminated mud–sand units in the Han River delta. Numbers indicate zones where such units are actively accumulating.

Table 9.1 Comparison of laminated heterolithic sediments, Han River delta

Depositional environment	Proximity to turbidity maximum	Gradational contacts	Cyclic tidal rhythmites	Mud layer thicknesses	Macerated organic debris
Tidal–fluvial point bar	Proximal ↑	Relatively common	Relatively common	1–10 mm	Common
Base of proximal parts of large tidal channels		Not observed	Not observed	1–10 mm	Rare
Open-coast tidal flats		Rare	Rare	1–10 mm	Rare
Flanks of large tidal bar	Distal ↓	Not observed	Not observed	1–5 mm	Rare

1. The frequency of **gradational contacts** between individual sand laminae and the mud laminae that overlie them decreases basinward from the turbidity maximum: well-developed normal grading was observed in the tidal–fluvial point bar and was best developed in the intertidal to shallow subtidal zone. Rare gradational contacts are also observed in open-coast tidal flats near the river mouth. By contrast, throughout the subaqueous delta, the contact between the sand and overlying mud is abrupt, with no transition between the two layers. The exact reason for this proximal-to-distal change is unclear, but several factors may contribute. First, the lower suspended-sediment concentrations that occur in more distal areas may mean that deposition from suspension is not as continuous as current speeds decrease toward slack water. Second, some areas such as the bottoms of the large tidal channels lack significant amounts of very fine sand, such that not all grain sizes are present and deposition is not continuous. Third, the greater influence of wave action in more exposed, distal areas may inhibit the development of a gradational contact between the sand and mud laminae.

2. The average **thickness of (nonamalgamated) mud layers** (i.e., of mud layers deposited during a single slack water period) decreases basinward from the turbidity maximum due to the basinward decrease in the amount of mud present in the water column, although this trend is complicated by the simultaneous tendency for mud layers to become thinner with increasing elevation above the base of the channels (cf. Dalrymple and Choi, 2007). For example, mud layers commonly reach 1 cm in thickness in the deposits of the tidal–fluvial point bar and at the base of the proximal part of the large tidal channel (e.g., Facies 4 in Figure 5.3), but are rarely thicker than several millimeters farther away from the turbidity maximum, such as on the fringing intertidal flats at Ganghwa Island and on the flanks of the large tidal bar (e.g., Facies 6 in Figure 5.3).

3. **Cyclic tidal rhythmites** were only observed to be accumulating proximally, beneath or near the turbidity maximum, on point bars in the tidal–fluvial transition and on the small

point bars of intertidal creeks that cut the open-coast tidal flats (e.g., Facies 5 in Figure 5.3). Three factors probably contribute to this: (1) the presence of abundant suspended sediment within the turbidity maximum, which allows for rapid sediment accumulation; (2) the sheltered nature of these settings and the resulting low levels of wave activity relative to the more open areas farther seaward on the large tidal bar, which allow for more regular accumulation of mud during slack water periods; and (3) the more strongly channelized nature of the proximal areas, which are more conducive to the development of true slack water periods than areas further seaward where some element of rotary tides exists (Figure 2.19) such that current speeds may reach zero less commonly.

4. The amount of **macerated organic debris** in tidal rhythmites decreases basinward away from the terrestrial source. Macerated organic debris is relatively common in the deposits of the tidal–fluvial point bar, but is rare further seaward.

If possible, these criteria should be used in combination to evaluate proximal–distal trends, because the use of a single criterion may lead to erroneous interpretations.

In addition to the nature of the interlaminated mud and sand units, some of the associated sandy facies may also help to determine one's position within the system. Perhaps the most obvious of these is the prevalence of dune-scale cross-bedding. Although cross-bedding is present in many different settings, including the base of the tidal–fluvial and large tidal channels along their entire lengths, in the swatchways and associated sandbars, on the distal part of the large tidal bar, and on the shelf, cross-bedded sand is most abundant in the more proximal areas and, particularly, in the part of the large tidal bar dominated by the swatchways and sandbars, where dunes abound (Figure 4.10). Here, the cross-bedding is oriented at a high angle to the onshore–offshore direction (Figure 4.7) because the flow passes across the large tidal bar through the swatchway channels.

The facies associated with cross-bedding also give important clues about the position in the system. As noted above, the presence of very thick mud drapes (i.e., individual mud layers many millimeters to several centimeters thick) in the bottomsets of the cross-beds is an excellent indication that deposition occurred in a proximal location, beneath the turbidity maximum and at a topographically low position (i.e., in the base of the large tidal channel that connects directly with the Han River). (The cross-beds that form in the relatively elevated swatchway–sandbar area appear to be free of mud drapes, or, if present, they will likely be much thinner because of the higher elevation and smaller water depths.) On the other hand, if the intervening mud beds are bioturbated, then they most likely accumulated in a more distal setting, such as in the seaward part of the large tidal channels. The dunes that are part of the transgressive sand sheet on the shelf are interspersed with bioturbated sand and perhaps also with finer sand containing *hummocky cross-stratification (HCS)* because of the exposure to large storm waves. In more distal areas where tidal currents are likely to be weaker and the dunes active more intermittently, the cross-bedding itself may be bioturbated, although this cannot be verified with the data available in this study.

Conceptual models propose that wave energy decreases landward in tide-dominated estuaries and deltas due to frictional dissipation in shallow water (e.g., Dalrymple and Choi, 2007). The presence of wave-formed sedimentary structures is therefore a potential proximal–distal criterion. Evidence of wave influence (e.g., small wave ripples, HCS) is lacking in the large tidal bar deposits formed early on during progradation (Figure 7.5), at a time when the tips of the bars were still well within the fetch-limited confines of Gyeonggi Bay. As the bar tips reached their current position and became exposed to large waves incoming from the Yellow Sea, it is possible that wave ripple cross-stratification was generated, but, again, this cannot be assessed due to the coring-induced liquefaction of the piston cores from the modern clinoform (Figure 5.2).

9.5 APPLICATIONS TO THE PETROLEUM INDUSTRY

The data set from Gyeonggi Bay provides us with a snapshot of a tide-dominated delta in early highstand. If the deposits were buried, consolidated, and infiltrated from below by hydrocarbons, the large tidal bars (sandbars and swatchways excluded) and the fringing muddy tidal flats would likely function as large-scale barriers to hydrocarbon movement. If progradation continued, the muddy tidal flats in particular could form a regional topseal. The main reservoirs in the strata beneath would likely be the sandbars and swatchways (Figures 4.14 and 7.4) and the lower parts of the tidal–fluvial point bars (Figure 7.2), as well as the ridge-shaped trangressive (estuarine) sandy deposits below the delta (Figure 6.7). Of these, the sandbar–swatchway reservoirs would likely be the easiest to produce because they contain the fewest baffles (fewer mud drapes on dune forests, and no intervening muddy tidal flat deposits) (Figure 7.4). The tidal–fluvial point bars and the ridge-shaped transgressive deposits would likely be more difficult to produce because they are hetero-lithic, with continuous mud layers on inclined surfaces (e.g., IHS) (Figure 7.2), as well as discontinuous mud drapes on the forests and in the troughs of ripples and dunes (Figure 8.10(g)). Sandy transgressive lag on the shelf would represent an additional, relatively easily produced reservoir, as would any sand that may have accumulated in the large tidal chan-nels, the thickness of which is difficult to ascertain given the data set. The hypothetical properties of these sandbodies as hydrocarbon reservoirs are detailed in Table 9.2.

Assuming the hydrocarbon source rocks are located in a more basinward position, the transgressive lag and the sand that lines the large tidal channels, along with incised valley fills in the strata below, would be potential pathways for updip hydrocarbon migration. Additional minor hydrocarbon source rocks might be present within the delta itself, including tidal flat and prodelta mudstones.

9.6 CONCLUDING REMARKS

Several decades ago, few tide-dominated deltas had been studied and, as a consequence, it was questioned whether they even existed (Walker, 1992). Today, it is understood that they are relatively common in the modern ocean. The key traits of several modern examples

Table 9.2 Han River delta sandbodies as reservoirs

| Sandbody type | Dimensions of individual sandbody* | | | Number of sandbodies present* | Reservoir characteristics | | | |
| | H (m) | L (m) | W (m) | | Porosity† | Volume of oil‡ | | Mud baffles |
						Per sandbody	Total	
Tidal–fluvial point bar	10	5000	2000	5	5%	5 million m^3 (0.03 billion US barrels)	25 million m^3 (0.16 billion US barrels)	Common
Sandbar in sandbar–swatch-way zone (i.e., mouth bar zone)	25	10,000	5000	10	5%	62.5 million m^3 (0.39 billion US barrels)	625 million m^3 (3.9 billion US barrels)	Rare
Transgressive ridge onto which the delta downlaps	25	25,000	5000	2	5%	156 million m^3 (0.98 billion US barrels)	312 million m^3 (1.96 billion US barrels)	Common
Transgressive lag on shelf	1	50,000	100,000	1	5%	250 million m^3 (1.57 billion US barrels)	250 million m^3 (1.57 billion US barrels)	Rare

H, height; L, length; W, width.
* Approximate.
† Hypothetical.
‡ For simplicity, assumes 100% oil saturation.

have been documented, and the physical processes that generate these traits are starting to be understood (Nittrouer et al., 1986; Wright and Nittrouer, 1995; Dalrymple et al., 2003; Friedrichs and Wright, 2004; Dalrymple and Choi, 2007; Goodbred and Saito, 2012). The Han River delta study contributes to this body of work because the Han is an end-member, the most tide-dominated of the tide-dominated deltas studied to date. Many of its key features are common to all tide-dominated deltas—the large subaqueous clino-form, the highly channelized mouth bar zone, the turbidity maximum with fluid muds at depth, the ubiquitous heterolithic sediment, and the open-coast tidal flats adjacent to the river mouth. The most obvious macroscopic difference is the degree of channelization: large tidal channels extend through the entire subaqueous delta, their position controlled by invaginations of the rocky coastline, and tidal channels are almost as pronounced off-axis as they are on-axis due to the macrotidal, fetch-limited setting, which results perhaps most notably in highly channelized open-coast tidal flats. Because of the degree of chan-nelization, lateral accretion of channel–barform pairs is an important generator of strati-graphic architecture in almost every subenvironment of the system (see Chapter 7).

Despite the advances made over the last several decades, many details of the nature and behavior of tide-dominated deltas remain poorly understood. This applies both to short-term, small-scale process–response relationships, and to long-term, large-scale ones. At the small scale, the behavior and dispersal of mud particles remains poorly understood, as does our ability to extract paleohydraulic information from mud deposits (cf. Mackay and Dalrymple, 2011). The question of whether a bedload convergence exists within these systems (e.g., Dalrymple et al., 2003) remains open. Fast currents make the shallow subtidal parts of these systems dangerous and difficult to study; these areas are therefore more poorly understood than the intertidal zone and the deeper shelf regions. There is also the issue of whether sufficient examples have been studied to allow proper "distilla-tion" of the facies, geomorphic and dynamic information to create a viable "facies model" for tide-dominated deltas, especially because the large tide-dominated deltas documented so far (Amazon, Changjiang, Ganges–Brahmaputra, Fly, Han) seem to bear little resem-blance to some of the ancient examples (e.g., Frewens system) that have been used to illustrate such deltas in textbooks (e.g., Bhattacharya, 2010). Are these differences real, or are they just an artifact of the scale of observation (e.g., was the Frewens system much larger and only a small part has been described)? Over longer timescales, the physics that cause tidal amplification as basin configurations change during sea level cycles is rela-tively well understood (Yoshida et al., 2007), but the application of this knowledge to the analysis of ancient sedimentary basins has rarely been performed. The question as to whether tide-dominated deltas, with their large muddy subaqueous clinoforms, deliver sediment to the deep ocean differently than wave- or fluvially dominanted deltas has not been explored. As with all scientific problems, the answers to these questions will require focused effort by multiple researchers over many years, and a desire to understand the complex interrelationships between all aspects of the system, at scales ranging from bed-form to basin fill.

APPENDIX 1A: YSDP-107 DRILL CORE

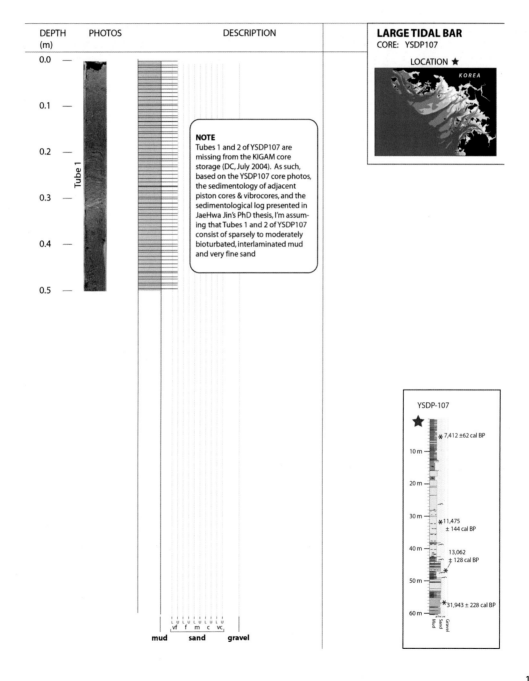

DEPTH (m)	PHOTOS	DESCRIPTION

NOTE
Tubes 1 and 2 of YSDP107 are missing from the KIGAM core storage (DC, July 2004). As such, based on the YSDP107 core photos, the sedimentology of adjacent piston cores & vibrocores, and the sedimentological log presented in JaeHwa Jin's PhD thesis, I'm assuming that Tubes 1 and 2 of YSDP107 consist of sparsely to moderately bioturbated, interlaminated mud and very fine sand

Tube 1

0.0
0.1
0.2
0.3
0.4
0.5

mud   sand   gravel

LARGE TIDAL BAR
CORE: YSDP107

LOCATION ★

KOREA

YSDP-107

★

7,412 ±62 cal BP

10 m

20 m

30 m
11,475 ± 144 cal BP

40 m
13,062 ± 128 cal BP

50 m

31,943 ± 228 cal BP

60 m
Mud Sand Gravel

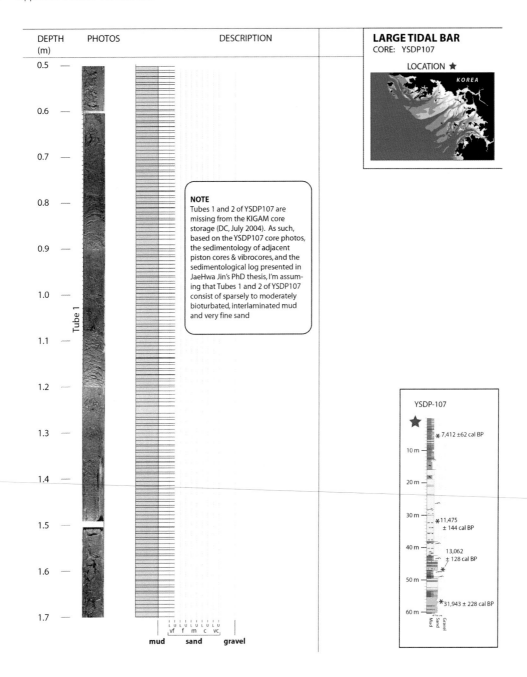

DEPTH (m)	PHOTOS	DESCRIPTION

NOTE

Tubes 1 and 2 of YSDP107 are missing from the KIGAM core storage (DC, July 2004). As such, based on the YSDP107 core photos, the sedimentology of adjacent piston cores & vibrocores, and the sedimentological log presented in JaeHwa Jin's PhD thesis, I'm assuming that Tubes 1 and 2 of YSDP107 consist of sparsely to moderately bioturbated, interlaminated mud and very fine sand

Tube 1

mud   sand   gravel

LARGE TIDAL BAR
CORE: YSDP107

LOCATION ★

KOREA

YSDP-107

★ 7,412 ±62 cal BP

10 m

20 m

30 m
★ 11,475 ± 144 cal BP

40 m
13,062 ± 128 cal BP
★

50 m

★ 31,943 ± 228 cal BP

60 m

Mud Sand Gravel

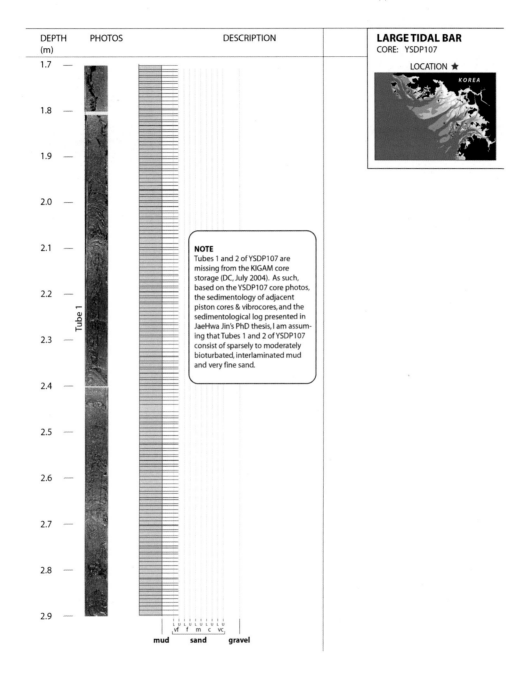

DEPTH (m) | PHOTOS | DESCRIPTION

LARGE TIDAL BAR
CORE: YSDP107

LOCATION ★

KOREA

NOTE
Tubes 1 and 2 of YSDP107 are missing from the KIGAM core storage (DC, July 2004). As such, based on the YSDP107 core photos, the sedimentology of adjacent piston cores & vibrocores, and the sedimentological log presented in JaeHwa Jin's PhD thesis, I am assuming that Tubes 1 and 2 of YSDP107 consist of sparsely to moderately bioturbated, interlaminated mud and very fine sand.

Tube 1

vf f m c vc
mud   sand   gravel

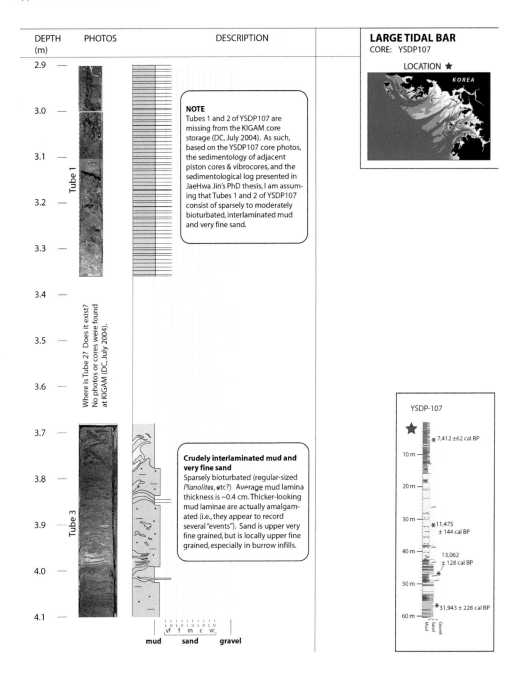

DEPTH (m)   PHOTOS   DESCRIPTION

Tube 1

NOTE
Tubes 1 and 2 of YSDP107 are missing from the KIGAM core storage (DC, July 2004). As such, based on the YSDP107 core photos, the sedimentology of adjacent piston cores & vibrocores, and the sedimentological log presented in JaeHwa Jin's PhD thesis, I am assuming that Tubes 1 and 2 of YSDP107 consist of sparsely to moderately bioturbated, interlaminated mud and very fine sand.

Where is Tube 2? Does it exist? No photos or cores were found at KIGAM (DC, July 2004).

Tube 3

Crudely interlaminated mud and very fine sand
Sparsely bioturbated (regular-sized *Planolites*, etc?) Average mud lamina thickness is ~0.4 cm. Thicker-looking mud laminae are actually amalgamated (i.e., they appear to record several "events"). Sand is upper very fine grained, but is locally upper fine grained, especially in burrow infills.

mud   sand   gravel

vf   f   m   c   vc

LARGE TIDAL BAR
CORE: YSDP107

LOCATION ★

KOREA

YSDP-107

★ 7,412 ±62 cal BP

10 m

20 m

30 m   *11,475 ± 144 cal BP

40 m   13,062 ± 128 cal BP *

50 m

*31,943 ± 228 cal BP

60 m

Mud   Sand   Gravel

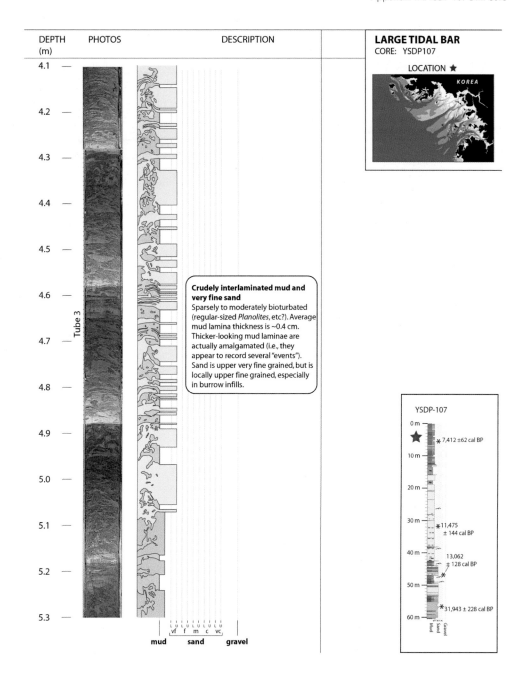

DEPTH (m) / PHOTOS / DESCRIPTION

Tube 3

Crudely interlaminated mud and very fine sand
Sparsely to moderately bioturbated (regular-sized *Planolites*, etc?). Average mud lamina thickness is ~0.4 cm. Thicker-looking mud laminae are actually amalgamated (i.e., they appear to record several "events"). Sand is upper very fine grained, but is locally upper fine grained, especially in burrow infills.

mud   sand   gravel

LARGE TIDAL BAR
CORE: YSDP107

LOCATION ★

KOREA

YSDP-107

0 m
★ * 7,412 ±62 cal BP
10 m
20 m
30 m
* 11,475 ± 144 cal BP
40 m
13,062 ± 128 cal BP
*
50 m
* 31,943 ± 228 cal BP
60 m
Mud Sand Gravel

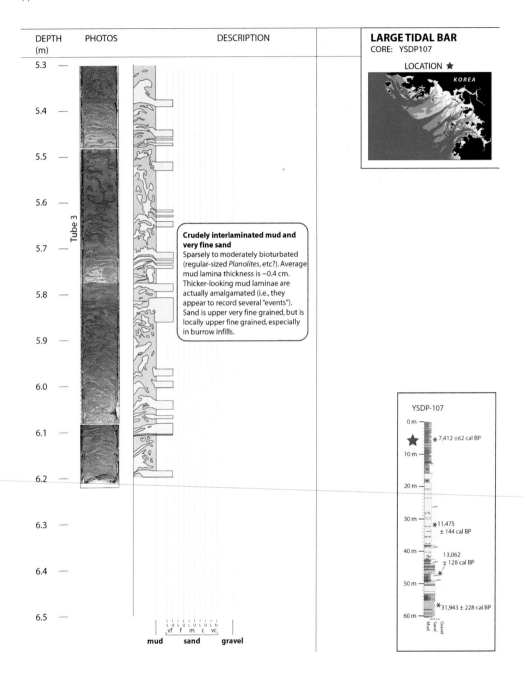

DEPTH (m)

PHOTOS

DESCRIPTION

Tube 3

Crudely interlaminated mud and very fine sand
Sparsely to moderately bioturbated (regular-sized *Planolites*, etc?). Average mud lamina thickness is ~0.4 cm. Thicker-looking mud laminae are actually amalgamated (i.e., they appear to record several "events"). Sand is upper very fine grained, but is locally upper fine grained, especially in burrow infills.

mud   sand   gravel

vf f m c vc

LARGE TIDAL BAR
CORE: YSDP107

LOCATION ★

KOREA

YSDP-107

0 m
★ ∗ 7,412 ±62 cal BP
10 m

20 m

30 m ∗ 11,475 ± 144 cal BP

40 m
13,062 ± 128 cal BP
∗
50 m

∗ 31,943 ± 228 cal BP
60 m

Mud Sand Gravel

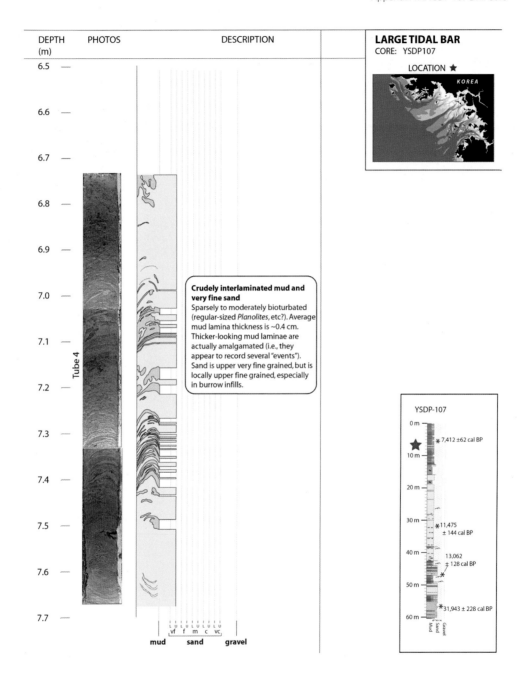

DEPTH (m)   PHOTOS   DESCRIPTION

Tube 4

Crudely interlaminated mud and very fine sand
Sparsely to moderately bioturbated (regular-sized *Planolites*, etc?). Average mud lamina thickness is ~0.4 cm. Thicker-looking mud laminae are actually amalgamated (i.e., they appear to record several "events"). Sand is upper very fine grained, but is locally upper fine grained, especially in burrow infills.

mud   sand   gravel
vf f m c vc

LARGE TIDAL BAR
CORE: YSDP107

LOCATION ★

KOREA

YSDP-107

0 m
★   ✱ 7,412 ±62 cal BP
10 m

20 m

30 m   ✱ 11,475 ± 144 cal BP
40 m
13,062 ± 128 cal BP
✱
50 m

✱ 31,943 ± 228 cal BP
60 m
Gravel
Sand
Mud

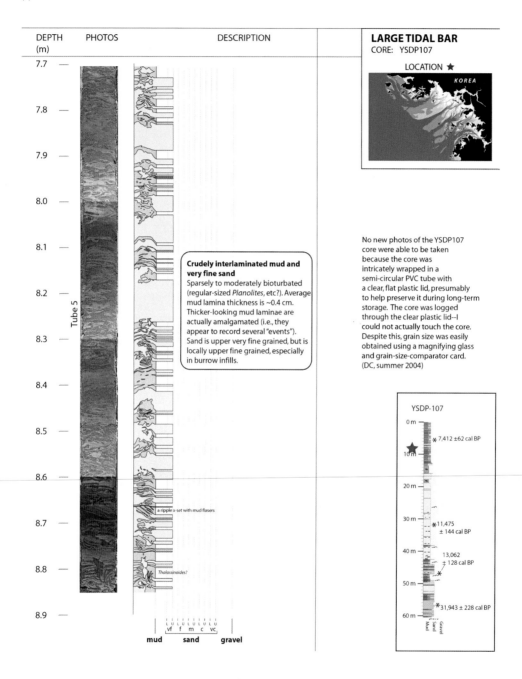

DEPTH (m)	PHOTOS	DESCRIPTION

LARGE TIDAL BAR
CORE: YSDP107

LOCATION ★

KOREA

Crudely interlaminated mud and very fine sand
Sparsely to moderately bioturbated (regular-sized *Planolites*, etc?). Average mud lamina thickness is ~0.4 cm. Thicker-looking mud laminae are actually amalgamated (i.e., they appear to record several "events"). Sand is upper very fine grained, but is locally upper fine grained, especially in burrow infills.

No new photos of the YSDP107 core were able to be taken because the core was intricately wrapped in a semi-circular PVC tube with a clear, flat plastic lid, presumably to help preserve it during long-term storage. The core was logged through the clear plastic lid--I could not actually touch the core. Despite this, grain size was easily obtained using a magnifying glass and grain-size-comparator card. (DC, summer 2004)

Tube 5

a ripple x-set with mud flasers

Thalassinoides?

mud   sand   gravel

vf f m c vc

YSDP-107

0 m

★ 7,412 ±62 cal BP

10 m

20 m

30 m

✳ 11,475 ± 144 cal BP

40 m

13,062 ± 128 cal BP
✳

50 m

✳ 31,943 ± 228 cal BP

60 m

Mud Sand Gravel

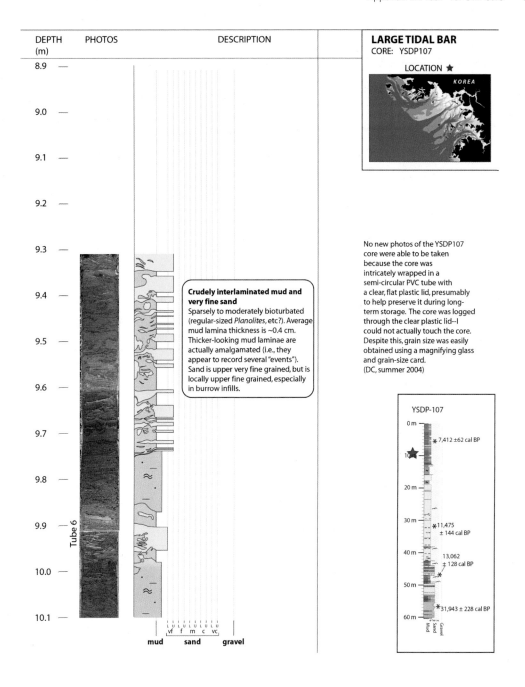

DEPTH (m)   PHOTOS   DESCRIPTION

Crudely interlaminated mud and very fine sand
Sparsely to moderately bioturbated (regular-sized *Planolites*, etc?). Average mud lamina thickness is ~0.4 cm. Thicker-looking mud laminae are actually amalgamated (i.e., they appear to record several "events"). Sand is upper very fine grained, but is locally upper fine grained, especially in burrow infills.

Tube 6

mud   sand   gravel

LARGE TIDAL BAR
CORE: YSDP107

LOCATION ★

KOREA

No new photos of the YSDP107 core were able to be taken because the core was intricately wrapped in a semi-circular PVC tube with a clear, flat plastic lid, presumably to help preserve it during long-term storage. The core was logged through the clear plastic lid--I could not actually touch the core. Despite this, grain size was easily obtained using a magnifying glass and grain-size card.
(DC, summer 2004)

YSDP-107

0 m
⚹ 7,412 ±62 cal BP
10 m
20 m
30 m
⚹ 11,475 ± 144 cal BP
40 m
13,062 ± 128 cal BP ⚹
50 m
⚹ 31,943 ± 228 cal BP
60 m

Mud Sand Gravel

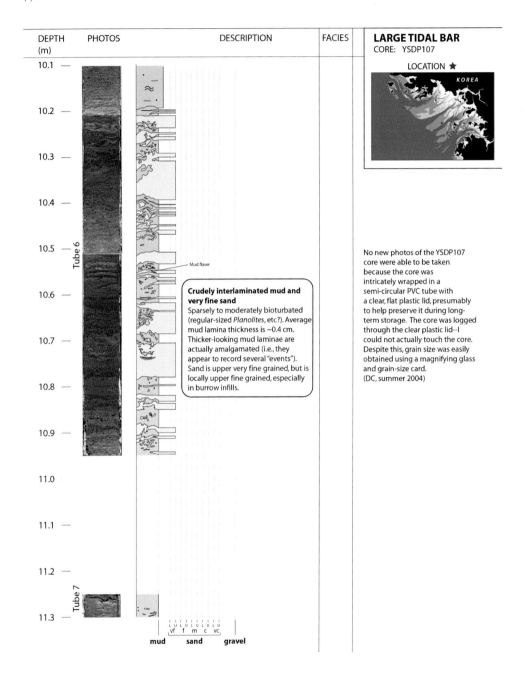

DEPTH (m)	PHOTOS	DESCRIPTION	FACIES

Crudely interlaminated mud and very fine sand
Sparsely to moderately bioturbated (regular-sized *Planolites*, etc?). Average mud lamina thickness is ~0.4 cm. Thicker-looking mud laminae are actually amalgamated (i.e., they appear to record several "events"). Sand is upper very fine grained, but is locally upper fine grained, especially in burrow infills.

Mud flaser

mud   sand   gravel
vf   f   m   c   vc

Tube 6

Tube 7

LARGE TIDAL BAR
CORE: YSDP107

LOCATION ★

KOREA

No new photos of the YSDP107 core were able to be taken because the core was intricately wrapped in a semi-circular PVC tube with a clear, flat plastic lid, presumably to help preserve it during long-term storage. The core was logged through the clear plastic lid--I could not actually touch the core. Despite this, grain size was easily obtained using a magnifying glass and grain-size card.
(DC, summer 2004)

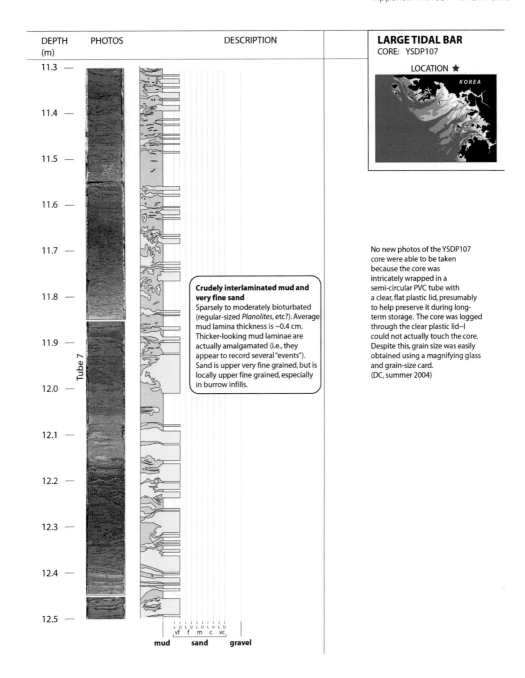

DEPTH (m)   PHOTOS   DESCRIPTION

Tube 7

Crudely interlaminated mud and very fine sand

Sparsely to moderately bioturbated (regular-sized *Planolites*, etc?). Average mud lamina thickness is ~0.4 cm. Thicker-looking mud laminae are actually amalgamated (i.e., they appear to record several "events"). Sand is upper very fine grained, but is locally upper fine grained, especially in burrow infills.

mud   sand   gravel

vf f m c vc

LARGE TIDAL BAR
CORE: YSDP107

LOCATION ★

KOREA

No new photos of the YSDP107 core were able to be taken because the core was intricately wrapped in a semi-circular PVC tube with a clear, flat plastic lid, presumably to help preserve it during long-term storage. The core was logged through the clear plastic lid--I could not actually touch the core. Despite this, grain size was easily obtained using a magnifying glass and grain-size card.
(DC, summer 2004)

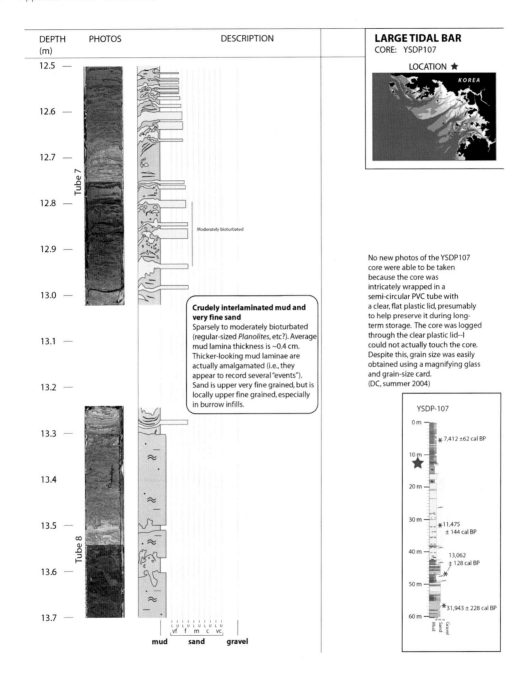

DEPTH (m)	PHOTOS	DESCRIPTION

LARGE TIDAL BAR
CORE: YSDP107

LOCATION ★

KOREA

Moderately bioturbated

Crudely interlaminated mud and very fine sand
Sparsely to moderately bioturbated (regular-sized *Planolites*, etc?). Average mud lamina thickness is ~0.4 cm. Thicker-looking mud laminae are actually amalgamated (i.e., they appear to record several "events"). Sand is upper very fine grained, but is locally upper fine grained, especially in burrow infills.

No new photos of the YSDP107 core were able to be taken because the core was intricately wrapped in a semi-circular PVC tube with a clear, flat plastic lid, presumably to help preserve it during long-term storage. The core was logged through the clear plastic lid--I could not actually touch the core. Despite this, grain size was easily obtained using a magnifying glass and grain-size card. (DC, summer 2004)

YSDP-107

0 m
★ 7,412 ±62 cal BP
10 m
20 m
30 m
★ 11,475 ± 144 cal BP
40 m
13,062 ± 128 cal BP
★
50 m
★ 31,943 ± 228 cal BP
60 m

Tube 7

Tube 8

L U L U L U L U L U
vf f   m   c   vc
mud   sand   gravel

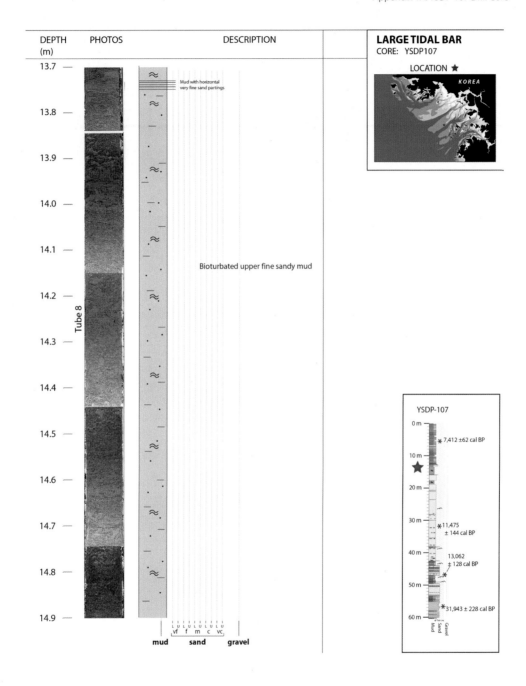

LARGE TIDAL BAR
CORE: YSDP107

LOCATION ★

KOREA

Mud with horizontal
very fine sand partings

Bioturbated upper fine sandy mud

Tube 8

mud   sand   gravel

YSDP-107

0 m — ∗ 7,412 ±62 cal BP
10 m — ★
20 m —
30 m — ∗ 11,475 ± 144 cal BP
40 m — 13,062 ± 128 cal BP ∗
50 m —
60 m — ∗ 31,943 ± 228 cal BP

Mud Sand Gravel

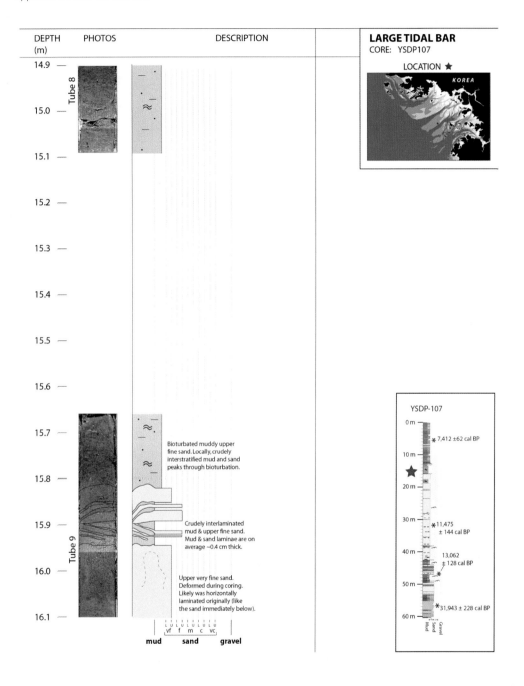

| DEPTH (m) | PHOTOS | DESCRIPTION |

LARGE TIDAL BAR
CORE: YSDP107

LOCATION ★

KOREA

Bioturbated muddy upper fine sand. Locally, crudely interstratified mud and sand peaks through bioturbation.

Crudely interlaminated mud & upper fine sand. Mud & sand laminae are on average ~0.4 cm thick.

Upper very fine sand. Deformed during coring. Likely was horizontally laminated originally (like the sand immediately below).

Tube 8

Tube 9

mud   sand   gravel

YSDP-107

0 m
    ✳ 7,412 ±62 cal BP
10 m

★

20 m

30 m
    ✳ 11,475 ± 144 cal BP
40 m
          13,062 ± 128 cal BP
50 m   ✳
    ✳ 31,943 ± 228 cal BP
60 m

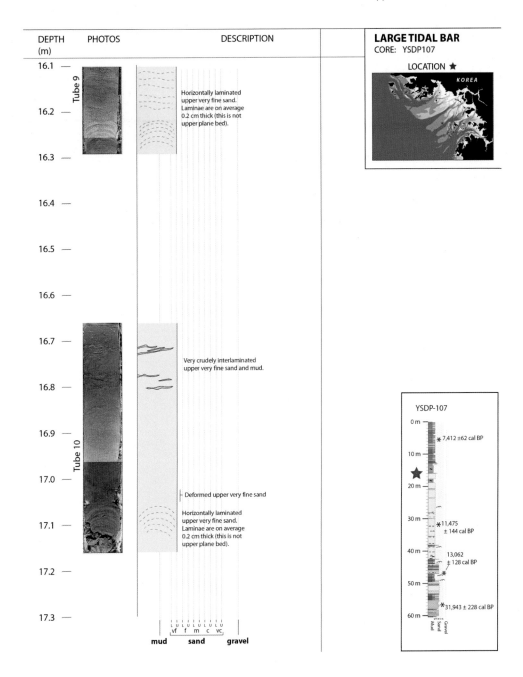

DEPTH (m) | PHOTOS | DESCRIPTION

Tube 9

Horizontally laminated upper very fine sand. Laminae are on average 0.2 cm thick (this is not upper plane bed).

Tube 10

Very crudely interlaminated upper very fine sand and mud.

Deformed upper very fine sand

Horizontally laminated upper very fine sand. Laminae are on average 0.2 cm thick (this is not upper plane bed).

mud   sand   gravel

LARGE TIDAL BAR
CORE: YSDP107

LOCATION ★

KOREA

YSDP-107

* 7,412 ±62 cal BP
* 11,475 ± 144 cal BP
13,062 ± 128 cal BP
* 31,943 ± 228 cal BP

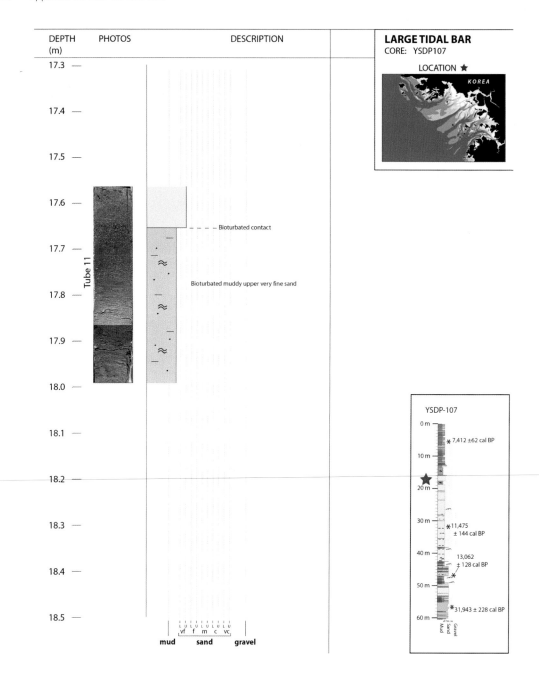

DEPTH (m)   PHOTOS   DESCRIPTION

17.3

17.4

17.5

17.6

17.7   — — — Bioturbated contact

Tube 11

17.8   Bioturbated muddy upper very fine sand

17.9

18.0

18.1

18.2

18.3

18.4

18.5

mud   sand   gravel

L U L U L U L U L U
vf   f   m   c   vc

LARGE TIDAL BAR
CORE: YSDP107

LOCATION ★

KOREA

YSDP-107

0 m
    ✳ 7,412 ±62 cal BP
10 m

20 m   ★

30 m
    ✳ 11,475
      ± 144 cal BP
40 m
        13,062
      ✳ ± 128 cal BP
50 m

        ✳ 31,943 ± 228 cal BP
60 m
    Mud Sand Gravel

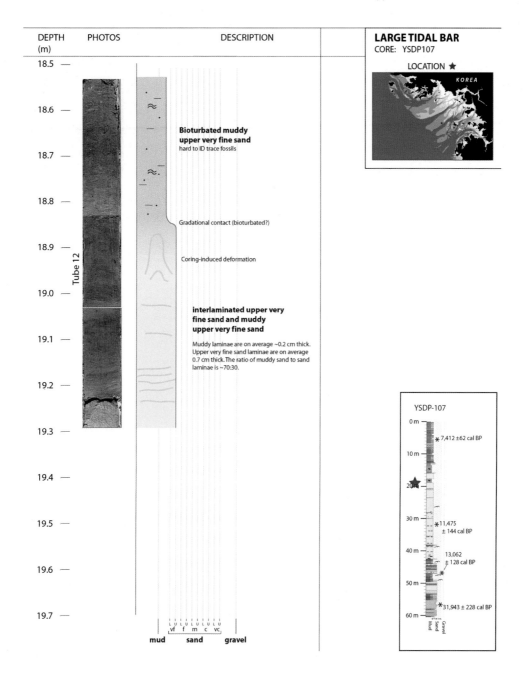

DEPTH (m)	PHOTOS	DESCRIPTION

Bioturbated muddy upper very fine sand
hard to ID trace fossils

Gradational contact (bioturbated?)

Coring-induced deformation

interlaminated upper very fine sand and muddy upper very fine sand

Muddy laminae are on average ~0.2 cm thick. Upper very fine sand laminae are on average 0.7 cm thick. The ratio of muddy sand to sand laminae is ~70:30.

Tube 12

18.5
18.6
18.7
18.8
18.9
19.0
19.1
19.2
19.3
19.4
19.5
19.6
19.7

vf f m c vc
mud   sand   gravel

LARGE TIDAL BAR
CORE: YSDP107

LOCATION ★

KOREA

YSDP-107

0 m — ✳ 7,412 ±62 cal BP
10 m
20 m ★
30 m — ✳ 11,475 ± 144 cal BP
40 m — 13,062 ± 128 cal BP ✳
50 m
60 m — ✳ 31,943 ± 228 cal BP

Mud Sand Gravel

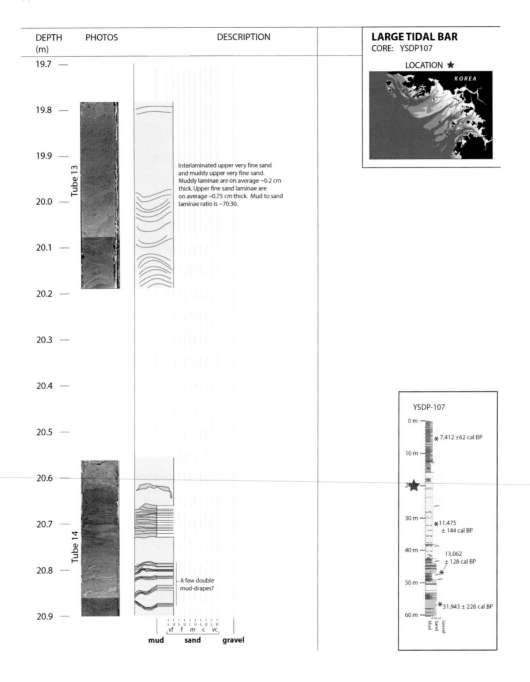

DEPTH (m)   PHOTOS   DESCRIPTION

Tube 13

Interlaminated upper very fine sand and muddy upper very fine sand. Muddy laminae are on average ~0.2 cm thick. Upper fine sand laminae are on average ~0.75 cm thick. Mud to sand laminae ratio is ~70:30.

Tube 14

A few double mud-drapes?

mud   sand   gravel

LARGE TIDAL BAR
CORE: YSDP107

LOCATION ★

KOREA

YSDP-107

0 m
★ 7,412 ±62 cal BP

10 m

20 m

30 m
★ 11,475 ± 144 cal BP

40 m
13,062 ± 128 cal BP
★

50 m

★ 31,943 ± 228 cal BP

60 m
Mud   Sand   Gravel

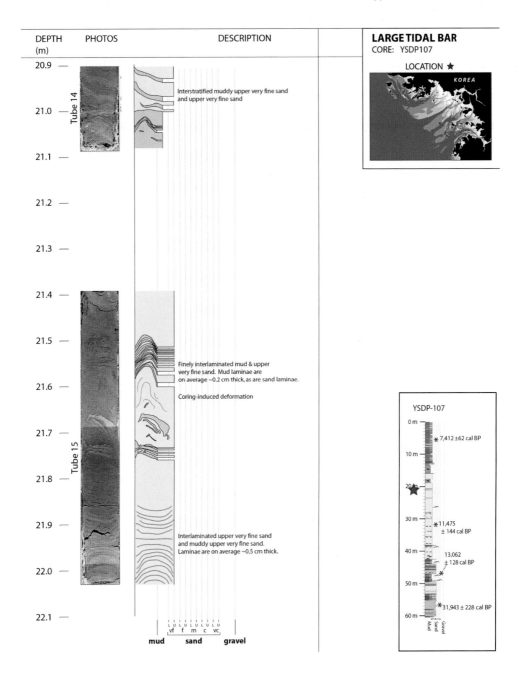

DEPTH (m)   PHOTOS   DESCRIPTION

Tube 14

Interstratified muddy upper very fine sand and upper very fine sand

Tube 15

Finely interlaminated mud & upper very fine sand. Mud laminae are on average ~0.2 cm thick, as are sand laminae.

Coring-induced deformation

Interlaminated upper very fine sand and muddy upper very fine sand. Laminae are on average ~0.5 cm thick.

mud   sand   gravel

vf   f   m   c   vc

LARGE TIDAL BAR
CORE: YSDP107

LOCATION ★

KOREA

YSDP-107

0 m
7,412 ±62 cal BP
10 m
20 m
30 m
11,475 ± 144 cal BP
40 m
13,062 ± 128 cal BP
50 m
31,943 ± 228 cal BP
60 m

Mud Sand Gravel

DEPTH (m)	PHOTOS	DESCRIPTION

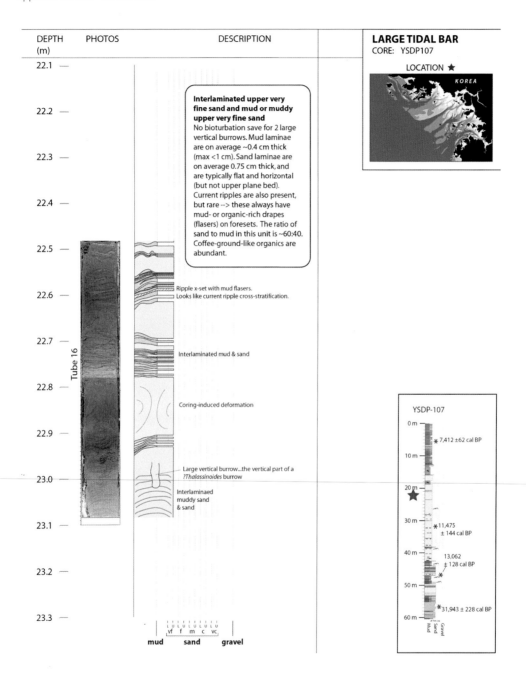

Interlaminated upper very fine sand and mud or muddy upper very fine sand
No bioturbation save for 2 large vertical burrows. Mud laminae are on average ~0.4 cm thick (max <1 cm). Sand laminae are on average 0.75 cm thick, and are typically flat and horizontal (but not upper plane bed). Current ripples are also present, but rare --> these always have mud- or organic-rich drapes (flasers) on foresets. The ratio of sand to mud in this unit is ~60:40. Coffee-ground-like organics are abundant.

Ripple x-set with mud flasers. Looks like current ripple cross-stratification.

Interlaminated mud & sand

Coring-induced deformation

Large vertical burrow...the vertical part of a ?*Thalassinoides* burrow

Interlaminaed muddy sand & sand

Tube 16

mud   sand   gravel

LARGE TIDAL BAR
CORE: YSDP107

LOCATION ★

KOREA

YSDP-107

0 m
∗ 7,412 ±62 cal BP
10 m
20 m
★
30 m
∗ 11,475 ± 144 cal BP
40 m
13,062 ± 128 cal BP
∗
50 m
∗ 31,943 ± 228 cal BP
60 m

Gravel
Sand
Mud

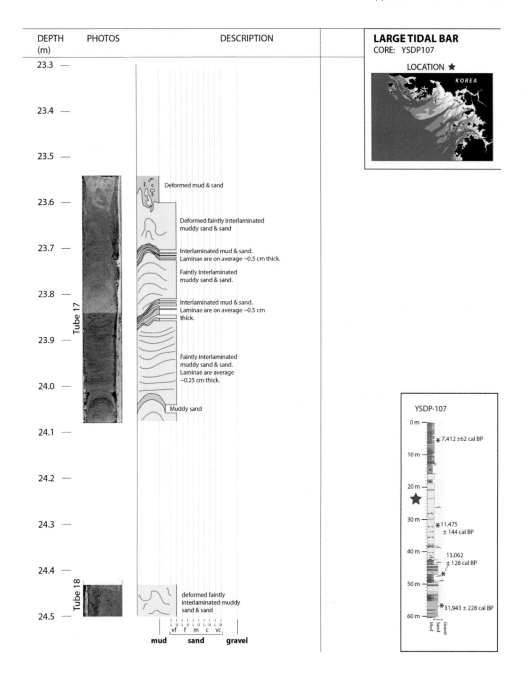

DEPTH (m) | PHOTOS | DESCRIPTION

Deformed mud & sand

Deformed faintly interlaminated muddy sand & sand

Interlaminated mud & sand. Laminae are on average ~0.5 cm thick.

Faintly interlaminated muddy sand & sand.

Interlaminated mud & sand. Laminae are on average ~0.5 cm thick.

Faintly interlaminated muddy sand & sand. Laminae are average ~0.25 cm thick.

Muddy sand

Tube 17

Tube 18

deformed faintly interlaminated muddy sand & sand

mud   sand   gravel

LARGE TIDAL BAR
CORE: YSDP107

LOCATION ★

KOREA

YSDP-107

0 m
* 7,412 ±62 cal BP
10 m
20 m
★
30 m
* 11,475 ± 144 cal BP
13,062 ± 128 cal BP
40 m
*
50 m
* 31,943 ± 228 cal BP
60 m
Mud   Sand   Gravel

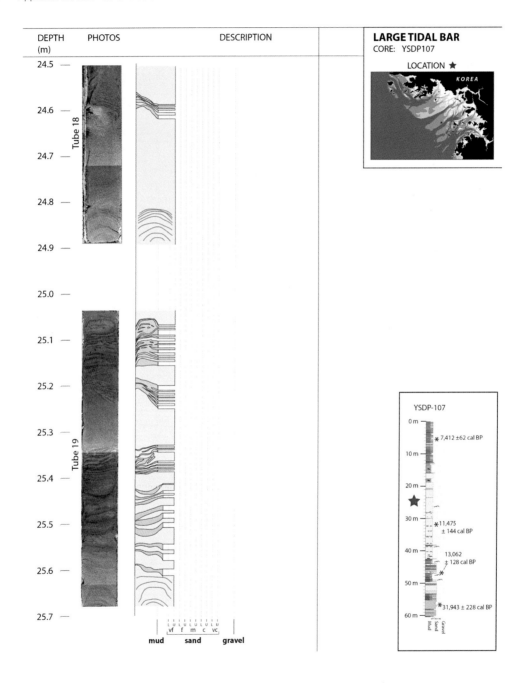

DEPTH (m)   PHOTOS   DESCRIPTION

LARGE TIDAL BAR
CORE: YSDP107

LOCATION ★

KOREA

YSDP-107

* 7,412 ±62 cal BP

*11,475 ± 144 cal BP

13,062 ± 128 cal BP *

*31,943 ± 228 cal BP

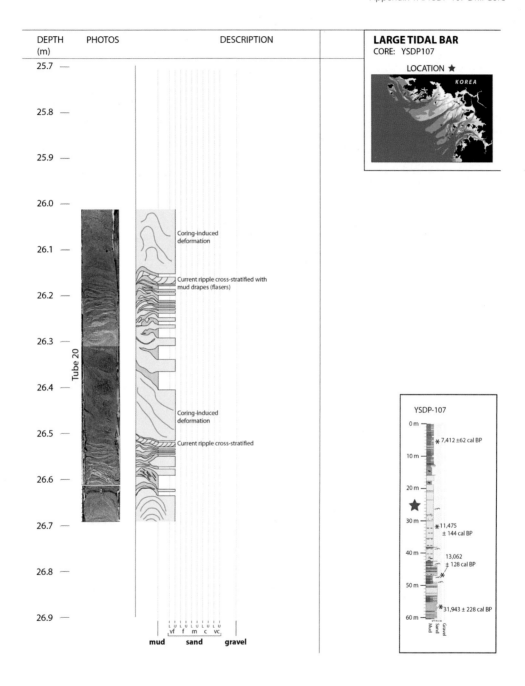

DEPTH (m) | PHOTOS | DESCRIPTION

Tube 20

Coring-induced deformation

Current ripple cross-stratified with mud drapes (flasers)

Coring-induced deformation

Current ripple cross-stratified

mud   sand   gravel
vf f m c vc

LARGE TIDAL BAR
CORE: YSDP107

LOCATION ★

KOREA

YSDP-107

0 m

✱ 7,412 ±62 cal BP

10 m

20 m

★

30 m

✱ 11,475 ± 144 cal BP

40 m

13,062 ± 128 cal BP
✱

50 m

✱ 31,943 ± 228 cal BP

60 m

Mud Sand Gravel

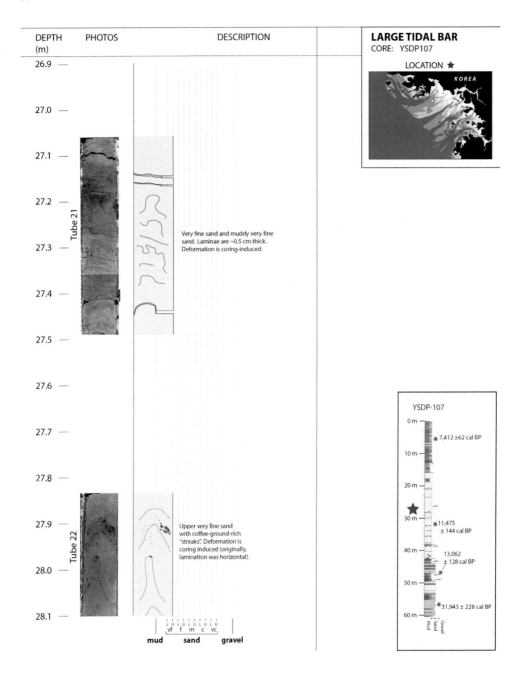

DEPTH (m)	PHOTOS	DESCRIPTION

Very fine sand and muddy very fine sand. Laminae are ~0.5 cm thick. Deformation is coring-induced.

Upper very fine sand with coffee-ground-rich "streaks". Deformation is coring induced (originally, lamination was horizontal).

Tube 21

Tube 22

mud   sand   gravel

LARGE TIDAL BAR
CORE: YSDP107

LOCATION ★

KOREA

YSDP-107

0 m
★ 7,412 ±62 cal BP
10 m
20 m
★
30 m ★ 11,475 ± 144 cal BP
40 m
13,062 ± 128 cal BP ★
50 m
★ 31,943 ± 228 cal BP
60 m

Mud   Sand   Gravel

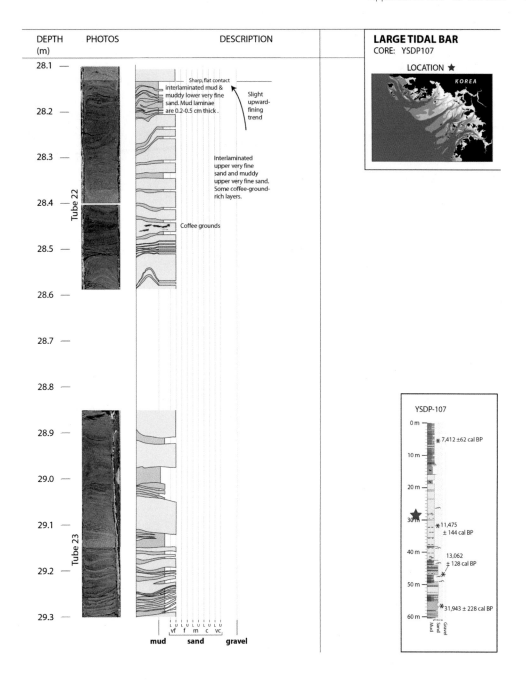

DEPTH (m)

PHOTOS

DESCRIPTION

Tube 22

Sharp, flat contact
interlaminated mud &
muddy lower very fine
sand. Mud laminae
are 0.2-0.5 cm thick .

Slight
upward-
fining
trend

Interlaminated
upper very fine
sand and muddy
upper very fine sand.
Some coffee-ground-
rich layers.

Coffee grounds

Tube 23

mud   sand   gravel

LARGE TIDAL BAR
CORE: YSDP107

LOCATION ★

KOREA

YSDP-107

0 m

*7,412 ±62 cal BP

10 m

20 m

30 m

*11,475
± 144 cal BP

40 m

13,062
± 128 cal BP

*

50 m

*31,943 ± 228 cal BP

60 m

Mud   Sand   Gravel

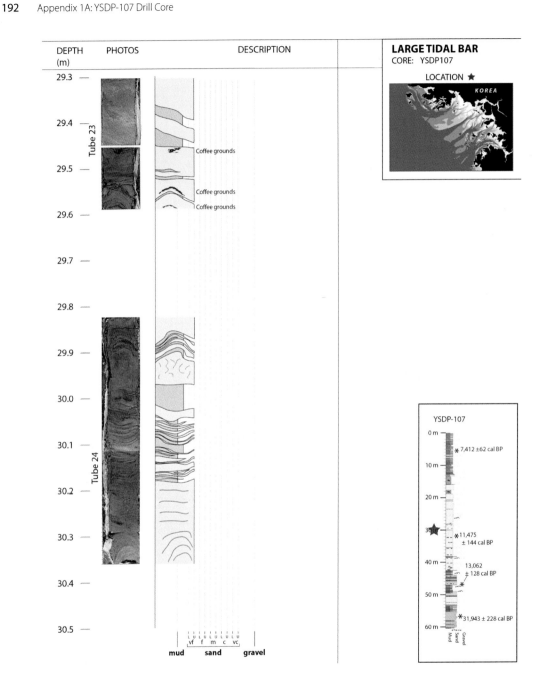

DEPTH (m)	PHOTOS	DESCRIPTION

Tube 23

Coffee grounds

Coffee grounds

Coffee grounds

Tube 24

mud   sand   gravel

vf f m c vc

LARGE TIDAL BAR
CORE: YSDP107

LOCATION ★

KOREA

YSDP-107

0 m

★ 7,412 ±62 cal BP

10 m

20 m

30 m   ✶11,475 ± 144 cal BP

40 m   13,062 ± 128 cal BP
          ✶

50 m

          ✶31,943 ± 228 cal BP

60 m

Mud Sand Gravel

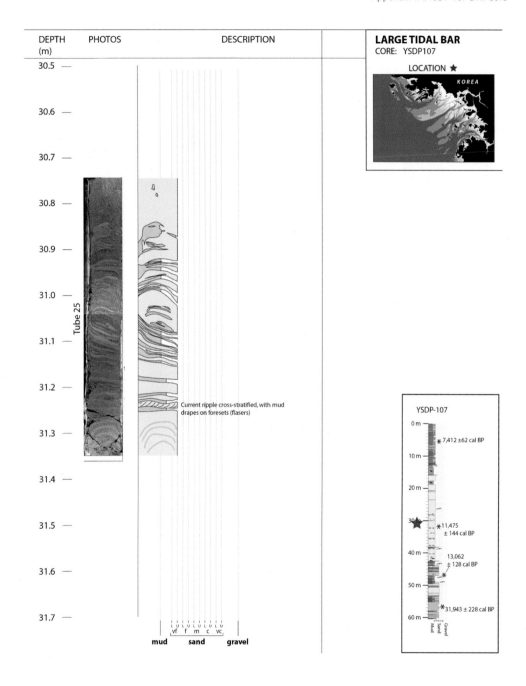

DEPTH (m)   PHOTOS   DESCRIPTION

30.5
30.6
30.7
30.8
30.9
31.0
31.1
31.2
31.3
31.4
31.5
31.6
31.7

Tube 25

Current ripple cross-stratified, with mud drapes on foresets (flasers)

vf f m c vc
mud   sand   gravel

LARGE TIDAL BAR
CORE: YSDP107

LOCATION ★

KOREA

YSDP-107

0 m
*7,412 ±62 cal BP
10 m
20 m
30
*11,475 ± 144 cal BP
40 m
13,062 ± 128 cal BP
*
50 m
*31,943 ± 228 cal BP
60 m
Mud Sand Gravel

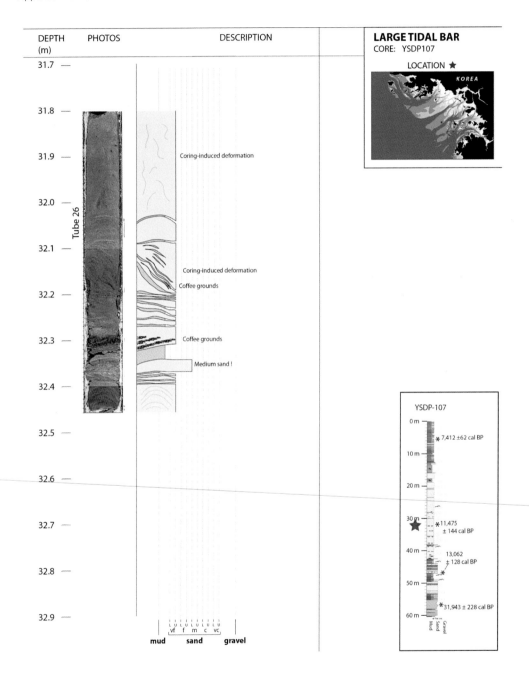

DEPTH (m)   PHOTOS   DESCRIPTION

31.7 —
31.8 —
31.9 —   Coring-induced deformation
32.0 —
32.1 —
32.2 —   Coring-induced deformation
         Coffee grounds
32.3 —   Coffee grounds
         Medium sand !
32.4 —
32.5 —
32.6 —
32.7 —
32.8 —
32.9 —

Tube 26

mud   sand   gravel

vf f m c vc

LARGE TIDAL BAR
CORE: YSDP107

LOCATION ★

KOREA

YSDP-107

0 m
         * 7,412 ±62 cal BP
10 m
20 m
30 m
         * 11,475 ± 144 cal BP
40 m
         13,062 ± 128 cal BP
         *
50 m
         * 31,943 ± 228 cal BP
60 m

Mud Sand Gravel

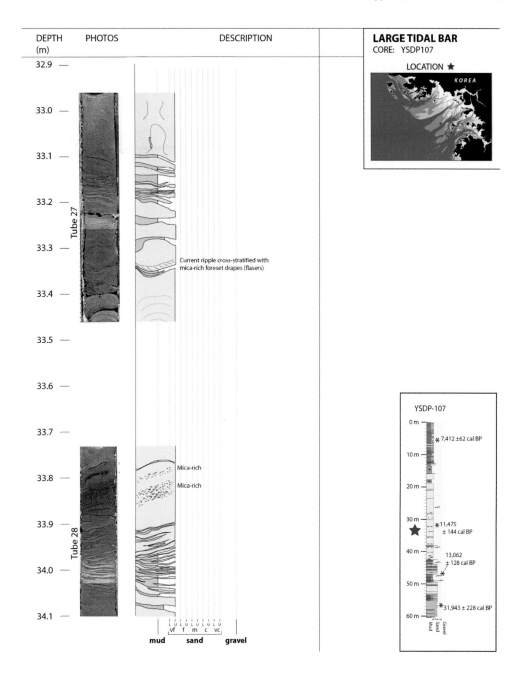

DEPTH (m) | PHOTOS | DESCRIPTION

Tube 27

Current ripple cross-stratified with mica-rich foreset drapes (flasers)

Tube 28

Mica-rich

Mica-rich

vf f m c vc
mud   sand   gravel

LARGE TIDAL BAR
CORE: YSDP107

LOCATION ★

KOREA

YSDP-107

0 m
✳ 7,412 ±62 cal BP
10 m
20 m
30 m
★ ✳ 11,475 ± 144 cal BP
40 m
13,062 ± 128 cal BP
✳
50 m
✳ 31,943 ± 228 cal BP
60 m
Mud Sand Gravel

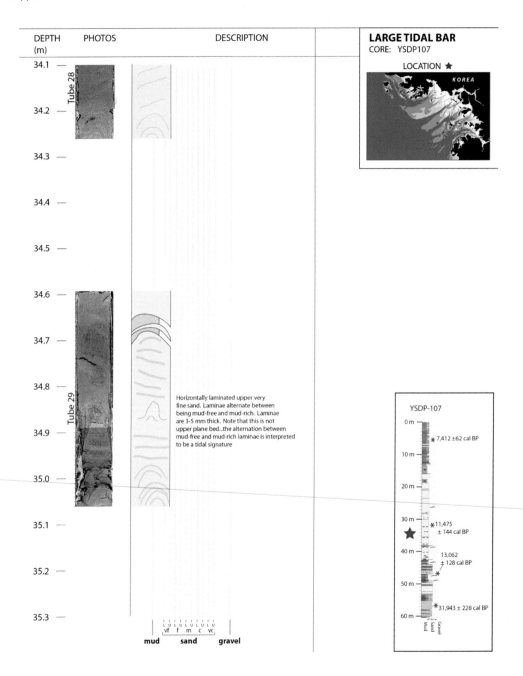

DEPTH (m)   PHOTOS   DESCRIPTION

Tube 28

Tube 29

Horizontally laminated upper very fine sand. Laminae alternate between being mud-free and mud-rich. Laminae are 3-5 mm thick. Note that this is not upper plane bed...the alternation between mud-free and mud-rich laminae is interpreted to be a tidal signature

mud   sand   gravel

vf f m c vc

LARGE TIDAL BAR
CORE: YSDP107

LOCATION ★

KOREA

YSDP-107

0 m
✱ 7,412 ±62 cal BP

10 m

20 m

30 m
✱ 11,475 ± 144 cal BP

40 m
13,062 ± 128 cal BP
✱

50 m

60 m
✱ 31,943 ± 228 cal BP

Mud Sand Gravel

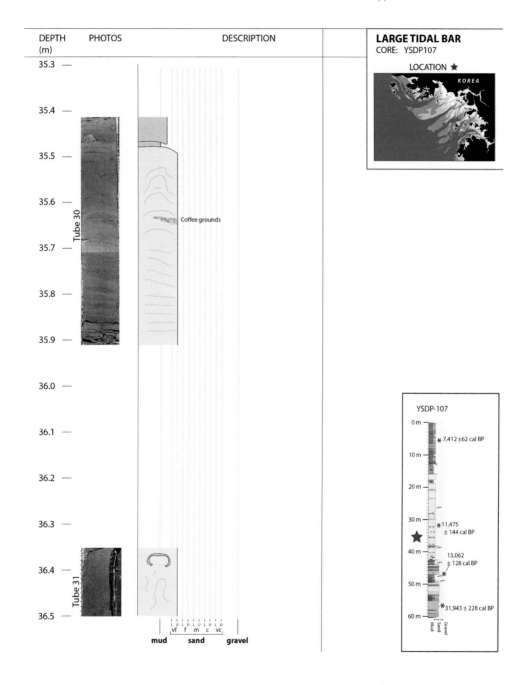

DEPTH (m)   PHOTOS   DESCRIPTION

Tube 30

Coffee grounds

Tube 31

mud   sand   gravel

LARGE TIDAL BAR
CORE: YSDP107

LOCATION ★

KOREA

YSDP-107

0 m
✳ 7,412 ±62 cal BP
10 m
20 m
30 m
✳ 11,475 ± 144 cal BP
★
40 m
13,062 ± 128 cal BP
✳
50 m
✳ 31,943 ± 228 cal BP
60 m

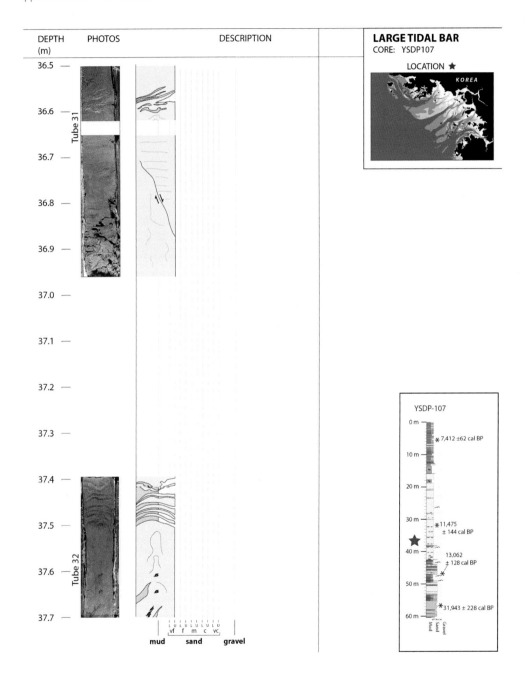

LARGE TIDAL BAR
CORE: YSDP107

LOCATION ★

KOREA

YSDP-107

* 7,412 ±62 cal BP

* 11,475 ± 144 cal BP

13,062 ± 128 cal BP *

* 31,943 ± 228 cal BP

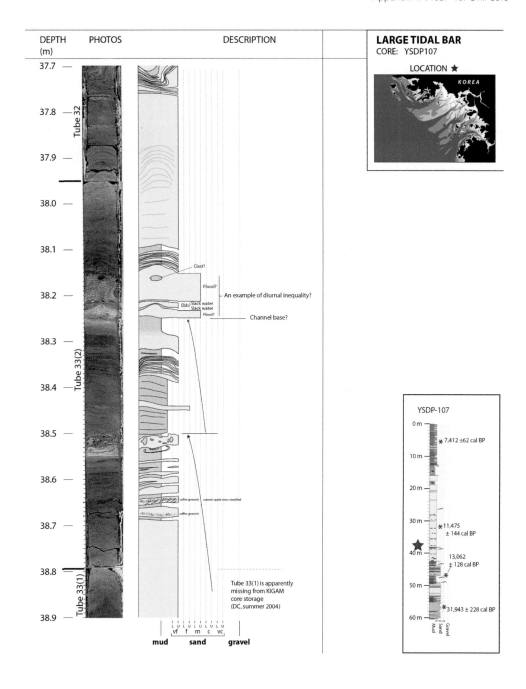

DEPTH (m) PHOTOS DESCRIPTION

LARGE TIDAL BAR
CORE: YSDP107

LOCATION ★

KOREA

37.7 — 37.8 — 37.9 — 38.0 — 38.1 — 38.2 — 38.3 — 38.4 — 38.5 — 38.6 — 38.7 — 38.8 — 38.9

Tube 32
Tube 33(2)
Tube 33(1)

Clast?
Flood?
An example of diurnal inequality?
Ebb? Slack water
Slack water
Flood?
Channel base?

coffee grounds   current ripple cross-stratified
coffee grounds

Tube 33(1) is apparently
missing from KIGAM
core storage
(DC, summer 2004)

L U L U L U L U L U
vf f m c vc
mud   sand   gravel

YSDP-107

0 m
10 m
20 m
30 m
40 m
50 m
60 m

* 7,412 ±62 cal BP

* 11,475
± 144 cal BP

13,062
± 128 cal BP
*

* 31,943 ± 228 cal BP

Mud Sand Gravel

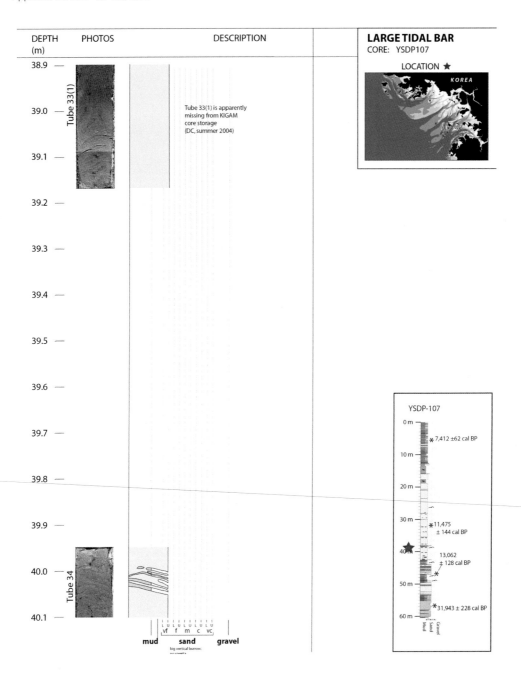

DEPTH (m)	PHOTOS	DESCRIPTION
	Tube 33(1)	Tube 33(1) is apparently missing from KIGAM core storage (DC, summer 2004)

LARGE TIDAL BAR
CORE: YSDP107

LOCATION ★

KOREA

YSDP-107

0 m
★ 7,412 ±62 cal BP
10 m
20 m
30 m
★ 11,475 ± 144 cal BP
★ 13,062 ± 128 cal BP
50 m
★ 31,943 ± 228 cal BP
60 m

mud   sand   gravel

big vertical burrow;

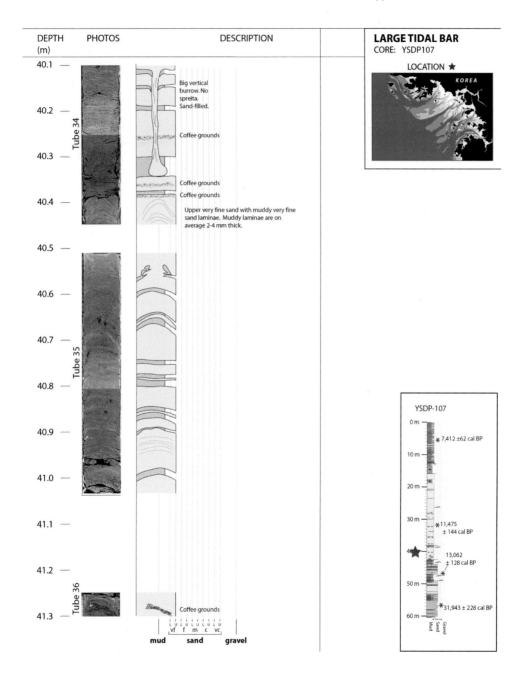

LARGE TIDAL BAR
CORE: YSDP107

LOCATION ★

KOREA

DEPTH (m) | PHOTOS | DESCRIPTION

Tube 34

Big vertical burrow. No spreita. Sand-filled.

Coffee grounds

Coffee grounds

Coffee grounds

Upper very fine sand with muddy very fine sand laminae. Muddy laminae are on average 2-4 mm thick.

Tube 35

Tube 36

Coffee grounds

mud   sand   gravel

YSDP-107

0 m
* 7,412 ±62 cal BP
10 m
20 m
30 m
*11,475 ± 144 cal BP
13,062 ± 128 cal BP
*
50 m
*31,943 ± 228 cal BP
60 m

Mud Sand Gravel

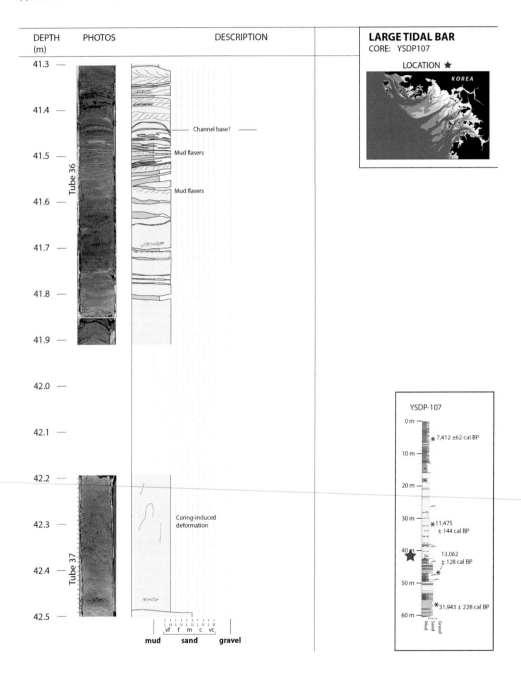

LARGE TIDAL BAR
CORE: YSDP107

LOCATION ★

KOREA

YSDP-107

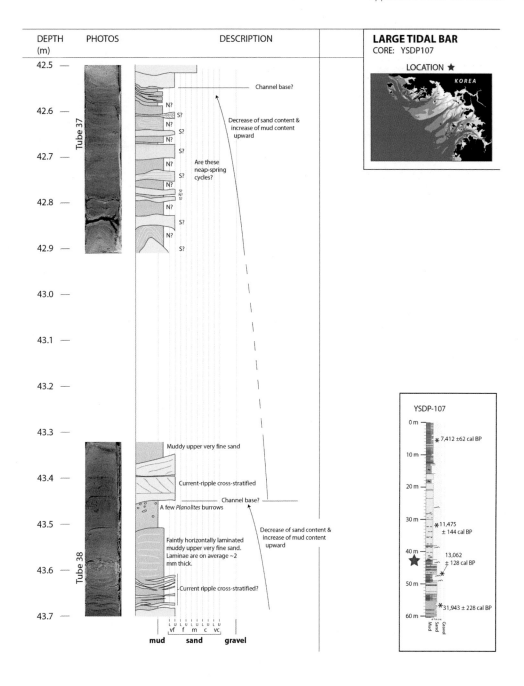

DEPTH (m) — PHOTOS — DESCRIPTION

Tube 37

Channel base?

Decrease of sand content & increase of mud content upward

Are these neap-spring cycles?

N? S? N? S? N? S? N? S? N? N? S? N? S? N? S?

Tube 38

Muddy upper very fine sand

Current-ripple cross-stratified

Channel base?

A few *Planolites* burrows

Faintly horizontally laminated muddy upper very fine sand. Laminae are on average ~2 mm thick.

Current ripple cross-stratified?

Decrease of sand content & increase of mud content upward

L U L U L U L U L U
vf   f   m   c   vc

mud   sand   gravel

LARGE TIDAL BAR
CORE: YSDP107

LOCATION ★

KOREA

YSDP-107

0 m
* 7,412 ±62 cal BP
10 m
20 m
30 m
* 11,475 ± 144 BP
40 m
★   13,062 ± 128 cal BP
*
50 m
* 31,943 ± 228 cal BP
60 m

Mud   Sand   Gravel

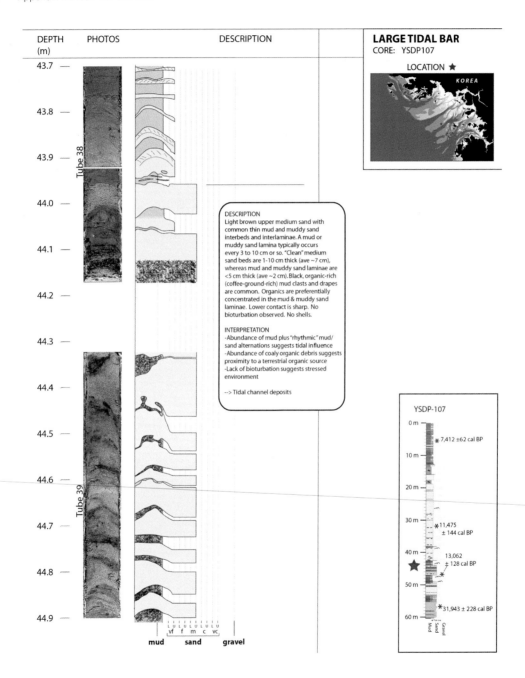

DEPTH (m)   PHOTOS   DESCRIPTION

Tube 38

Tube 39

DESCRIPTION
Light brown upper medium sand with common thin mud and muddy sand interbeds and interlaminae. A mud or muddy sand lamina typically occurs every 3 to 10 cm or so. "Clean" medium sand beds are 1-10 cm thick (ave ~7 cm), whereas mud and muddy sand laminae are <5 cm thick (ave ~2 cm). Black, organic-rich (coffee-ground-rich) mud clasts and drapes are common. Organics are preferentially concentrated in the mud & muddy sand laminae. Lower contact is sharp. No bioturbation observed. No shells.

INTERPRETATION
-Abundance of mud plus "rhythmic" mud/sand alternations suggests tidal influence
-Abundance of coaly organic debris suggests proximity to a terrestrial organic source
-Lack of bioturbation suggests stressed environment

--> Tidal channel deposits

mud   sand   gravel

LARGE TIDAL BAR
CORE: YSDP107

LOCATION ★

KOREA

YSDP-107

0 m
★ 7,412 ±62 cal BP
10 m
20 m
30 m
★ 11,475 ± 144 cal BP
40 m
13,062 ± 128 cal BP
★
50 m
★ 31,943 ± 228 cal BP
60 m
Mud Sand Gravel

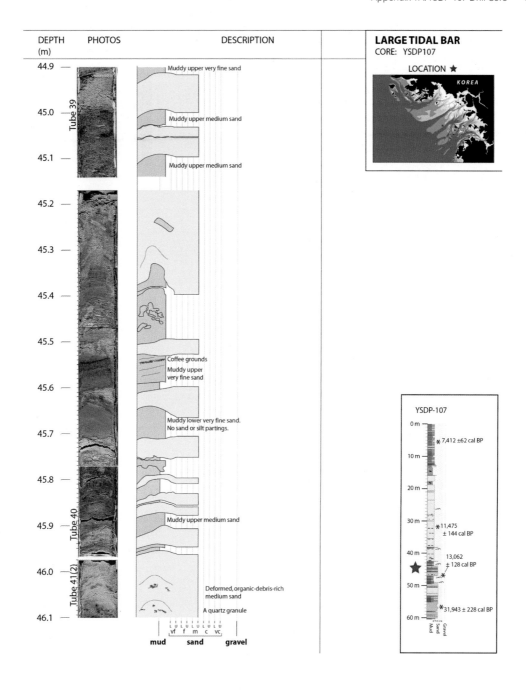

DEPTH (m) | PHOTOS | DESCRIPTION

44.9 — Muddy upper very fine sand

Tube 39

45.0 — Muddy upper medium sand

45.1 — Muddy upper medium sand

45.2 —

45.3 —

45.4 —

45.5 —

Coffee grounds
Muddy upper very fine sand

45.6 —

Muddy lower very fine sand. No sand or silt partings.

45.7 —

45.8 —

Tube 40

45.9 — Muddy upper medium sand

Tube 41(2)

46.0 —

Deformed, organic-debris-rich medium sand

A quartz granule

46.1 —

vf f m c vc
mud   sand   gravel

LARGE TIDAL BAR
CORE: YSDP107

LOCATION ★

KOREA

YSDP-107

0 m

✱ 7,412 ±62 cal BP

10 m

20 m

30 m

✱ 11,475 ± 144 cal BP

40 m

13,062 ± 128 cal BP

★   ✱

50 m

✱ 31,943 ± 228 cal BP

60 m

Mud Sand Gravel

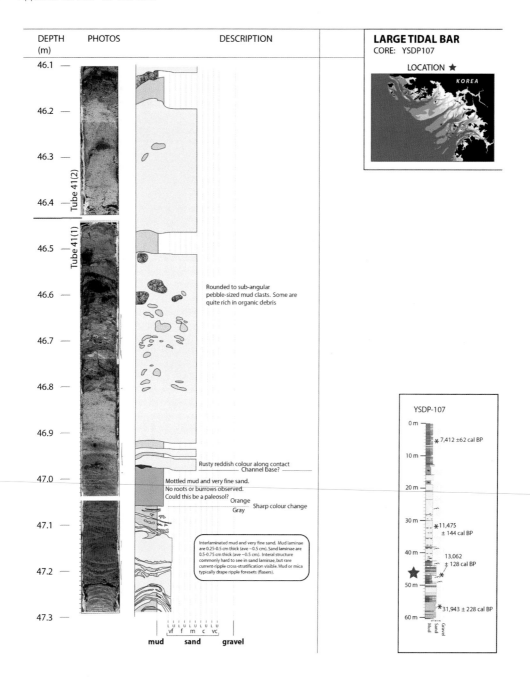

DEPTH (m)   PHOTOS   DESCRIPTION

LARGE TIDAL BAR
CORE: YSDP107

LOCATION ★

KOREA

Tube 41(2)

Tube 41(1)

Rounded to sub-angular pebble-sized mud clasts. Some are quite rich in organic debris

Rusty reddish colour along contact
Channel base?

Mottled mud and very fine sand.
No roots or burrows observed.
Could this be a paleosol?   Orange

Gray   Sharp colour change

Interlaminated mud and very fine sand. Mud laminae are 0.25-0.5 cm thick (ave ~0.5 cm). Sand laminae are 0.5-0.75 cm thick (ave ~0.5 cm). Interal structure commonly hard to see in sand laminae, but rare current-ripple cross-stratification visible. Mud or mica typically drape ripple foresets (flasers).

mud   sand   gravel

vf f m c vc

YSDP-107

0 m

★ 7,412 ±62 cal BP

10 m

20 m

30 m

★ 11,475 ± 144 cal BP

40 m

13,062 ± 128 cal BP

★ 31,943 ± 228 cal BP

50 m

60 m

Mud Sand Gravel

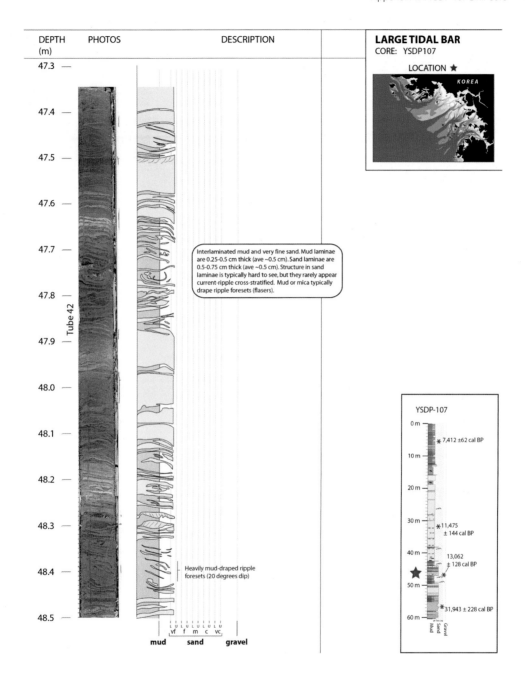

DEPTH (m) / PHOTOS / DESCRIPTION

LARGE TIDAL BAR
CORE: YSDP107

LOCATION ★

KOREA

Interlaminated mud and very fine sand. Mud laminae are 0.25-0.5 cm thick (ave ~0.5 cm). Sand laminae are 0.5-0.75 cm thick (ave ~0.5 cm). Structure in sand laminae is typically hard to see, but they rarely appear current-ripple cross-stratified. Mud or mica typically drape ripple foresets (flasers).

Heavily mud-draped ripple foresets (20 degrees dip)

Tube 42

mud   sand   gravel
vf f m c vc

YSDP-107

0 m
✳ 7,412 ±62 cal BP
10 m
20 m
30 m
✳ 11,475 ± 144 cal BP
40 m
13,062 ± 128 cal BP
★   ✳
50 m
✳ 31,943 ± 228 cal BP
60 m
Mud Sand Gravel

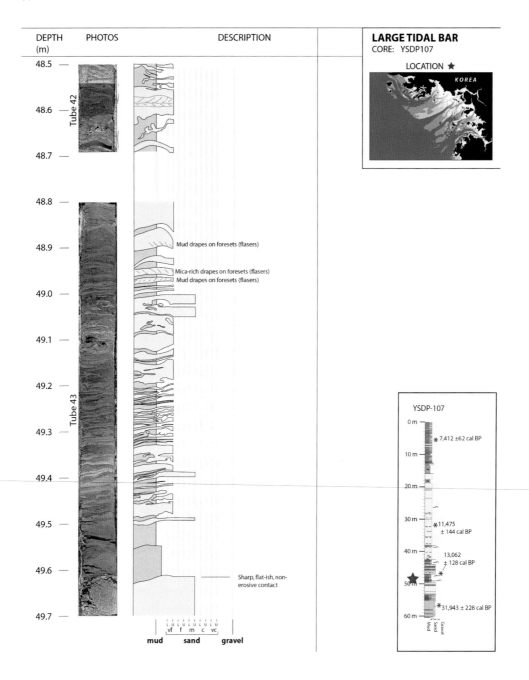

DEPTH (m) | PHOTOS | DESCRIPTION

Tube 42

Tube 43

Mud drapes on foresets (flasers)

Mica-rich drapes on foresets (flasers)
Mud drapes on foresets (flasers)

Sharp, flat-ish, non-erosive contact

mud   sand   gravel
vf f m c vc

LARGE TIDAL BAR
CORE: YSDP107

LOCATION ★

KOREA

YSDP-107

0 m

★ 7,412 ±62 cal BP

10 m

20 m

30 m

★ 11,475 ± 144 cal BP

40 m

13,062 ± 128 cal BP

★

50 m

★ 31,943 ± 228 cal BP

60 m

Mud Sand Gravel

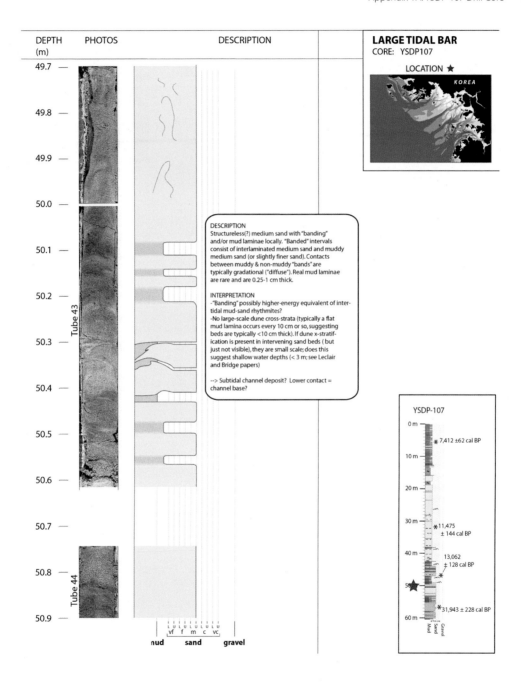

DEPTH (m) / PHOTOS / DESCRIPTION

Tube 43

Tube 44

LARGE TIDAL BAR
CORE: YSDP107

LOCATION ★

KOREA

DESCRIPTION
Structureless(?) medium sand with "banding" and/or mud laminae locally. "Banded" intervals consist of interlaminated medium sand and muddy medium sand (or slightly finer sand). Contacts between muddy & non-muddy "bands" are typically gradational ("diffuse"). Real mud laminae are rare and are 0.25-1 cm thick.

INTERPRETATION
- "Banding" possibly higher-energy equivalent of inter-tidal mud-sand rhythmites?
- No large-scale dune cross-strata (typically a flat mud lamina occurs every 10 cm or so, suggesting beds are typically <10 cm thick). If dune x-stratification is present in intervening sand beds (but just not visible), they are small scale; does this suggest shallow water depths (< 3 m; see Leclair and Bridge papers)

--> Subtidal channel deposit? Lower contact = channel base?

mud   sand   gravel
vf f m c vc

YSDP-107

0 m
※ 7,412 ±62 cal BP
10 m
20 m
30 m
※ 11,475 ± 144 cal BP
40 m
13,062 ± 128 cal BP
※
50 m
※ 31,943 ± 228 cal BP
60 m
Mud Sand Gravel

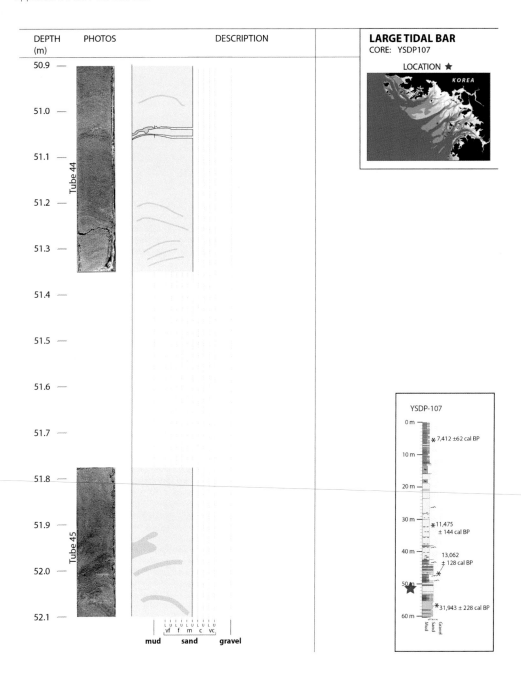

DEPTH (m)	PHOTOS	DESCRIPTION

LARGE TIDAL BAR
CORE: YSDP107

LOCATION ★

KOREA

YSDP-107

0 m
✱ 7,412 ±62 cal BP
10 m
20 m
30 m
✱ 11,475 ± 144 cal BP
40 m
13,062 ± 128 cal BP
✱
50 m
★
✱ 31,943 ± 228 cal BP
60 m

mud   sand   gravel

vf f m c vc

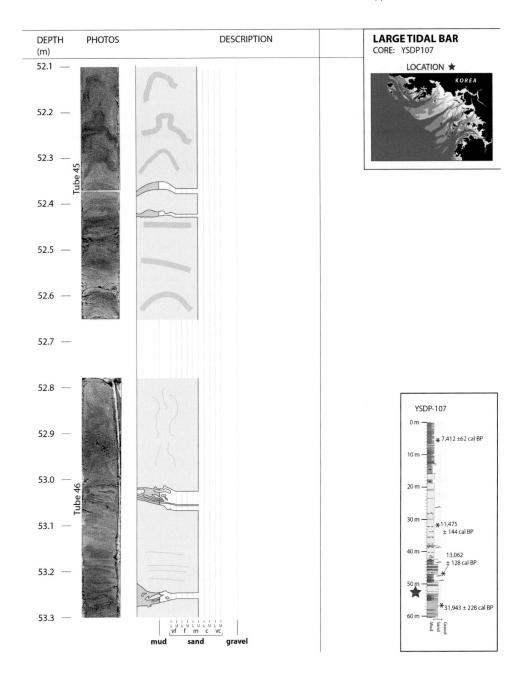

DEPTH (m)   PHOTOS   DESCRIPTION

Tube 45

Tube 46

mud   sand   gravel

vf f m c vc

LARGE TIDAL BAR
CORE: YSDP107

LOCATION ★

KOREA

YSDP-107

0 m

* 7,412 ±62 cal BP

10 m

20 m

30 m

* 11,475 ± 144 cal BP

40 m

13,062 ± 128 cal BP

*

50 m

★

* 31,943 ± 228 cal BP

60 m

Mud   Sand   Gravel

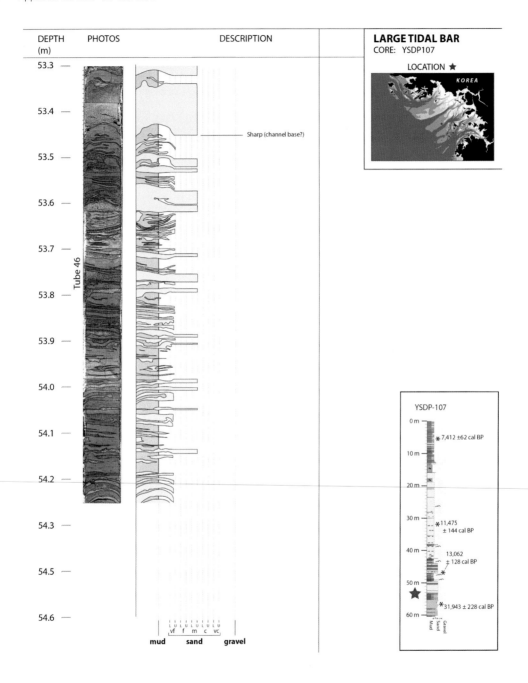

LARGE TIDAL BAR
CORE: YSDP107

LOCATION ★

DEPTH (m)   PHOTOS   DESCRIPTION

Tube 46

Sharp (channel base?)

mud   sand   gravel

vf f m c vc

YSDP-107

0 m
*7,412 ±62 cal BP
10 m
20 m
30 m
*11,475 ± 144 cal BP
40 m
13,062 ± 128 cal BP
*
50 m
★
*31,943 ± 228 cal BP
60 m

Mud Sand Gravel

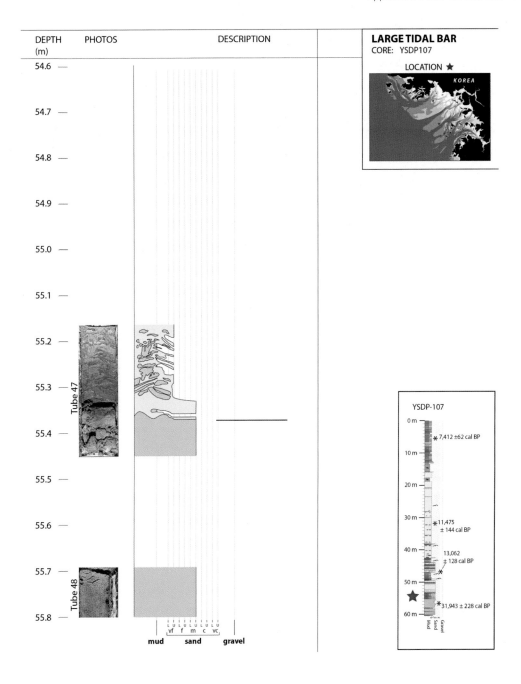

DEPTH
(m)

PHOTOS

DESCRIPTION

54.6

54.7

54.8

54.9

55.0

55.1

55.2

55.3

55.4

55.5

55.6

55.7

55.8

Tube 47

Tube 48

mud   sand   gravel

vf f m c vc

LARGE TIDAL BAR
CORE: YSDP107

LOCATION ★

KOREA

YSDP-107

0 m

*7,412 ±62 cal BP

10 m

20 m

30 m

*11,475
± 144 cal BP

40 m

13,062
± 128 cal BP
*

50 m

*31,943 ± 228 cal BP

60 m

Mud Sand Gravel

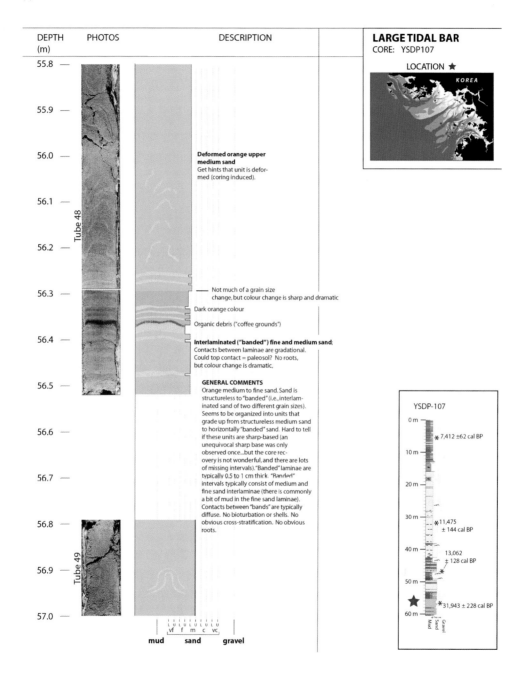

DEPTH (m)	PHOTOS	DESCRIPTION

Deformed orange upper medium sand
Get hints that unit is deformed (coring induced).

— Not much of a grain size change, but colour change is sharp and dramatic

Dark orange colour

Organic debris ("coffee grounds")

Interlaminated ("banded") fine and medium sand;
Contacts between laminae are gradational.
Could top contact = paleosol? No roots, but colour change is dramatic.

GENERAL COMMENTS
Orange medium to fine sand. Sand is structureless to "banded" (i.e., interlaminated sand of two different grain sizes). Seems to be organized into units that grade up from structureless medium sand to horizontally "banded" sand. Hard to tell if these units are sharp-based (an unequivocal sharp base was only observed once...but the core recovery is not wonderful, and there are lots of missing intervals). "Banded" laminae are typically 0.5 to 1 cm thick. "Banded" intervals typically consist of medium and fine sand interlaminae (there is commonly a bit of mud in the fine sand laminae). Contacts between "bands" are typically diffuse. No bioturbation or shells. No obvious cross-stratification. No obvious roots.

Tube 48

Tube 49

mud   sand   gravel

LARGE TIDAL BAR
CORE: YSDP107

LOCATION ★

KOREA

YSDP-107

0 m
✱ 7,412 ±62 cal BP
10 m
20 m
30 m
✱ 11,475 ± 144 cal BP
40 m
13,062 ± 128 cal BP
✱
50 m
★
✱ 31,943 ± 228 cal BP
60 m
Mud Sand Gravel

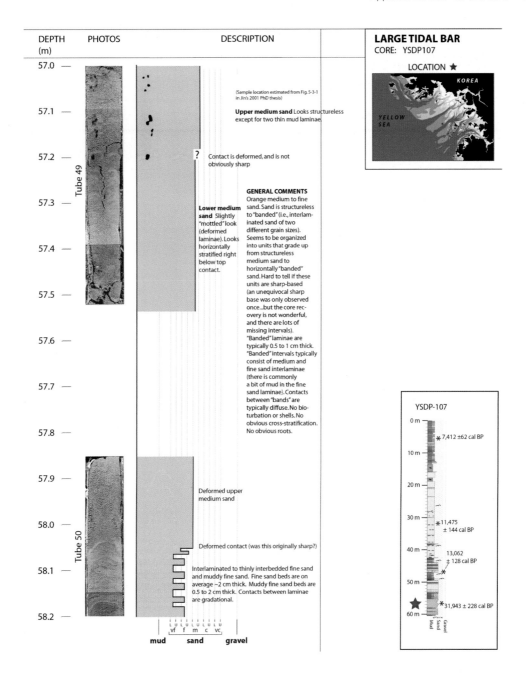

DEPTH (m) PHOTOS DESCRIPTION

Tube 49

(Sample location estimated from Fig. 5-3-1 in Jin's 2001 PhD thesis)

Upper medium sand Looks structureless except for two thin mud laminae.

? Contact is deformed, and is not obviously sharp

Lower medium sand Slightly "mottled" look (deformed laminae). Looks horizontally stratified right below top contact.

GENERAL COMMENTS
Orange medium to fine sand. Sand is structureless to "banded" (i.e., interlaminated sand of two different grain sizes). Seems to be organized into units that grade up from structureless medium sand to horizontally "banded" sand. Hard to tell if these units are sharp-based (an unequivocal sharp base was only observed once...but the core recovery is not wonderful, and there are lots of missing intervals). "Banded" laminae are typically 0.5 to 1 cm thick. "Banded" intervals typically consist of medium and fine sand interlaminae (there is commonly a bit of mud in the fine sand laminae). Contacts between "bands" are typically diffuse. No bioturbation or shells. No obvious cross-stratification. No obvious roots.

Tube 50

Deformed upper medium sand

Deformed contact (was this originally sharp?)

Interlaminated to thinly interbedded fine sand and muddy fine sand. Fine sand beds are on average ~2 cm thick. Muddy fine sand beds are 0.5 to 2 cm thick. Contacts between laminae are gradational.

mud   sand   gravel
vf f m c vc

LARGE TIDAL BAR
CORE: YSDP107

LOCATION ★

KOREA

YELLOW SEA

YSDP-107

0 m
✳ 7,412 ±62 cal BP

10 m

20 m

30 m
✳ 11,475 ± 144 cal BP

40 m
13,062 ± 128 cal BP
✳

50 m
✳ 31,943 ± 228 cal BP

60 m

Mud Sand Gravel

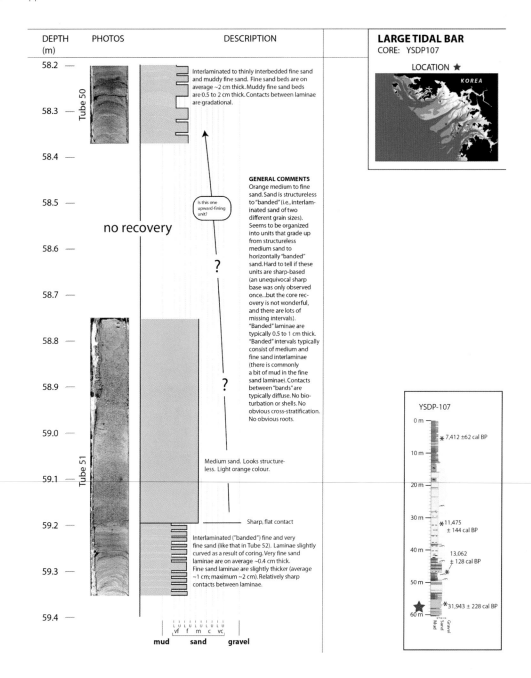

DEPTH (m) | **PHOTOS** | **DESCRIPTION**

58.2 — Tube 50 — Interlaminated to thinly interbedded fine sand and muddy fine sand. Fine sand beds are on average ~2 cm thick. Muddy fine sand beds are 0.5 to 2 cm thick. Contacts between laminae are gradational.

58.3 —

58.4 —

58.5 —

Is this one upward-fining unit?

no recovery

58.6 —

GENERAL COMMENTS
Orange medium to fine sand. Sand is structureless to "banded" (i.e., interlaminated sand of two different grain sizes). Seems to be organized into units that grade up from structureless medium sand to horizontally "banded" sand. Hard to tell if these units are sharp-based (an unequivocal sharp base was only observed once...but the core recovery is not wonderful, and there are lots of missing intervals). "Banded" laminae are typically 0.5 to 1 cm thick. "Banded" intervals typically consist of medium and fine sand interlaminae (there is commonly a bit of mud in the fine sand laminae). Contacts between "bands" are typically diffuse. No bioturbation or shells. No obvious cross-stratification. No obvious roots.

58.7 —

?

58.8 —

58.9 —

?

59.0 —

Tube 51

59.1 —

Medium sand. Looks structureless. Light orange colour.

59.2 —

Sharp, flat contact

Interlaminated ("banded") fine and very fine sand (like that in Tube 52). Laminae slightly curved as a result of coring. Very fine sand laminae are on average ~0.4 cm thick. Fine sand laminae are slightly thicker (average ~1 cm; maximum ~2 cm). Relatively sharp contacts between laminae.

59.3 —

59.4 —

L U L U L U L U L U
vf f m c vc
mud   sand   gravel

LARGE TIDAL BAR
CORE: YSDP107

LOCATION ★

KOREA

YSDP-107

0 m
*7,412 ±62 cal BP
10 m
20 m
30 m
*11,475 ± 144 cal BP
40 m
13,062 ± 128 cal BP
*
50 m
*31,943 ± 228 cal BP
60 m
Mud Sand Gravel

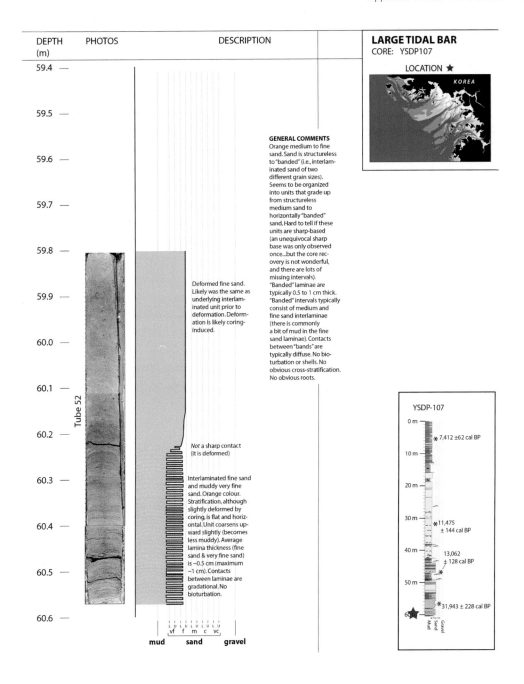

DEPTH (m)   PHOTOS   DESCRIPTION

LARGE TIDAL BAR
CORE: YSDP107

LOCATION ★

KOREA

GENERAL COMMENTS
Orange medium to fine sand. Sand is structureless to "banded" (i.e., interlaminated sand of two different grain sizes). Seems to be organized into units that grade up from structureless medium sand to horizontally "banded" sand. Hard to tell if these units are sharp-based (an unequivocal sharp base was only observed once...but the core recovery is not wonderful, and there are lots of missing intervals). "Banded" laminae are typically 0.5 to 1 cm thick. "Banded" intervals typically consist of medium and fine sand interlaminae (there is commonly a bit of mud in the fine sand laminae). Contacts between "bands" are typically diffuse. No bioturbation or shells. No obvious cross-stratification. No obvious roots.

Tube 52

Deformed fine sand. Likely was the same as underlying interlaminated unit prior to deformation. Deformation is likely coring-induced.

Not a sharp contact (it is deformed)

Interlaminated fine sand and muddy very fine sand. Orange colour. Stratification, although slightly deformed by coring, is flat and horizontal. Unit coarsens upward slightly (becomes less muddy). Average lamina thickness (fine sand & very fine sand) is ~0.5 cm (maximum ~1 cm). Contacts between laminae are gradational. No bioturbation.

L U L U L U L U L U
vf   f   m   c   vc
mud   sand   gravel

YSDP-107

0 m

✳ 7,412 ±62 cal BP

10 m

20 m

30 m

✳ 11,475 ± 144 cal BP

40 m

13,062 ± 128 cal BP
✳

50 m

✳ 31,943 ± 228 cal BP

60
★
                vf   f   m   c
                Sand   Gravel
        Mud

APPENDIX 1B: NYSDP-101 DRILL CORE

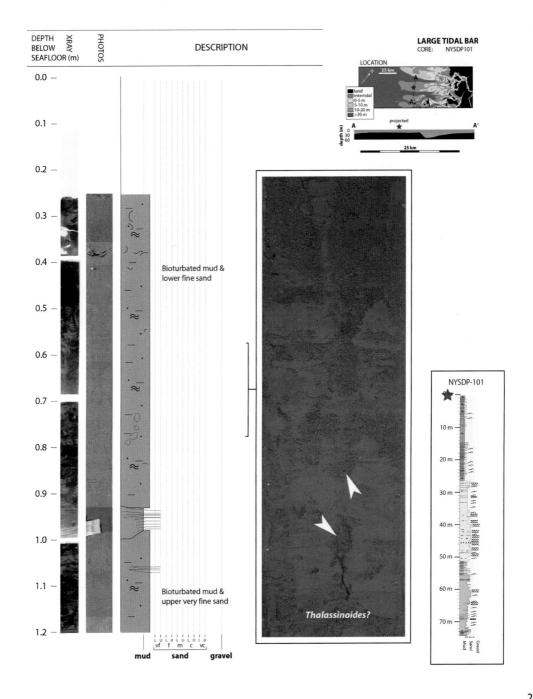

DEPTH BELOW SEAFLOOR (m)

XRAY

PHOTOS

DESCRIPTION

0.0

0.1

0.2

0.3

0.4

0.5

0.6

0.7

0.8

0.9

1.0

1.1

1.2

Bioturbated mud & lower fine sand

Bioturbated mud & upper very fine sand

mud   sand   gravel

L U L U L U L U L U
vf f m c vc

LARGE TIDAL BAR
CORE: NYSDP101

LOCATION

25 km

land
intertidal
0-5 m
5-10 m
10-20 m
>20 m

projected

A   A'
depth (m) 0
30
60

25 km

Thalassinoides?

NYSDP-101

10 m

20 m

30 m

40 m

50 m

60 m

70 m

Gravel
Sand
Mud

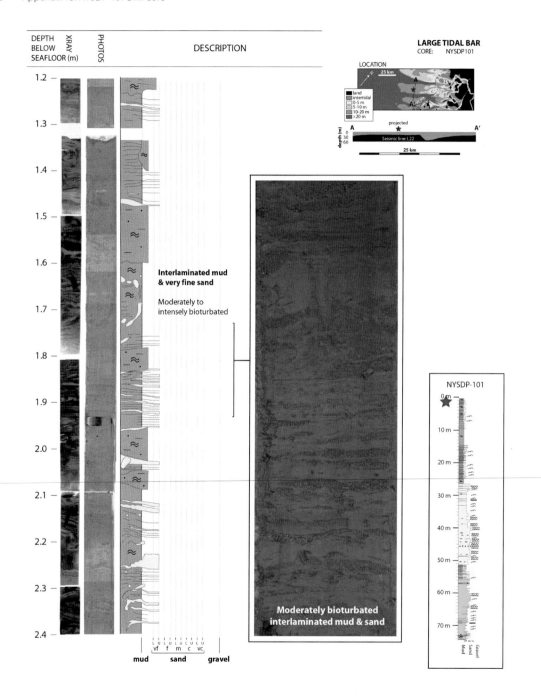

DEPTH BELOW SEAFLOOR (m)

XRAY

PHOTOS

DESCRIPTION

Interlaminated mud & very fine sand

Moderately to intensely bioturbated

mud   sand   gravel

vf f m c vc

LARGE TIDAL BAR
CORE: NYSDP101

LOCATION

25 km

land
intertidal
0–5 m
5–10 m
10–20 m
>20 m

A   projected   A'

depth (m)   0   30   60

Seismic line: L22

25 km

Moderately bioturbated interlaminated mud & sand

NYSDP-101

0 m

10 m

20 m

30 m

40 m

50 m

60 m

70 m

Gravel
Sand
Mud

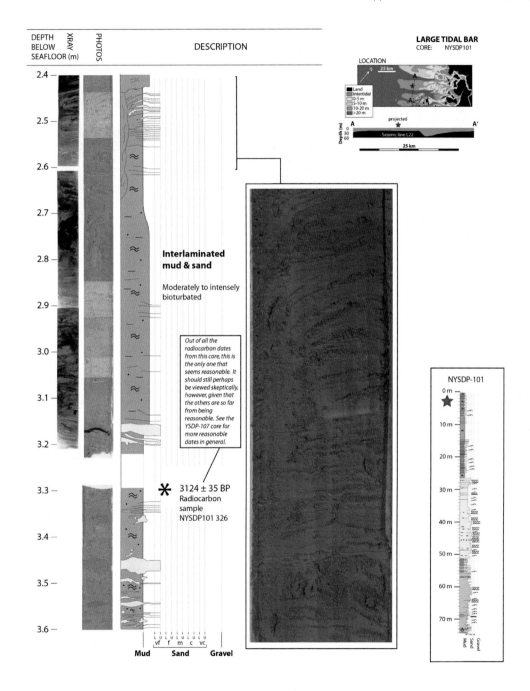

DEPTH BELOW SEAFLOOR (m)

XRAY

PHOTOS

DESCRIPTION

Interlaminated mud & sand

Moderately to intensely bioturbated

Out of all the radiocarbon dates from this core, this is the only one that seems reasonable. It should still perhaps be viewed skeptically, however, given that the others are so far from being reasonable. See the YSDP-107 core for more reasonable dates in general.

✱ 3124 ± 35 BP Radiocarbon sample NYSDP101 326

vf f m c vc

Mud   Sand   Gravel

LARGE TIDAL BAR
CORE: NYSDP101

LOCATION

25 km

Land
Intertidal
0–5 m
5–10 m
10–20 m
> 20 m

projected

A   A'

Depth (m) 0 30 60

Seismic line: L22

25 km

NYSDP-101

0 m

10 m

20 m

30 m

40 m

50 m

60 m

70 m

Gravel
Sand
Mud

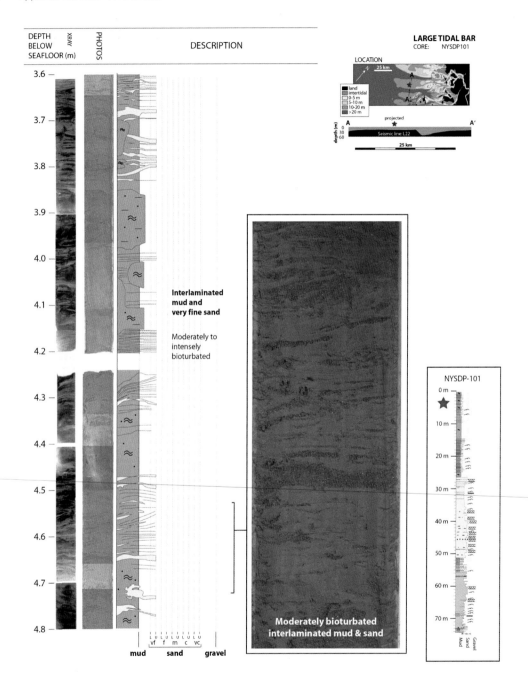

DEPTH BELOW SEAFLOOR (m)

XRAY

PHOTOS

DESCRIPTION

LARGE TIDAL BAR
CORE: NYSDP101

LOCATION
25 km

land
intertidal
0-5 m
5-10 m
10-20 m
>20 m

A
depth (m)
0
30
60
projected
Seismic line: L22
A'
25 km

Interlaminated mud and very fine sand

Moderately to intensely bioturbated

mud   sand   gravel

vf f m c vc

Moderately bioturbated interlaminated mud & sand

NYSDP-101
0 m
10 m
20 m
30 m
40 m
50 m
60 m
70 m
Mud Sand Gravel

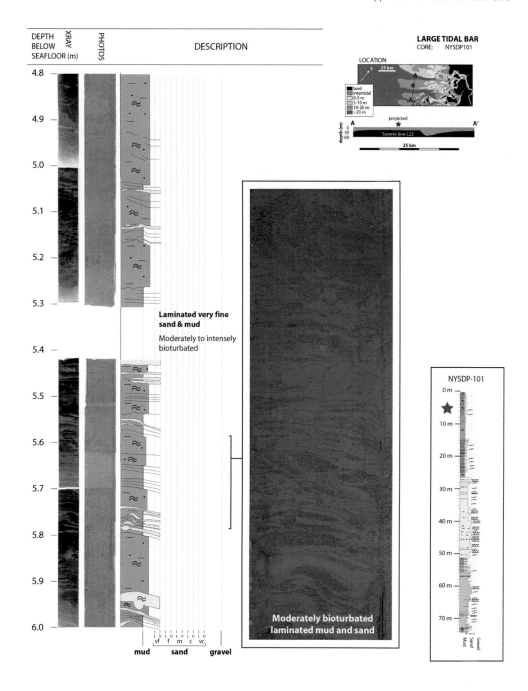

DEPTH BELOW SEAFLOOR (m)

XRAY

PHOTOS

DESCRIPTION

LARGE TIDAL BAR
CORE: NYSDP101

LOCATION

25 km

land
intertidal
0-5 m
5-10 m
10-20 m
>20 m

projected

A   A'

Seismic line:L22

25 km

**Laminated very fine
sand & mud**

Moderately to intensely
bioturbated

mud   sand   gravel

L U L U L U L U L U
vf   f   m   c   vc

**Moderately bioturbated
laminated mud and sand**

NYSDP-101

0 m

10 m

20 m

30 m

40 m

50 m

60 m

70 m

Gravel
Sand
Mud

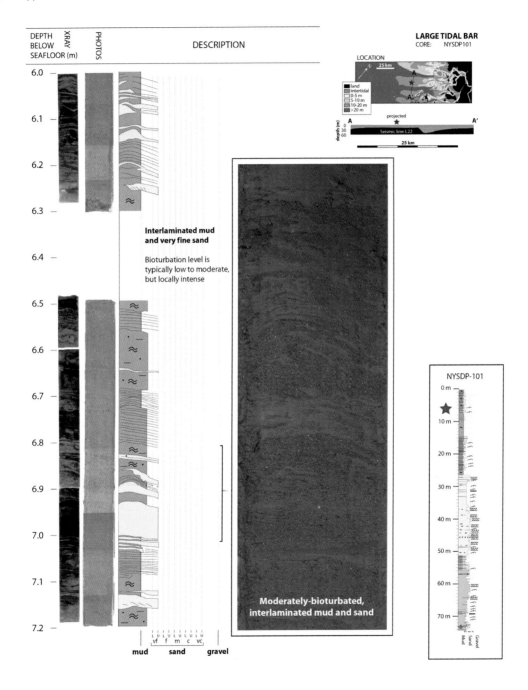

LARGE TIDAL BAR
CORE: NYSDP101

DEPTH BELOW SEAFLOOR (m)

XRAY

PHOTOS

DESCRIPTION

Interlaminated mud and very fine sand

Bioturbation level is typically low to moderate, but locally intense

Moderately-bioturbated, interlaminated mud and sand

LOCATION

25 km

land
intertidal
0–5 m
5–10 m
10–20 m
>20 m

projected

depth (m)   A   Seismic line: L22   A'
0
30
60

25 km

NYSDP-101

0 m

10 m

20 m

30 m

40 m

50 m

60 m

70 m

Mud Sand Gravel

mud   sand   gravel

vf f m c vc

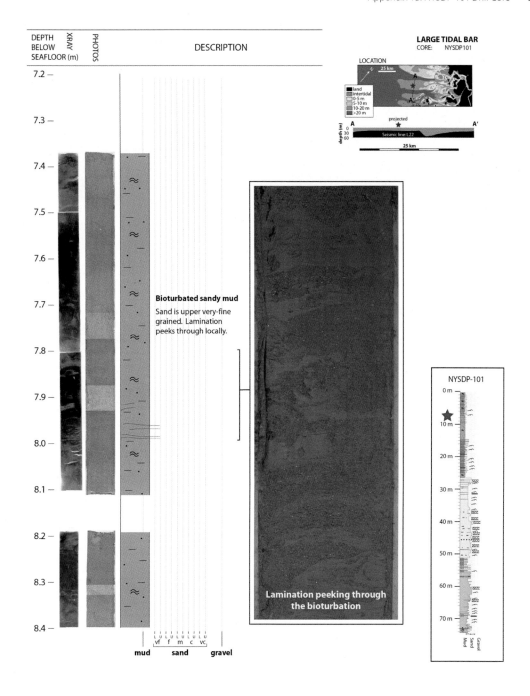

DEPTH BELOW SEAFLOOR (m)

XRAY

PHOTOS

DESCRIPTION

Bioturbated sandy mud
Sand is upper very-fine grained. Lamination peeks through locally.

Lamination peeking through the bioturbation

LARGE TIDAL BAR
CORE: NYSDP101

LOCATION

land
intertidal
0-5 m
5-10 m
10-20 m
>20 m

projected
Seismic line:L22

mud   sand   gravel

NYSDP-101

Gravel
Sand
Mud

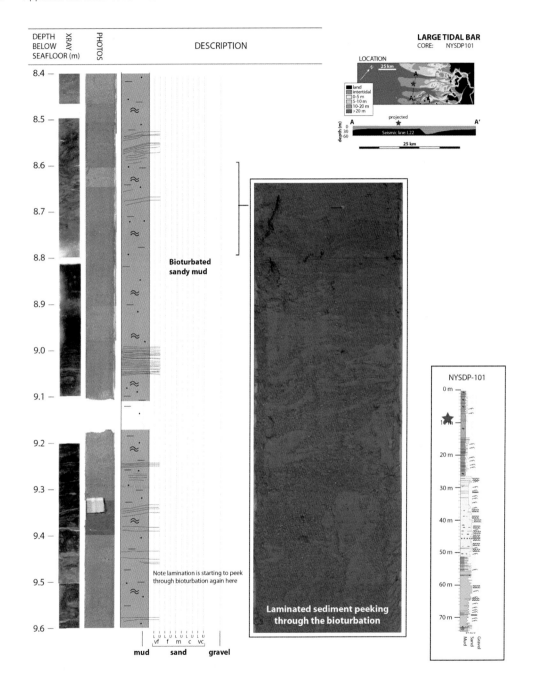

LARGE TIDAL BAR
CORE: NYSDP101

DEPTH BELOW SEAFLOOR (m)

XRAY

PHOTOS

DESCRIPTION

Bioturbated sandy mud

Note lamination is starting to peek through bioturbation again here

mud   sand   gravel

Laminated sediment peeking through the bioturbation

NYSDP-101

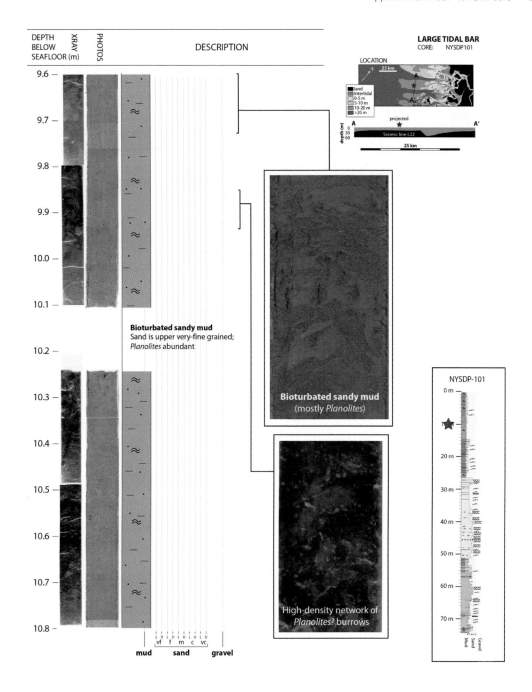

DEPTH BELOW SEAFLOOR (m)

XRAY

PHOTOS

DESCRIPTION

LARGE TIDAL BAR
CORE: NYSDP101

LOCATION

Bioturbated sandy mud
Sand is upper very-fine grained;
Planolites abundant

Bioturbated sandy mud
(mostly *Planolites*)

High-density network of
Planolites? burrows

mud   sand   gravel

NYSDP-101

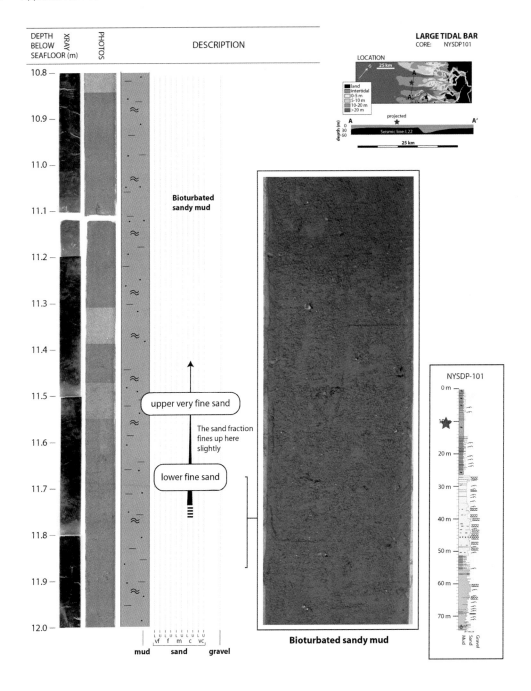

DEPTH BELOW SEAFLOOR (m)

XRAY

PHOTOS

DESCRIPTION

Bioturbated sandy mud

upper very fine sand

The sand fraction fines up here slightly

lower fine sand

mud   sand   gravel

LARGE TIDAL BAR
CORE: NYSDP101

LOCATION

land
intertidal
0-5 m
5-10 m
10-20 m
>20 m

projected

depth (m)

Seismic line: L22

25 km

Bioturbated sandy mud

NYSDP-101

0 m

20 m

30 m

40 m

50 m

60 m

70 m

Gravel
Sand
Mud

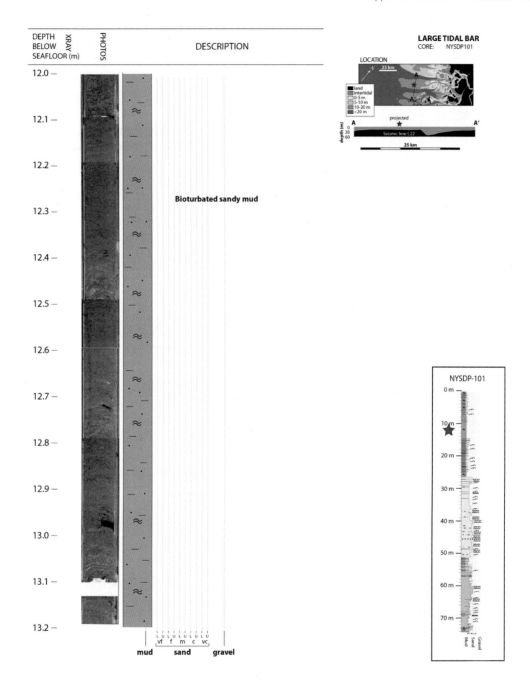

DEPTH BELOW SEAFLOOR (m)

XRAY

PHOTOS

DESCRIPTION

Bioturbated sandy mud

L U L U L U L U L U
vf   f   m   c   vc
mud   sand   gravel

LARGE TIDAL BAR
CORE: NYSDP101

LOCATION

land
intertidal
0-5 m
5-10 m
10-20 m
>20 m

projected

A   A'
Seismic line: L22

25 km

NYSDP-101

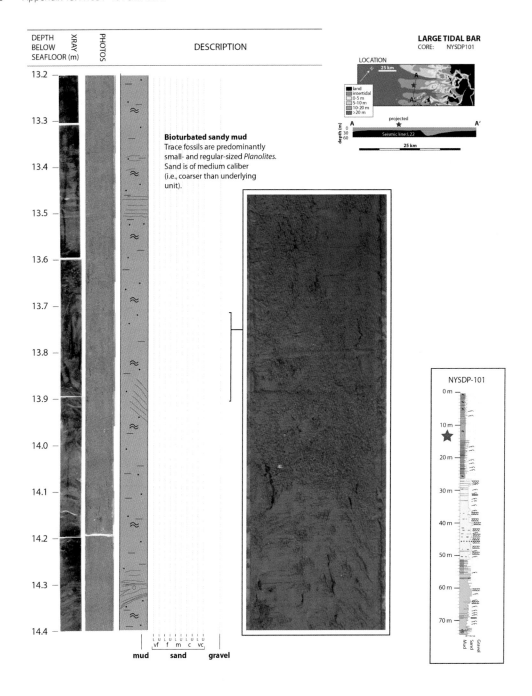

DEPTH BELOW SEAFLOOR (m)

XRAY

PHOTOS

DESCRIPTION

Bioturbated sandy mud
Trace fossils are predominantly small- and regular-sized *Planolites*. Sand is of medium caliber (i.e., coarser than underlying unit).

mud   sand   gravel

vf f m c vc

LARGE TIDAL BAR
CORE: NYSDP101

LOCATION

25 km

land
intertidal
0–5 m
5–10 m
10–20 m
>20 m

projected

A   A'

depth (m)   0 30 60

Seismic line: L22

25 km

NYSDP-101

0 m

10 m

20 m

30 m

40 m

50 m

60 m

70 m

Mud Sand Gravel

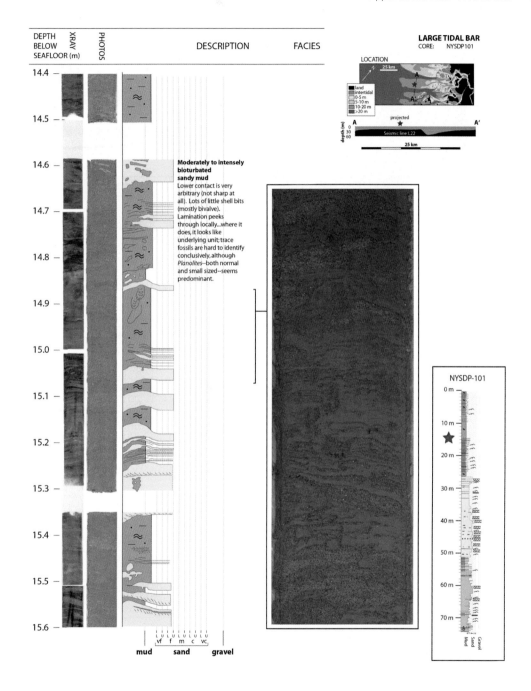

DEPTH BELOW SEAFLOOR (m)

XRAY

PHOTOS

DESCRIPTION

FACIES

Moderately to intensely bioturbated sandy mud
Lower contact is very arbitrary (not sharp at all). Lots of little shell bits (mostly bivalve). Lamination peeks through locally...where it does, it looks like underlying unit; trace fossils are hard to identify conclusively, although *Planolites*--both normal and small sized--seems predominant.

mud   sand   gravel

LARGE TIDAL BAR
CORE: NYSDP101

LOCATION

land
intertidal
0-5 m
5-10 m
10-20 m
>20 m

projected

A   A'
Seismic line L22
25 km

NYSDP-101

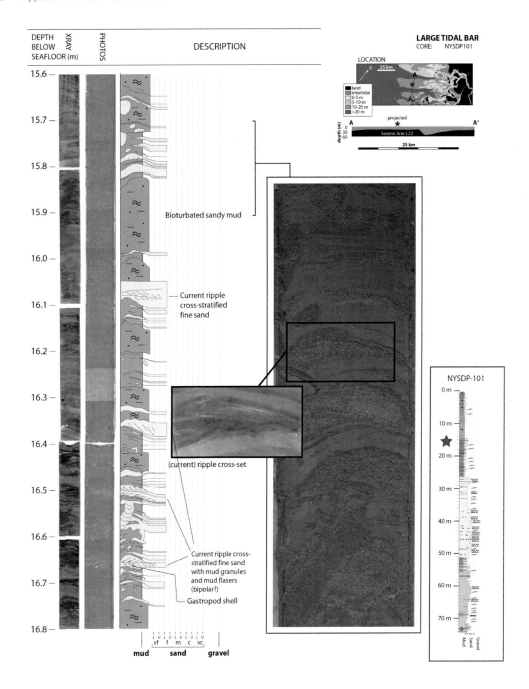

DEPTH BELOW SEAFLOOR (m)

XRAY

PHOTOS

DESCRIPTION

Bioturbated sandy mud

Current ripple cross-stratified fine sand

(current) ripple cross-set

Current ripple cross-stratified fine sand with mud granules and mud flasers (bipolar?)

Gastropod shell

mud   sand   gravel

LARGE TIDAL BAR
CORE: NYSDP101

LOCATION

land
intertidal
0-5 m
5-10 m
10-20 m
>20 m

projected

A   A'
depth (m)
0
30
60
Seismic line: L22

25 km

NYSDP-101

0 m

10 m

20 m

30 m

40 m

50 m

60 m

70 m

Gravel
Sand
Mud

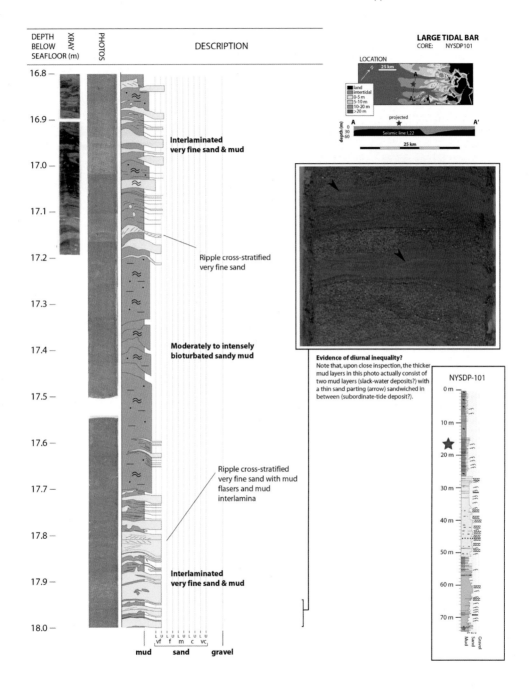

LARGE TIDAL BAR
CORE: NYSDP101

LOCATION

Evidence of diurnal inequality?
Note that, upon close inspection, the thicker mud layers in this photo actually consist of two mud layers (slack-water deposits?) with a thin sand parting (arrow) sandwiched in between (subordinate-tide deposit?).

DEPTH BELOW SEAFLOOR (m)

XRAY

PHOTOS

DESCRIPTION

Interlaminated very fine sand & mud

Ripple cross-stratified very fine sand

Moderately to intensely bioturbated sandy mud

Ripple cross-stratified very fine sand with mud flasers and mud interlamina

Interlaminated very fine sand & mud

mud   sand   gravel

NYSDP-101

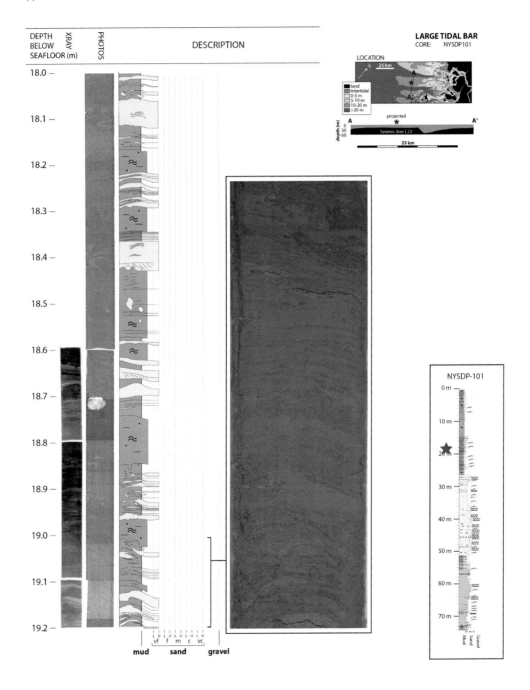

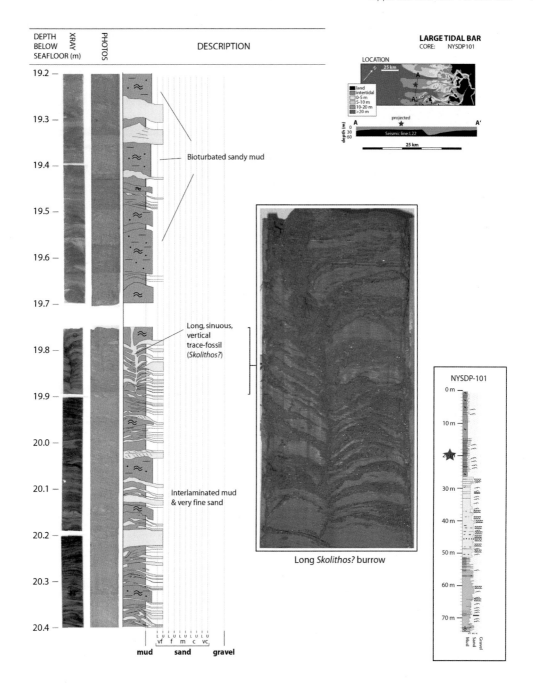

DEPTH BELOW SEAFLOOR (m)

XRAY

PHOTOS

DESCRIPTION

Bioturbated sandy mud

Long, sinuous, vertical trace-fossil (*Skolithos?*)

Interlaminated mud & very fine sand

mud   sand   gravel

vf f m c vc

LARGE TIDAL BAR
CORE: NYSDP101

LOCATION

25 km

land
intertidal
0-5 m
5-10 m
10-20 m
>20 m

projected

depth (m) A   A'
0
30
60   Seismic line: L22

25 km

Long *Skolithos?* burrow

NYSDP-101

0 m

10 m

30 m

40 m

50 m

60 m

70 m

Mud Sand Gravel

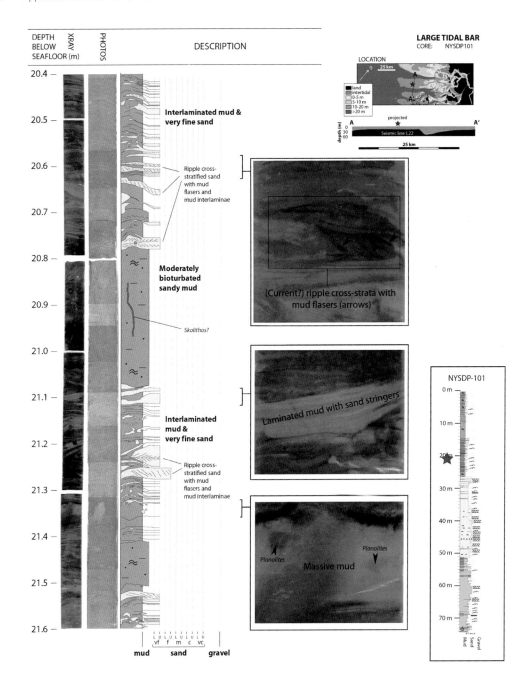

LARGE TIDAL BAR
CORE: NYSDP101

LOCATION

DEPTH BELOW SEAFLOOR (m)

XRAY

PHOTOS

DESCRIPTION

Interlaminated mud & very fine sand

Ripple cross-stratified sand with mud flasers and mud interlaminae

Moderately bioturbated sandy mud

Skolithos?

Interlaminated mud & very fine sand

Ripple cross-stratified sand with mud flasers and mud interlaminae

(Current?) ripple cross-strata with mud flasers (arrows)

Laminated mud with sand stringers

Planolites   *Planolites*

Massive mud

mud   sand   gravel

NYSDP-101

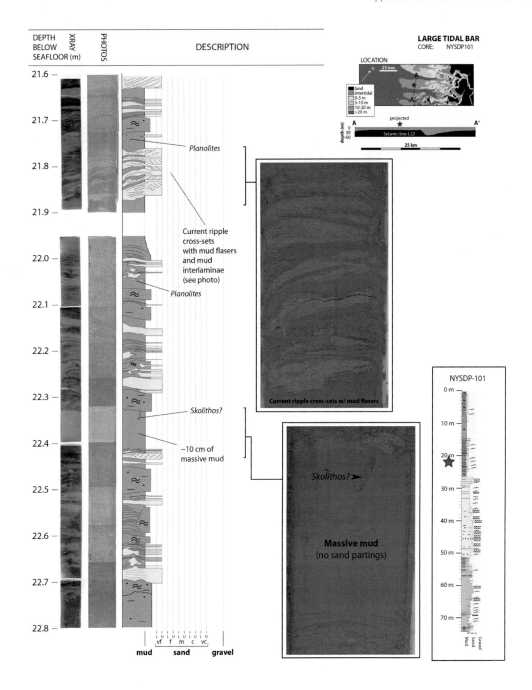

DEPTH BELOW SEAFLOOR (m)

XRAY

PHOTOS

DESCRIPTION

Planolites

Current ripple cross-sets with mud flasers and mud interlaminae (see photo)

Planolites

Skolithos?

~10 cm of massive mud

mud   sand   gravel

LARGE TIDAL BAR
CORE: NYSDP101

LOCATION

25 km

land
intertidal
0–5 m
5–10 m
10–20 m
>20 m

projected

A   A'
depth (m)   Seismic line: L22
25 km

Current ripple cross-sets w/ mud flasers

Skolithos?

Massive mud
(no sand partings)

NYSDP-101

0 m

10 m

20 m

30 m

40 m

50 m

60 m

70 m

Mud Sand Gravel

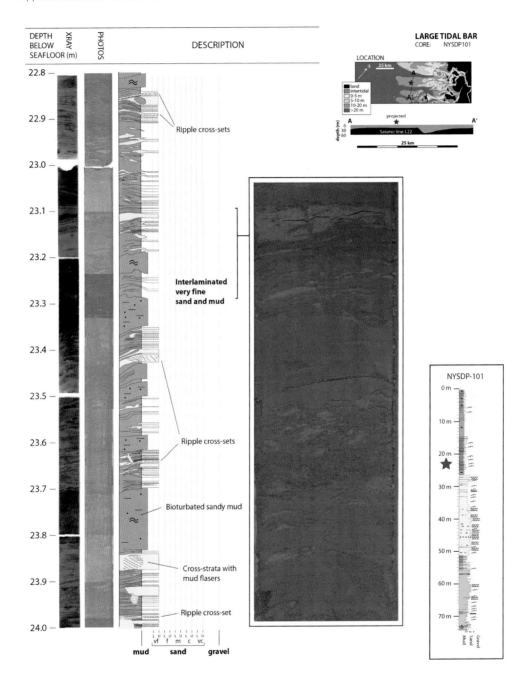

DEPTH BELOW SEAFLOOR (m)

XRAY

PHOTOS

DESCRIPTION

22.8

22.9 — Ripple cross-sets

23.0

23.1

23.2

23.3 — **Interlaminated very fine sand and mud**

23.4

23.5

23.6 — Ripple cross-sets

23.7

23.8 — Bioturbated sandy mud

23.9 — Cross-strata with mud flasers

24.0 — Ripple cross-set

L U L U L U L U L U
vf   f   m   c   vc
mud   sand   gravel

LARGE TIDAL BAR
CORE: NYSDP101

LOCATION
25 km

land
intertidal
0–5 m
5–10 m
10–20 m
>20 m

projected

depth (m)
A   0
          30
          60   A'
Seismic line: L22

25 km

NYSDP-101

0 m

10 m

20 m

30 m

40 m

50 m

60 m

70 m

Mud Sand Gravel

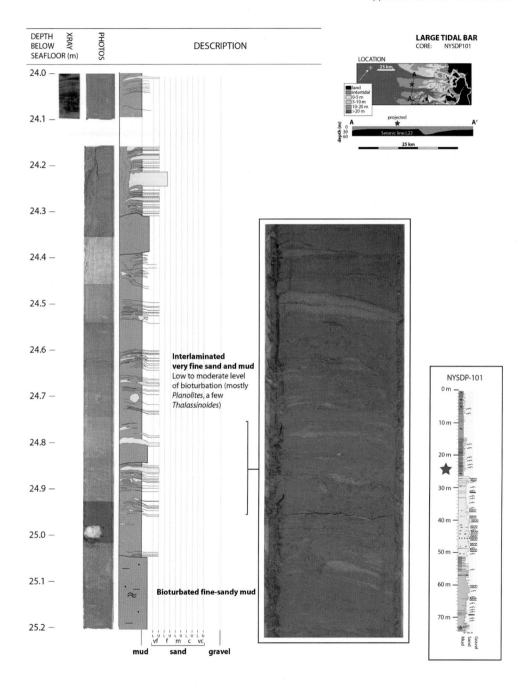

DEPTH BELOW SEAFLOOR (m)

XRAY

PHOTOS

DESCRIPTION

24.0

24.1

24.2

24.3

24.4

24.5

24.6

24.7

24.8

24.9

25.0

25.1

25.2

Interlaminated very fine sand and mud
Low to moderate level of bioturbation (mostly *Planolites*, a few *Thalassinoides*)

Bioturbated fine-sandy mud

L U L U L U L U L U
vf   f   m   c   vc

mud   sand   gravel

LARGE TIDAL BAR
CORE: NYSDP101

LOCATION

25 km

land
intertidal
0-5 m
5-10 m
10-20 m
>20 m

projected

A   A'
depth (m)
0
30
60
Seismic line:L22

25 km

NYSDP-101

0 m

10 m

20 m

30 m

40 m

50 m

60 m

70 m

Gravel
Sand
Mud

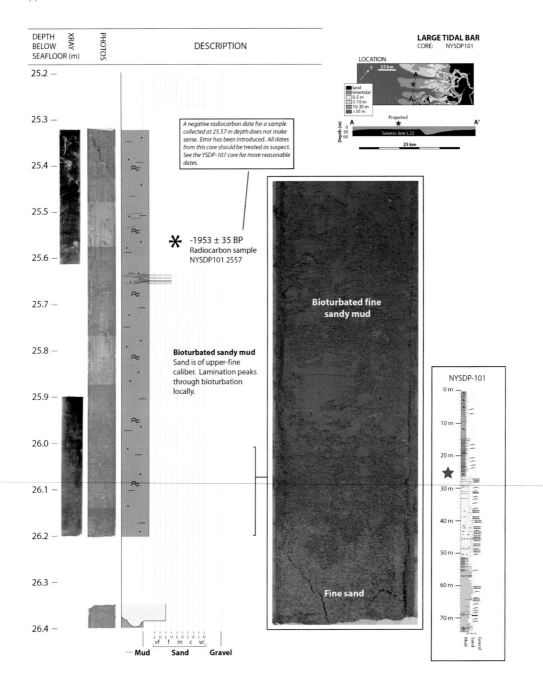

DEPTH BELOW SEAFLOOR (m)
XRAY
PHOTOS

DESCRIPTION

A negative radiocarbon date for a sample collected at 25.57 m depth does not make sense. Error has been introduced. All dates from this core should be treated as suspect. See the YSDP-107 core for more reasonable dates.

✳ -1953 ± 35 BP
Radiocarbon sample
NYSDP101 2557

Bioturbated sandy mud
Sand is of upper-fine caliber. Lamination peaks through bioturbation locally.

Mud   Sand   Gravel

L U L U L U L U L U
vf f   m   c vc

LARGE TIDAL BAR
CORE: NYSDP101

LOCATION
25 km

land
intertidal
0-5 m
5-10 m
10-20 m
>20 m

Projected
★
Depth (m)   A   A'
Seismic line: L22
25 km

Bioturbated fine sandy mud

Fine sand

NYSDP-101
0 m
10 m
20 m
★
30 m
40 m
50 m
60 m
70 m
Mud Sand Gravel

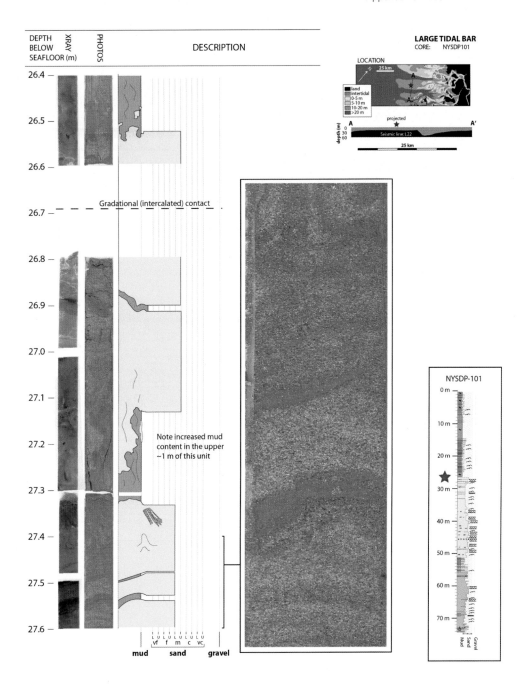

LARGE TIDAL BAR
CORE: NYSDP101

DEPTH BELOW SEAFLOOR (m) | XRAY | PHOTOS | DESCRIPTION

Gradational (intercalated) contact

Note increased mud content in the upper ~1 m of this unit

mud   sand   gravel

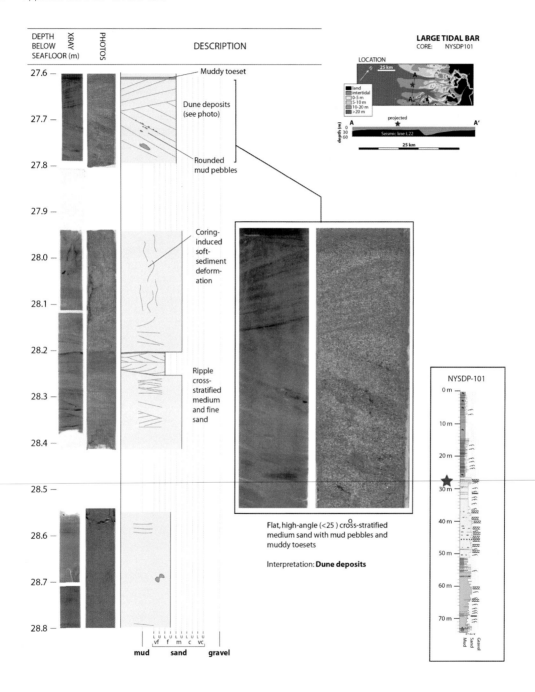

Flat, high-angle (<25°) cross-stratified medium sand with mud pebbles and muddy toesets

Interpretation: **Dune deposits**

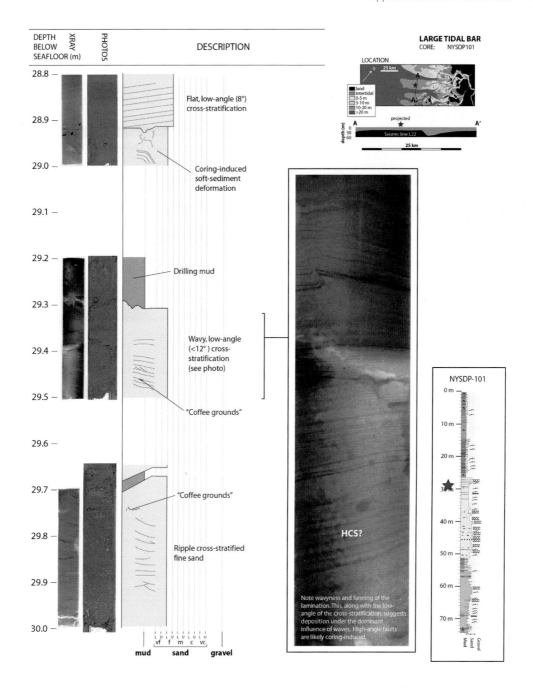

DEPTH BELOW SEAFLOOR (m)

XRAY

PHOTOS

DESCRIPTION

28.8

28.9

29.0

Flat, low-angle (8°) cross-stratification

Coring-induced soft-sediment deformation

29.1

29.2

29.3

Drilling mud

29.4

Wavy, low-angle (<12°) cross-stratification (see photo)

29.5

"Coffee grounds"

29.6

29.7

"Coffee grounds"

29.8

Ripple cross-stratified fine sand

29.9

30.0

L U L U L U L U L U
vf   f   m   c   vc

mud   sand   gravel

LARGE TIDAL BAR
CORE: NYSDP101

LOCATION

25 km

land
intertidal
0-5 m
5-10 m
10-20 m
>20 m

projected

A   A′
depth (m)
0
30
60
Seismic line: L22

25 km

HCS?

Note wavyness and fanning of the lamination. This, along with the low-angle of the cross-stratification, suggests deposition under the dominant influence of waves. High-angle faults are likely coring-induced.

NYSDP-101

0 m

10 m

20 m

30 m

40 m

50 m

60 m

70 m

Mud Sand Gravel

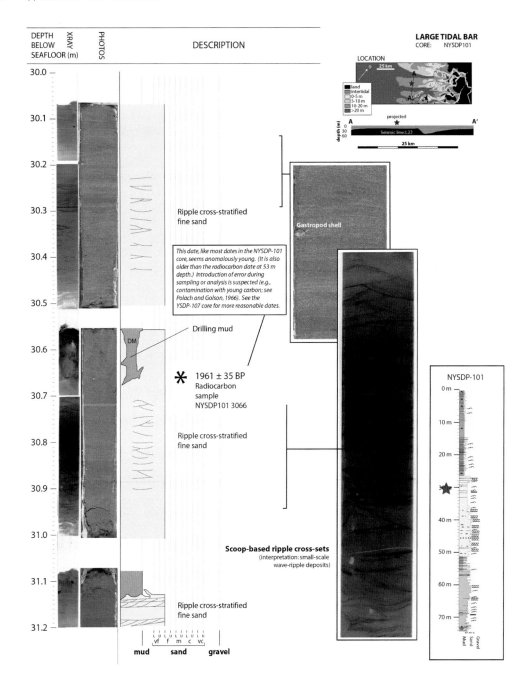

DEPTH BELOW SEAFLOOR (m)

XRAY

PHOTOS

DESCRIPTION

LARGE TIDAL BAR
CORE: NYSDP101

LOCATION

Ripple cross-stratified fine sand

Gastropod shell

This date, like most dates in the NYSDP-101 core, seems anomalously young. (It is also older than the radiocarbon date at 53 m depth.) Introduction of error during sampling or analysis is suspected (e.g., contamination with young carbon; see Polach and Golson, 1966). See the YSDP-107 core for more reasonable dates.

Drilling mud

DM

✱   1961 ± 35 BP
Radiocarbon sample
NYSDP101 3066

Ripple cross-stratified fine sand

Scoop-based ripple cross-sets
(interpretation: small-scale wave-ripple deposits)

Ripple cross-stratified fine sand

vf f m c vc

mud   sand   gravel

NYSDP-101

0 m

10 m

20 m

40 m

50 m

60 m

70 m

Gravel
Sand
Mud

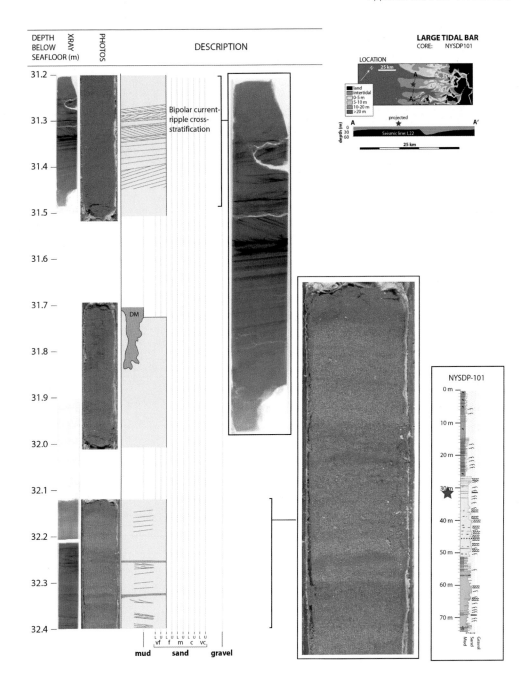

DEPTH BELOW SEAFLOOR (m)

XRAY

PHOTOS

DESCRIPTION

31.2
31.3
31.4
31.5
31.6
31.7
31.8
31.9
32.0
32.1
32.2
32.3
32.4

Bipolar current-ripple cross-stratification

DM

L U L U L U L U L U
vf f   m   c   vc
mud   sand   gravel

LARGE TIDAL BAR
CORE: NYSDP101

LOCATION
25 km
land
intertidal
0-5 m
5-10 m
10-20 m
>20 m
projected
A   A'
depth (m) 0 30 60
Seismic line: L22
25 km

NYSDP-101
0 m
10 m
20 m
30 m
40 m
50 m
60 m
70 m
Mud Sand Gravel

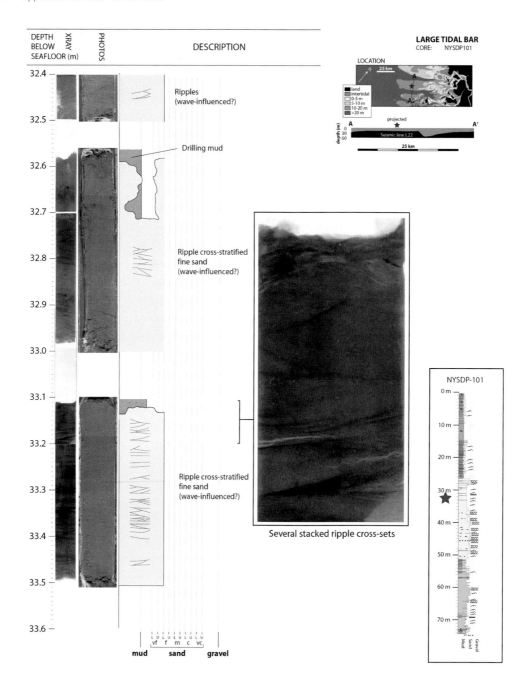

Several stacked ripple cross-sets

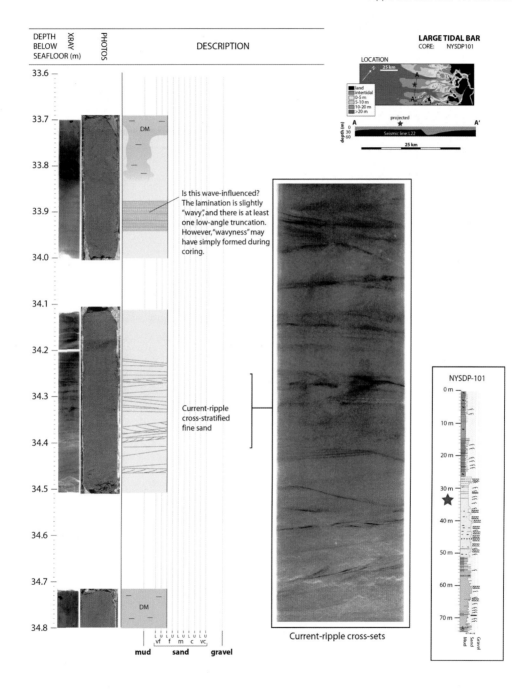

DEPTH BELOW SEAFLOOR (m)

XRAY

PHOTOS

DESCRIPTION

LARGE TIDAL BAR
CORE: NYSDP101

LOCATION

25 km

land
intertidal
0-5 m
5-10 m
10-20 m
>20 m

projected

A   depth (m) 0 30 60   Seismic line L22   A'

25 km

DM

Is this wave-influenced? The lamination is slightly "wavy", and there is at least one low-angle truncation. However, "wavyness" may have simply formed during coring.

Current-ripple cross-stratified fine sand

DM

L U L U L U L U L U
vf   f   m   c   vc

mud   sand   gravel

Current-ripple cross-sets

NYSDP-101

0 m

10 m

20 m

30 m

40 m

50 m

60 m

70 m

Gravel
Sand
Mud

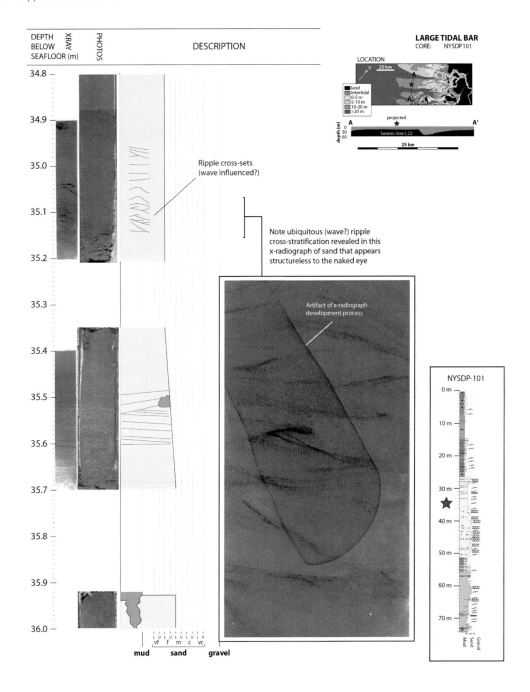

LARGE TIDAL BAR
CORE: NYSDP101

LOCATION

Ripple cross-sets
(wave influenced?)

Note ubiquitous (wave?) ripple
cross-stratification revealed in this
x-radiograph of sand that appears
structureless to the naked eye

Artifact of x-radiograph
development process

NYSDP-101

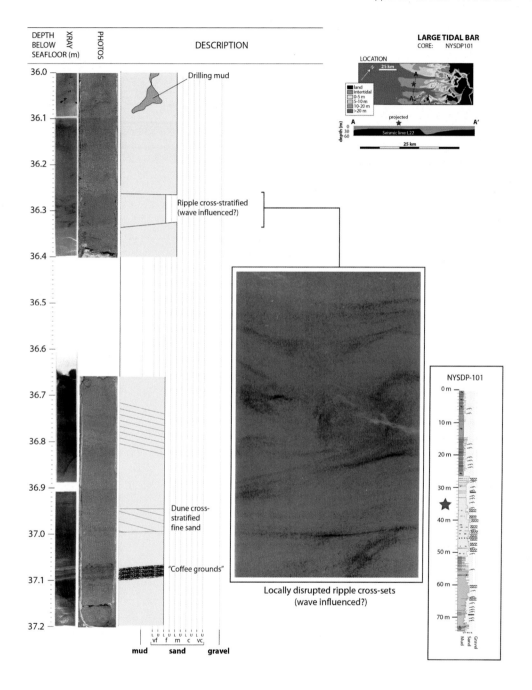

DEPTH BELOW SEAFLOOR (m)

XRAY

PHOTOS

DESCRIPTION

Drilling mud

Ripple cross-stratified (wave influenced?)

Dune cross-stratified fine sand

"Coffee grounds"

mud   sand   gravel

Locally disrupted ripple cross-sets (wave influenced?)

LARGE TIDAL BAR
CORE: NYSDP101

LOCATION

25 km

land
intertidal
0-5 m
5-10 m
10-20 m
>20 m

projected

A   A′

depth (m)
0
30
60

Seismic line: L22

25 km

NYSDP-101

0 m

10 m

20 m

30 m

40 m

50 m

60 m

70 m

Mud Sand Gravel

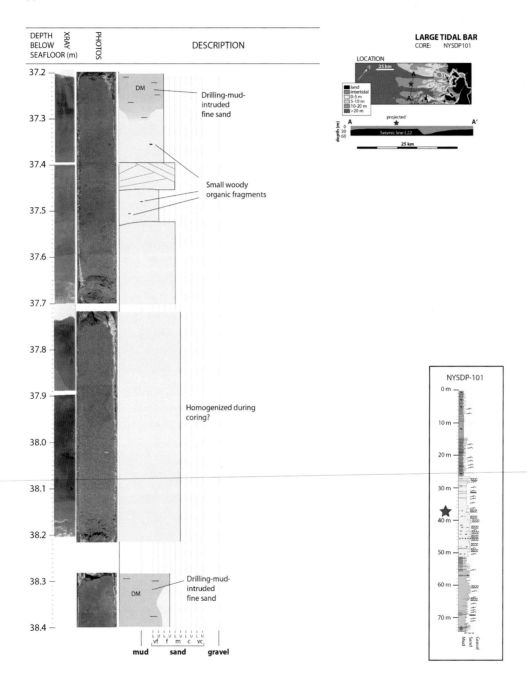

DEPTH BELOW SEAFLOOR (m)

XRAY

PHOTOS

DESCRIPTION

DM — Drilling-mud-intruded fine sand

Small woody organic fragments

Homogenized during coring?

DM — Drilling-mud-intruded fine sand

mud   sand   gravel

LARGE TIDAL BAR
CORE: NYSDP101

LOCATION
25 km
land
intertidal
0-5 m
5-10 m
10-20 m
>20 m

projected
depth (m)   A   A'
Seismic line: L22
25 km

NYSDP-101
0 m
10 m
20 m
30 m
40 m
50 m
60 m
70 m
Mud   Sand   Gravel

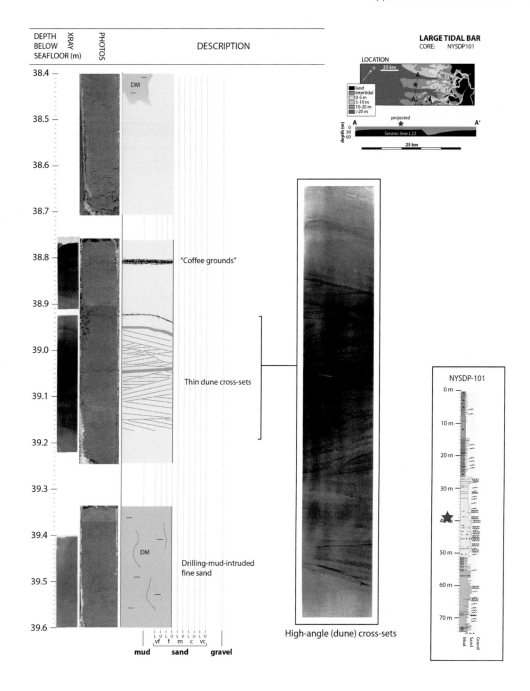

DEPTH BELOW SEAFLOOR (m)

XRAY

PHOTOS

DESCRIPTION

DM

"Coffee grounds"

Thin dune cross-sets

Drilling-mud-intruded fine sand

DM

mud   sand   gravel

vf   f   m   c   vc

High-angle (dune) cross-sets

LARGE TIDAL BAR
CORE: NYSDP101

LOCATION

25 km

land
intertidal
0-5 m
5-10 m
10-20 m
>20 m

projected

A   A'

depth (m)

Seismic line: L22

25 km

NYSDP-101

0 m

10 m

20 m

30 m

40 m

50 m

60 m

70 m

Gravel
Sand
Mud

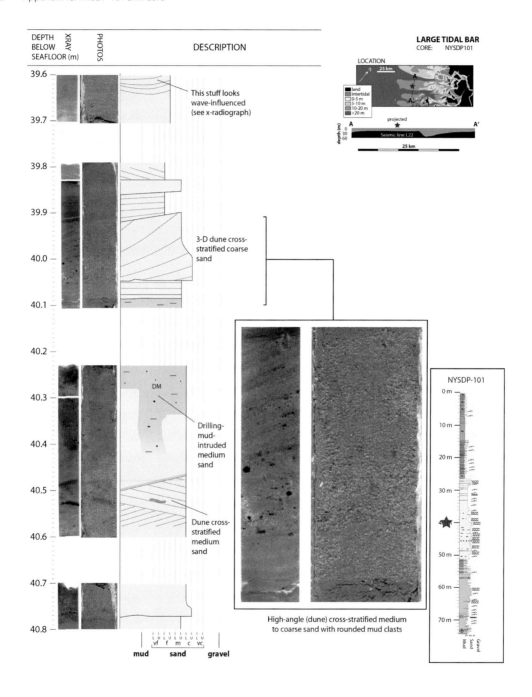

High-angle (dune) cross-stratified medium to coarse sand with rounded mud clasts

DEPTH BELOW SEAFLOOR (m)

XRAY

PHOTOS

DESCRIPTION

This stuff looks wave-influenced (see x-radiograph)

3-D dune cross-stratified coarse sand

DM

Drilling-mud-intruded medium sand

Dune cross-stratified medium sand

mud   sand   gravel

vf f m c vc

LARGE TIDAL BAR
CORE: NYSDP101

LOCATION
25 km

land
intertidal
0–5 m
5–10 m
10–20 m
>20 m

projected

A   A'
depth (m) 0
30
60
Seismic line: L22
25 km

NYSDP-101
0 m
10 m
20 m
30 m
40 m
50 m
60 m
70 m
Mud Sand Gravel

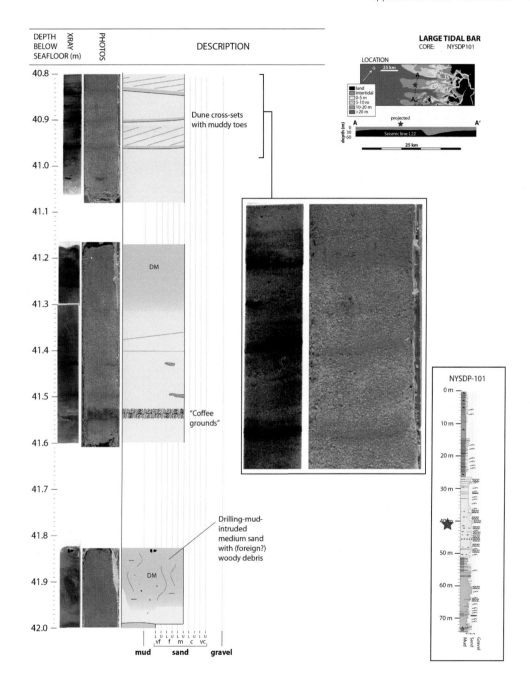

DEPTH BELOW SEAFLOOR (m)

XRAY

PHOTOS

DESCRIPTION

Dune cross-sets with muddy toes

DM

"Coffee grounds"

Drilling-mud-intruded medium sand with (foreign?) woody debris

DM

mud   sand   gravel

vf f m c vc

LARGE TIDAL BAR
CORE: NYSDP101

LOCATION
25 km

land
intertidal
0-5 m
5-10 m
10-20 m
>20 m

projected

A   A'
depth (m)
0
30
60
Seismic line: L22
25 km

NYSDP-101
0 m
10 m
20 m
30 m
40 m
50 m
60 m
70 m

Gravel
Sand
Mud

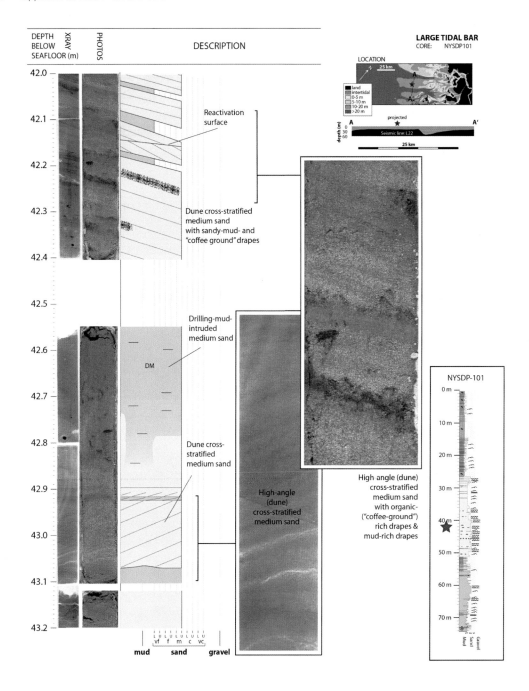

DEPTH BELOW SEAFLOOR (m)

XRAY

PHOTOS

DESCRIPTION

Reactivation surface

Dune cross-stratified medium sand with sandy-mud- and "coffee ground" drapes

Drilling-mud-intruded medium sand

DM

Dune cross-stratified medium sand

High-angle (dune) cross-stratified medium sand

High-angle (dune) cross-stratified medium sand with organic-("coffee-ground") rich drapes & mud-rich drapes

mud   sand   gravel

LARGE TIDAL BAR
CORE: NYSDP101

LOCATION
25 km

land
intertidal
0-5 m
5-10 m
10-20 m
>20 m

projected

A   A'
depth (m)
Seismic line: L22
25 km

NYSDP-101
0 m
10 m
20 m
30 m
40 m
50 m
60 m
70 m
Mud Sand Gravel

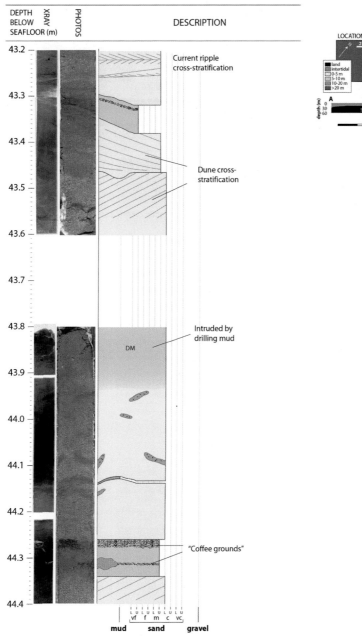

DEPTH BELOW SEAFLOOR (m)

XRAY

PHOTOS

DESCRIPTION

43.2 — Current ripple cross-stratification

43.3

43.4 — Dune cross-stratification

43.5

43.6

43.7

43.8 — Intruded by drilling mud

DM

43.9

44.0

44.1

44.2

44.3 — "Coffee grounds"

44.4

mud   sand   gravel

vf f m c vc

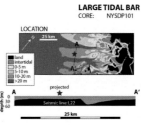

LARGE TIDAL BAR
CORE: NYSDP101

LOCATION

25 km

land
intertidal
0–5 m
5–10 m
10–20 m
>20 m

projected

A   A'
depth (m)   0   30   60
Seismic line: L22

25 km

NYSDP-101

0 m

10 m

20 m

30 m

40 m

50 m

60 m

70 m

Mud   Sand   Gravel

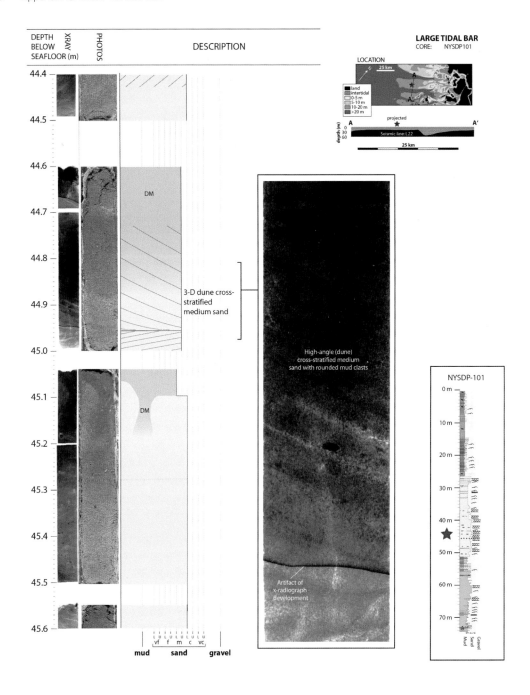

LARGE TIDAL BAR
CORE: NYSDP101

DEPTH BELOW SEAFLOOR (m)

XRAY

PHOTOS

DESCRIPTION

LOCATION

land
intertidal
0-5 m
5-10 m
10-20 m
>20 m

projected

A   A'
depth (m)
0
30
60   Seismic line: L.22

25 km

DM

3-D dune cross-stratified medium sand

High-angle (dune) cross-stratified medium sand with rounded mud clasts

DM

Artifact of x-radiograph development

NYSDP-101

0 m

10 m

20 m

30 m

40 m

50 m

60 m

70 m

Mud Sand Gravel

L U L U L U L U L U
vf   f   m   c   vc
mud   sand   gravel

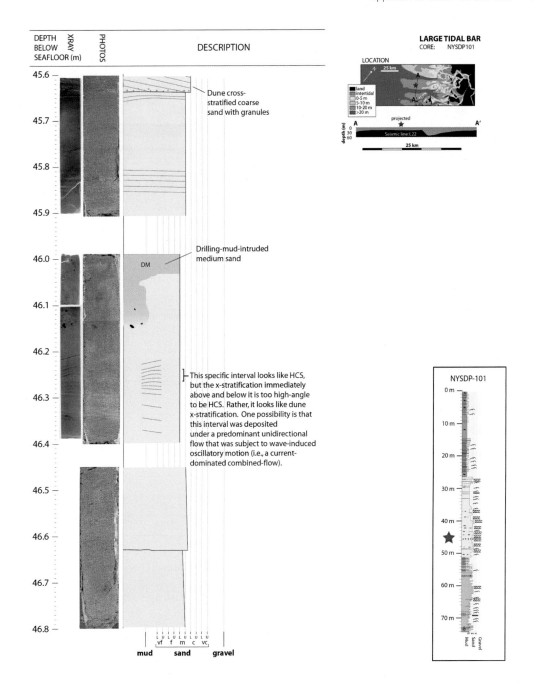

DEPTH BELOW SEAFLOOR (m)

XRAY PHOTOS

DESCRIPTION

Dune cross-stratified coarse sand with granules

Drilling-mud-intruded medium sand

DM

This specific interval looks like HCS, but the x-stratification immediately above and below it is too high-angle to be HCS. Rather, it looks like dune x-stratification. One possibility is that this interval was deposited under a predominant unidirectional flow that was subject to wave-induced oscillatory motion (i.e., a current-dominated combined-flow).

mud   sand   gravel

LARGE TIDAL BAR
CORE: NYSDP101

LOCATION

25 km

land
intertidal
0–5 m
5–10 m
10–20 m
>20 m

A   A'

projected

depth (m)
0
30
60

Seismic line: L.22

25 km

NYSDP-101

0 m

10 m

20 m

30 m

40 m

50 m

60 m

70 m

Mud Sand Gravel

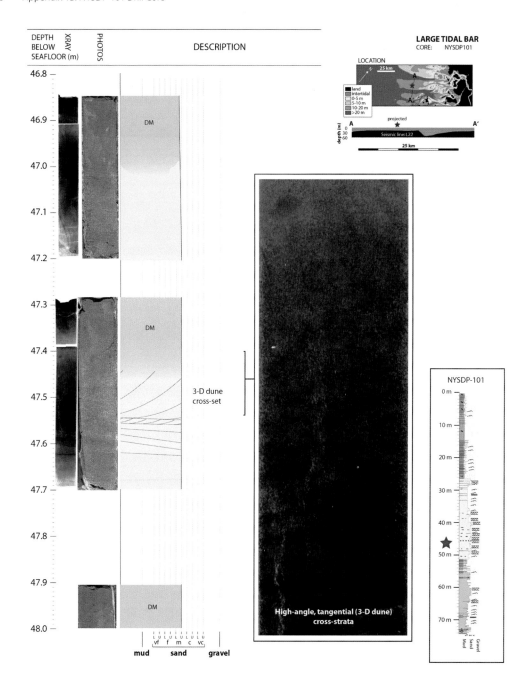

High-angle, tangential (3-D dune) cross-strata

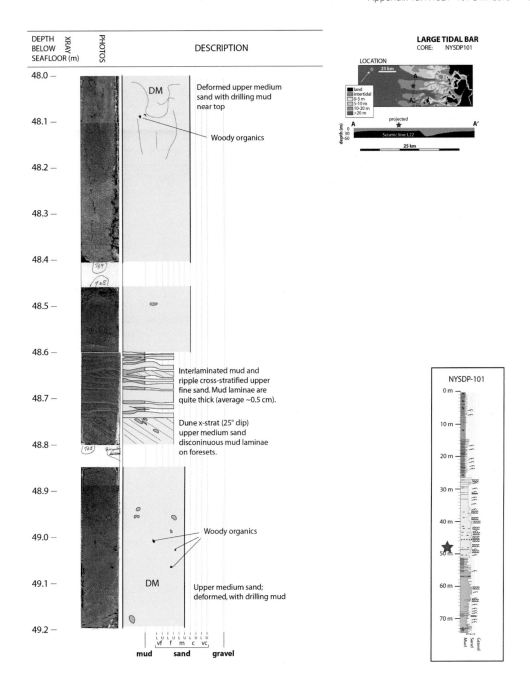

LARGE TIDAL BAR
CORE: NYSDP101

LOCATION

DEPTH BELOW SEAFLOOR (m) **XRAY** **PHOTOS** **DESCRIPTION**

DM — Deformed upper medium sand with drilling mud near top

Woody organics

Interlaminated mud and ripple cross-stratified upper fine sand. Mud laminae are quite thick (average ~0.5 cm).

Dune x-strat (25° dip) upper medium sand disconinuous mud laminae on foresets.

Woody organics

DM — Upper medium sand; deformed, with drilling mud

mud   sand   gravel

NYSDP-101

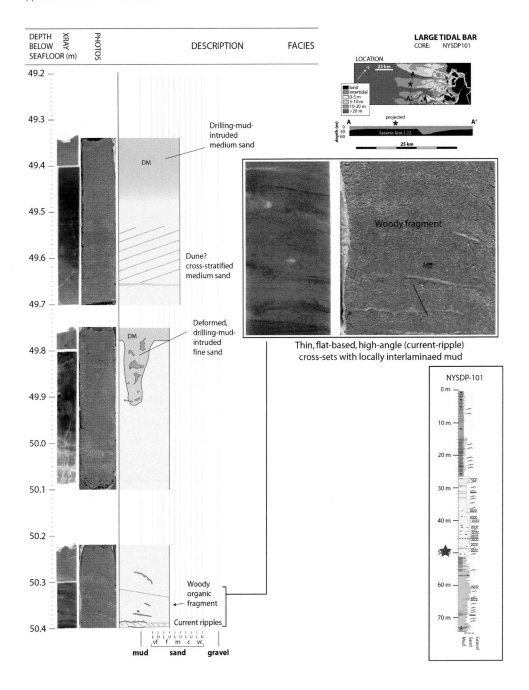

LARGE TIDAL BAR
CORE: NYSDP101

LOCATION

DEPTH BELOW SEAFLOOR (m)
XRAY
PHOTOS
DESCRIPTION
FACIES

49.2
49.3
49.4
49.5
49.6
49.7
49.8
49.9
50.0
50.1
50.2
50.3
50.4

DM — Drilling-mud-intruded medium sand

Dune? cross-stratified medium sand

DM — Deformed, drilling-mud-intruded fine sand

Woody organic fragment

Current ripples

mud   sand   gravel

vf f m c vc

Woody fragment

Thin, flat-based, high-angle (current-ripple) cross-sets with locally interlaminaed mud

NYSDP-101

0 m
10 m
20 m
30 m
40 m
50 m
60 m
70 m

Mud   Sand   Gravel

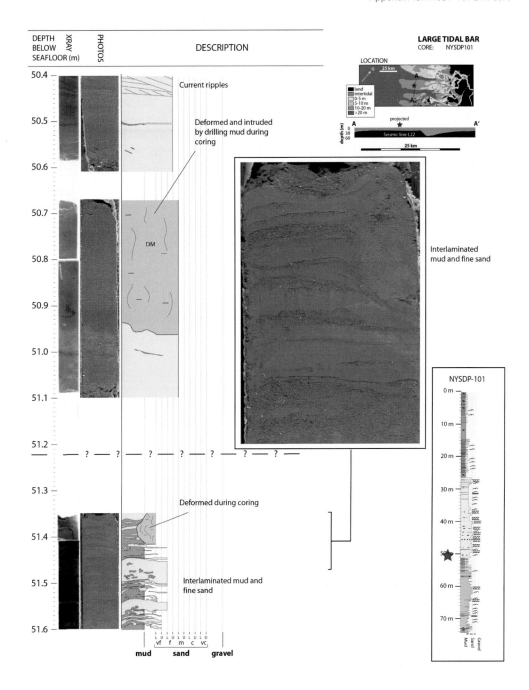

DEPTH BELOW SEAFLOOR (m)

XRAY

PHOTOS

DESCRIPTION

50.4 — Current ripples

50.5 — Deformed and intruded by drilling mud during coring

50.6

50.7

50.8 — DM

50.9

51.0

51.1

51.2 — ? — ? — ? — ? — ? — ? — ? —

51.3

51.4 — Deformed during coring

51.5 — Interlaminated mud and fine sand

51.6

L U L U L U L U L U
vf f m c vc
mud   sand   gravel

LARGE TIDAL BAR
CORE: NYSDP101

LOCATION

25 km

- land
- intertidal
- 0-5 m
- 5-10 m
- 10-20 m
- >20 m

projected

A _____ A'
depth (m)
0
30
60
Seismic line: L22

25 km

Interlaminated mud and fine sand

NYSDP-101

0 m

10 m

20 m

30 m

40 m

50 m

60 m

70 m

Gravel
Sand
Mud

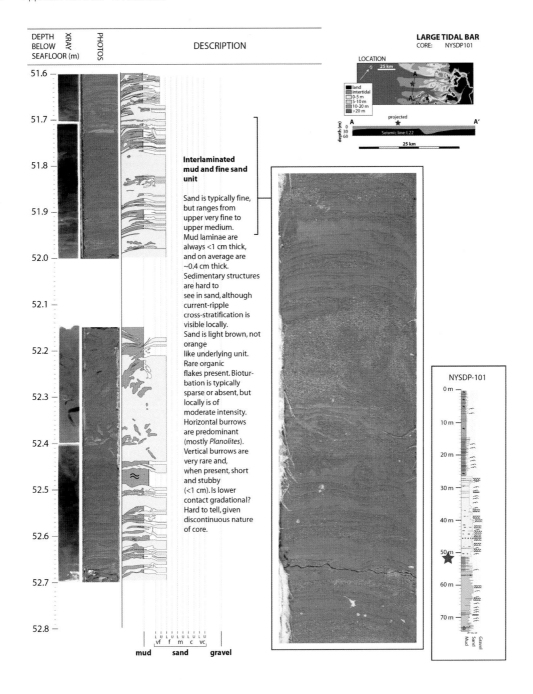

DEPTH BELOW SEAFLOOR (m)

XRAY

PHOTOS

DESCRIPTION

Interlaminated mud and fine sand unit

Sand is typically fine, but ranges from upper very fine to upper medium. Mud laminae are always <1 cm thick, and on average are ~0.4 cm thick. Sedimentary structures are hard to see in sand, although current-ripple cross-stratification is visible locally. Sand is light brown, not orange like underlying unit. Rare organic flakes present. Bioturbation is typically sparse or absent, but locally is of moderate intensity. Horizontal burrows are predominant (mostly *Planolites*). Vertical burrows are very rare and, when present, short and stubby (<1 cm). Is lower contact gradational? Hard to tell, given discontinuous nature of core.

mud   sand   gravel

LARGE TIDAL BAR
CORE: NYSDP101

LOCATION

land
intertidal
0-5 m
5-10 m
10-20 m
>20 m

projected

A   A'

depth (m)

Seismic line: L22

25 km

NYSDP-101

0 m

10 m

20 m

30 m

40 m

50 m

60 m

70 m

Mud Sand Gravel

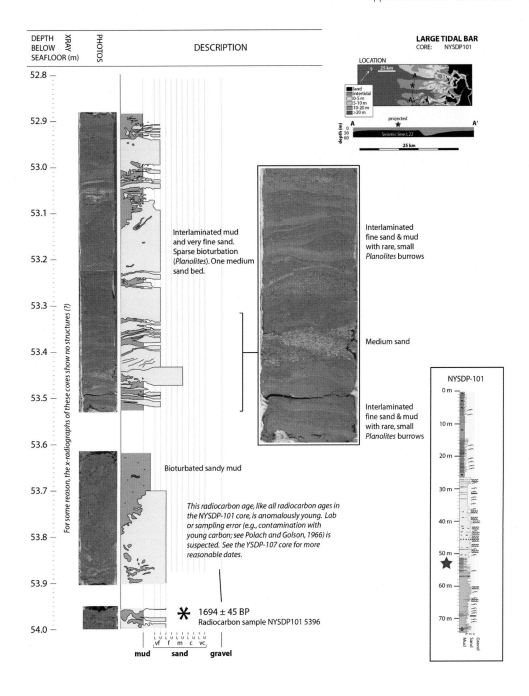

DEPTH
BELOW
SEAFLOOR (m)

XRAY

PHOTOS

DESCRIPTION

LARGE TIDAL BAR
CORE: NYSDP101

LOCATION

land
intertidal
0–5 m
5–10 m
10–20 m
>20 m

projected

A   A'
Seismic line: L22

25 km

For some reason, the x-radiographs of these cores show no structures (?)

Interlaminated mud
and very fine sand.
Sparse bioturbation
(*Planolites*). One medium
sand bed.

Interlaminated
fine sand & mud
with rare, small
Planolites burrows

Medium sand

Interlaminated
fine sand & mud
with rare, small
Planolites burrows

Bioturbated sandy mud

*This radiocarbon age, like all radiocarbon ages in
the NYSDP-101 core, is anomalously young. Lab
or sampling error (e.g., contamination with
young carbon; see Polach and Golson, 1966) is
suspected. See the YSDP-107 core for more
reasonable dates.*

* 1694 ± 45 BP
Radiocarbon sample NYSDP101 5396

vf f m c vc

mud   sand   gravel

NYSDP-101

0 m

10 m

20 m

30 m

40 m

50 m

60 m

70 m

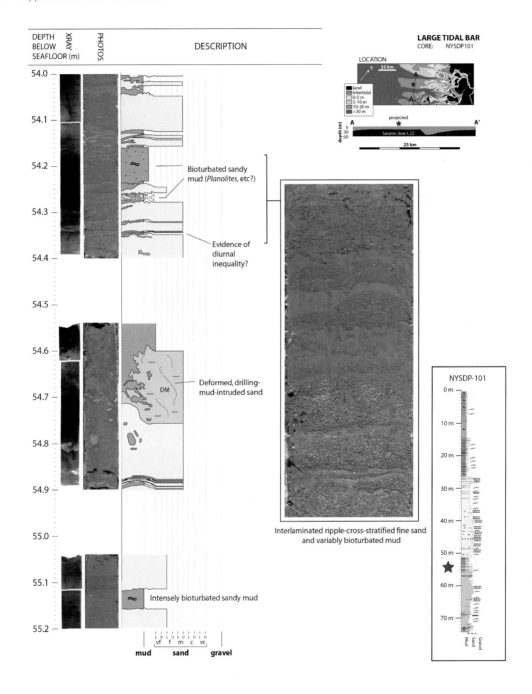

LARGE TIDAL BAR
CORE: NYSDP101

Interlaminated ripple-cross-stratified fine sand
and variably bioturbated mud

NYSDP-101

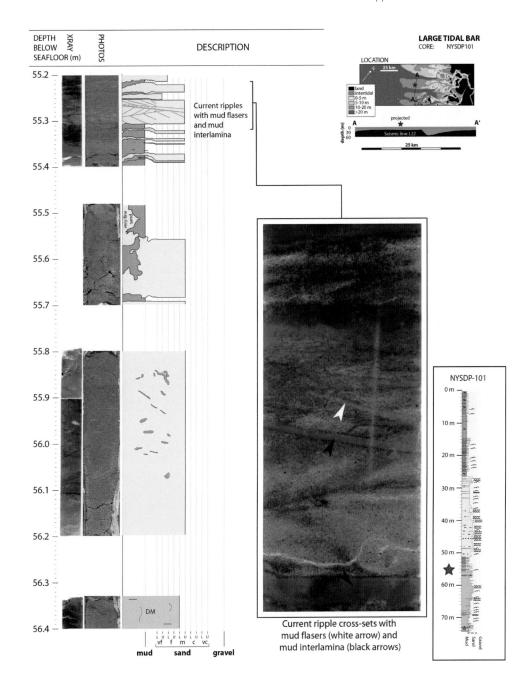

Current ripple cross-sets with
mud flasers (white arrow) and
mud interlamina (black arrows)

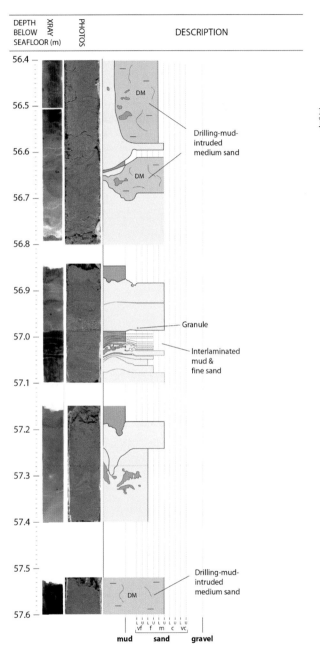

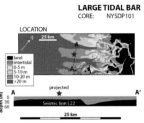

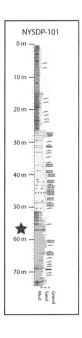

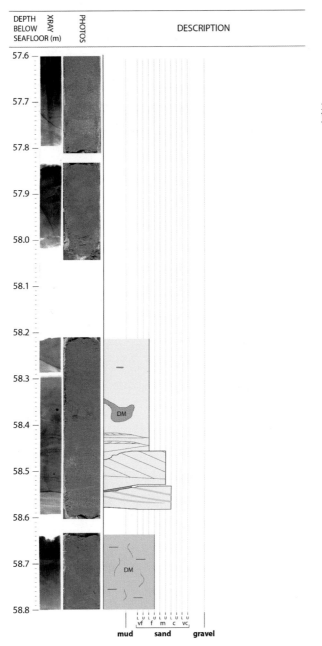

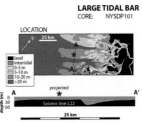

LARGE TIDAL BAR
CORE: NYSDP101

LOCATION

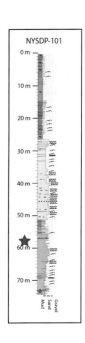

NYSDP-101

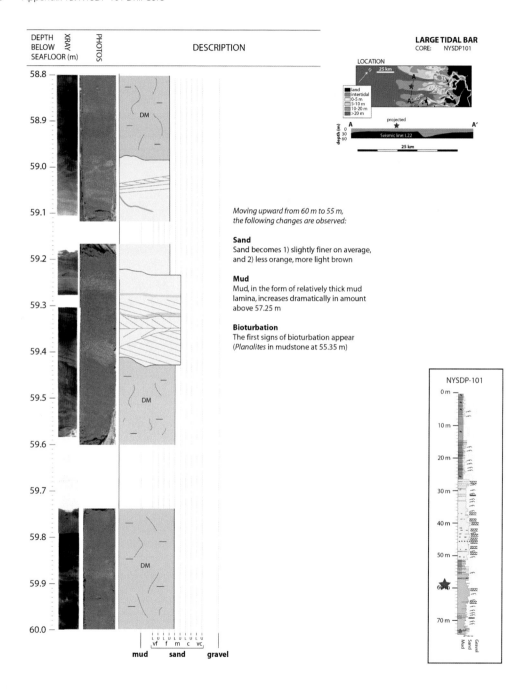

DEPTH BELOW SEAFLOOR (m) — **XRAY** — **PHOTOS** — **DESCRIPTION**

LARGE TIDAL BAR
CORE: NYSDP101

LOCATION

Moving upward from 60 m to 55 m, the following changes are observed:

Sand
Sand becomes 1) slightly finer on average, and 2) less orange, more light brown

Mud
Mud, in the form of relatively thick mud lamina, increases dramatically in amount above 57.25 m

Bioturbation
The first signs of bioturbation appear (*Planolites* in mudstone at 55.35 m)

mud   sand   gravel

NYSDP-101

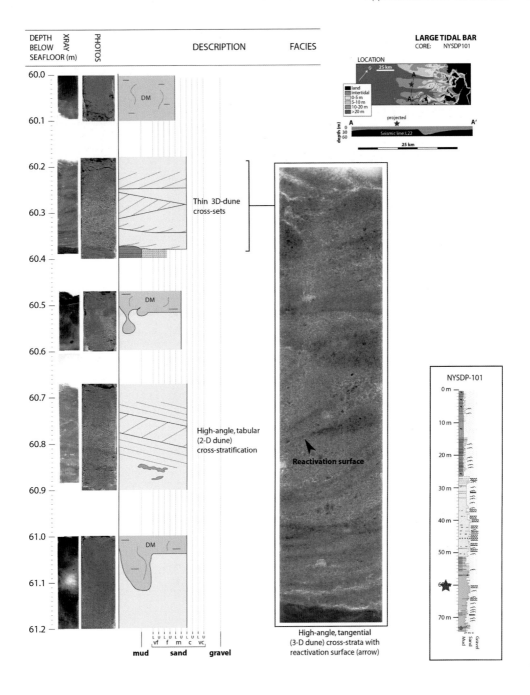

DEPTH BELOW SEAFLOOR (m)

XRAY

PHOTOS

DESCRIPTION

FACIES

DM

Thin 3D-dune cross-sets

DM

High-angle, tabular (2-D dune) cross-stratification

DM

60.0
60.1
60.2
60.3
60.4
60.5
60.6
60.7
60.8
60.9
61.0
61.1
61.2

L U L U L U L U L U
vf f   m   c   vc
mud   sand   gravel

LARGE TIDAL BAR
CORE: NYSDP101

LOCATION

25 km

land
intertidal
0-5 m
5-10 m
10-20 m
>20 m

A   A'

projected

depth (m)
0
30
60

Seismic line:L22

25 km

Reactivation surface

High-angle, tangential (3-D dune) cross-strata with reactivation surface (arrow)

NYSDP-101

0 m
10 m
20 m
30 m
40 m
50 m
60 m
70 m

Mud
Sand
Gravel

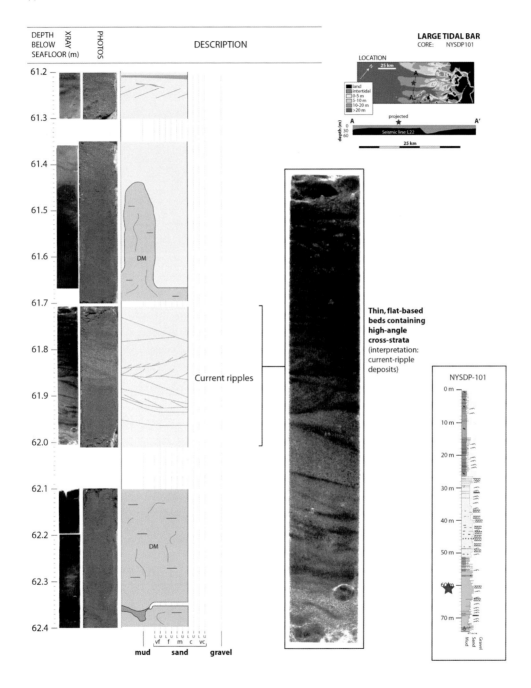

LARGE TIDAL BAR
CORE: NYSDP101

Current ripples

Thin, flat-based beds containing high-angle cross-strata (interpretation: current-ripple deposits)

NYSDP-101

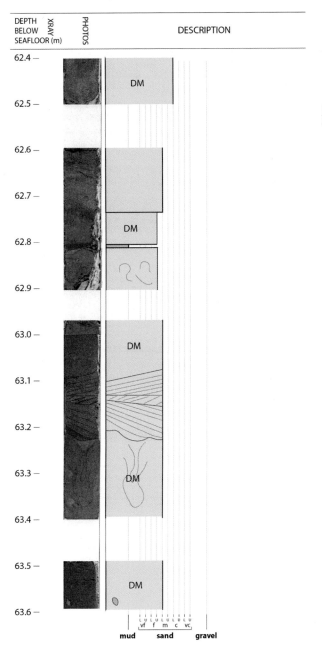

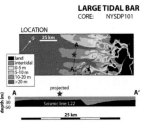

LARGE TIDAL BAR
CORE: NYSDP101

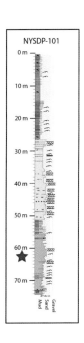

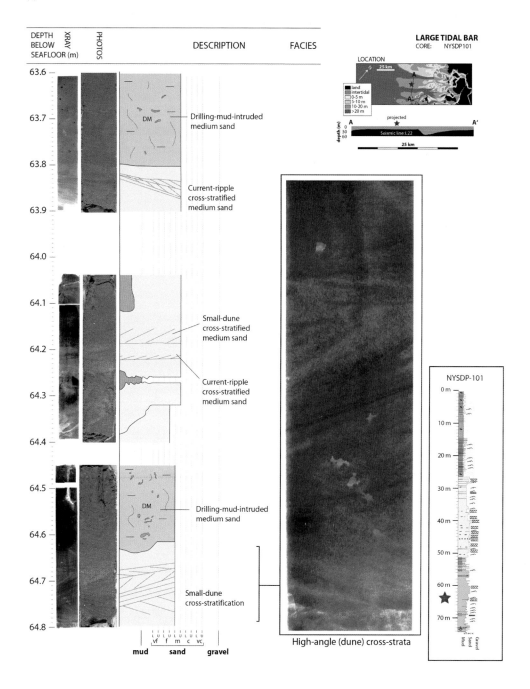

LARGE TIDAL BAR
CORE: NYSDP101

DEPTH BELOW SEAFLOOR (m)

XRAY

PHOTOS

DESCRIPTION

FACIES

LOCATION

Drilling-mud-intruded medium sand

Current-ripple cross-stratified medium sand

Small-dune cross-stratified medium sand

Current-ripple cross-stratified medium sand

Drilling-mud-intruded medium sand

Small-dune cross-stratification

mud   sand   gravel

High-angle (dune) cross-strata

NYSDP-101

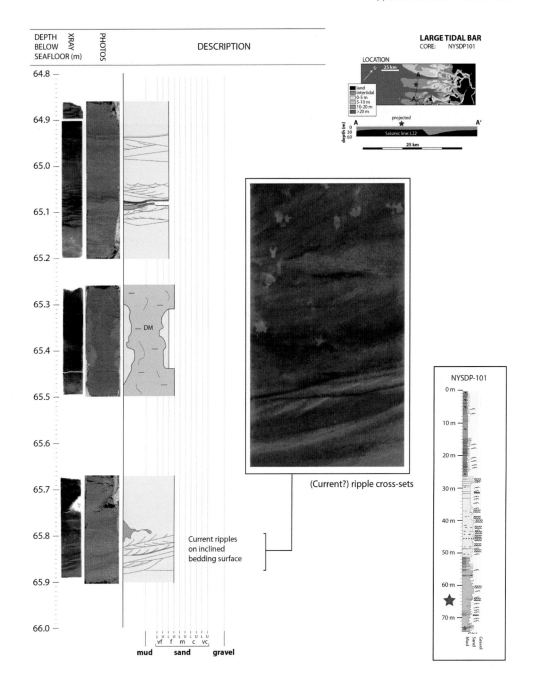

(Current?) ripple cross-sets

Current ripples on inclined bedding surface

LARGE TIDAL BAR
CORE: NYSDP101

LOCATION

NYSDP-101

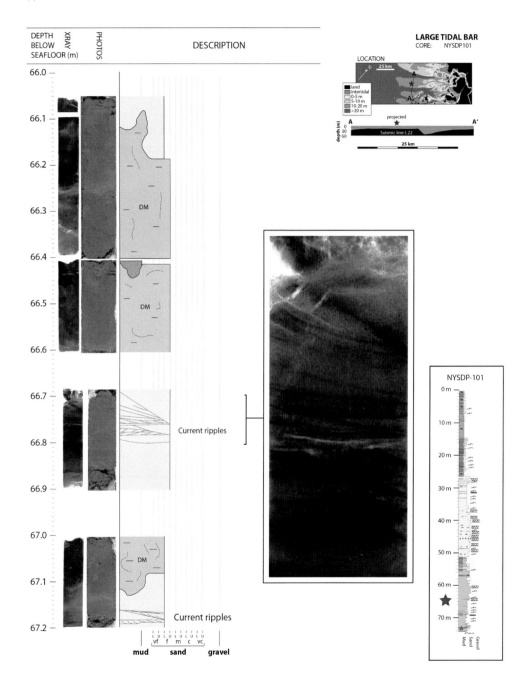

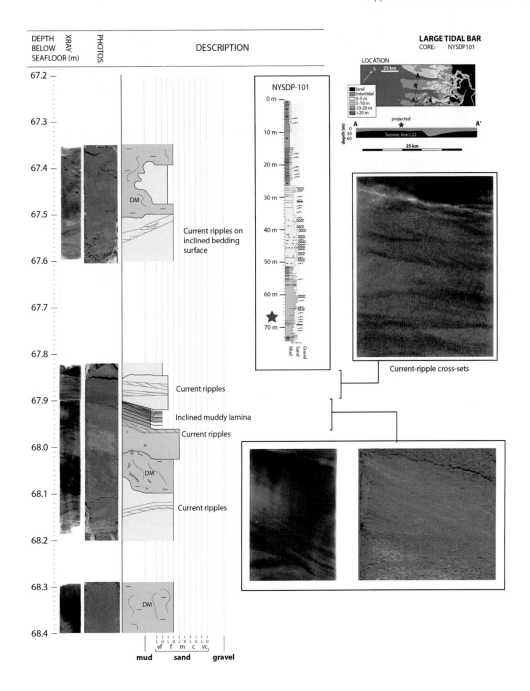

DEPTH BELOW SEAFLOOR (m)

XRAY

PHOTOS

DESCRIPTION

DM

Current ripples on inclined bedding surface

Current ripples

Inclined muddy lamina

Current ripples

DM

Current ripples

DM

mud   sand   gravel

L U L U L U L U L U
vf f m c vc

NYSDP-101

0 m
10 m
20 m
30 m
40 m
50 m
60 m
70 m

Mud
Sand
Gravel

LARGE TIDAL BAR
CORE: NYSDP101

LOCATION

25 km

land
intertidal
0–5 m
5–10 m
10–20 m
>20 m

A   A'

projected

depth (m)
0
30
60

Seismic line: L22

25 km

Current-ripple cross-sets

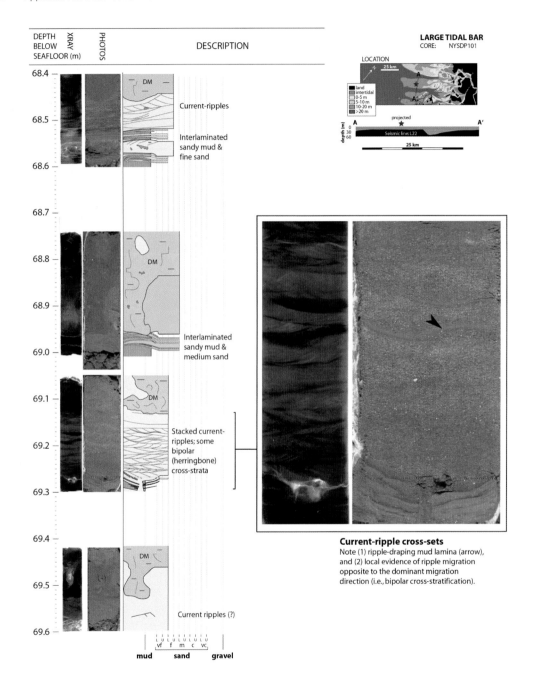

Current-ripple cross-sets
Note (1) ripple-draping mud lamina (arrow), and (2) local evidence of ripple migration opposite to the dominant migration direction (i.e., bipolar cross-stratification).

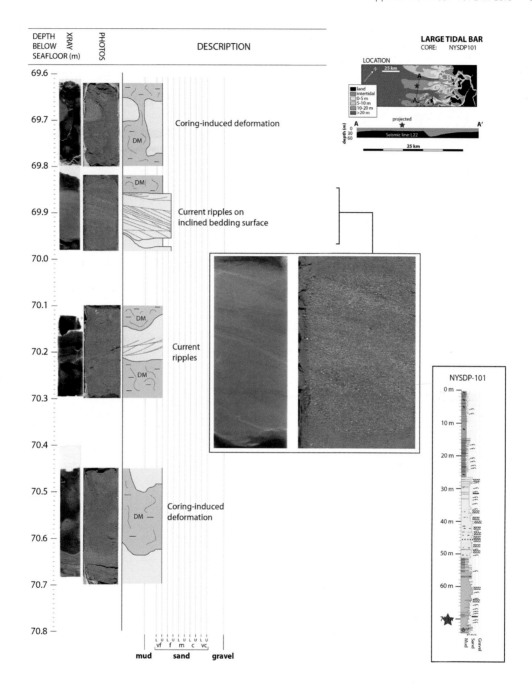

DEPTH
BELOW
SEAFLOOR (m)

XRAY

PHOTOS

DESCRIPTION

69.6

69.7   Coring-induced deformation

DM

69.8

DM

69.9   Current ripples on
inclined bedding surface

70.0

70.1   DM

70.2   Current
ripples

DM

70.3

70.4

70.5   Coring-induced
deformation

DM

70.6

70.7

70.8

L U L U L U L U L U
vf f m c vc
mud   sand   gravel

LARGE TIDAL BAR
CORE: NYSDP101

LOCATION

25 km

land
Intertidal
0–5 m
5–10 m
10–20 m
>20 m

A   projected   A'

depth (m)
0
30
60

Seismic line: L22

25 km

NYSDP-101

0 m

10 m

20 m

30 m

40 m

50 m

60 m

Gravel
Sand
Mud

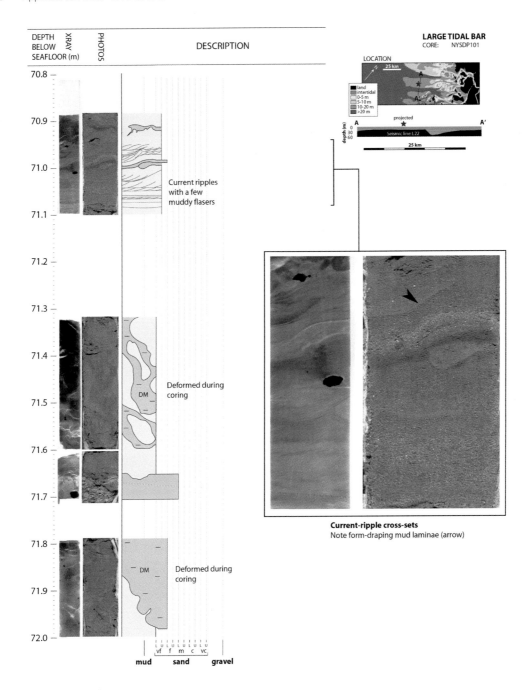

Current-ripple cross-sets
Note form-draping mud laminae (arrow)

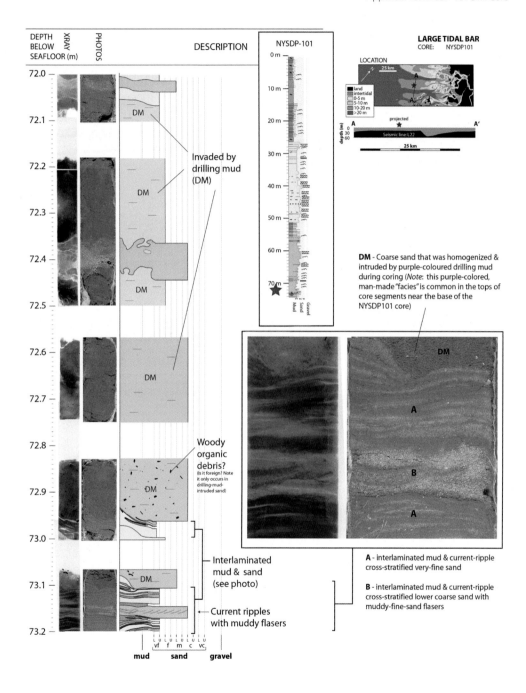

DEPTH BELOW SEAFLOOR (m)

XRAY

PHOTOS

DESCRIPTION

NYSDP-101

0 m

Invaded by drilling mud (DM)

Woody organic debris?
(Is it foreign? Note it only occurs in drilling-mud-intruded sand)

Interlaminated mud & sand (see photo)

Current ripples with muddy flasers

mud   sand   gravel
vf f m c vc

LARGE TIDAL BAR
CORE: NYSDP101

LOCATION

25 km

land
intertidal
0–5 m
5–10 m
10–20 m
>20 m

projected

A   A'
depth (m)
0
30
60
Seismic line: L22

25 km

DM – Coarse sand that was homogenized & intruded by purple-coloured drilling mud during coring (*Note*: this purple-colored, man-made "facies" is common in the tops of core segments near the base of the NYSDP101 core)

DM

A

B

A

A – interlaminated mud & current-ripple cross-stratified very-fine sand

B – interlaminated mud & current-ripple cross-stratified lower coarse sand with muddy-fine-sand flasers

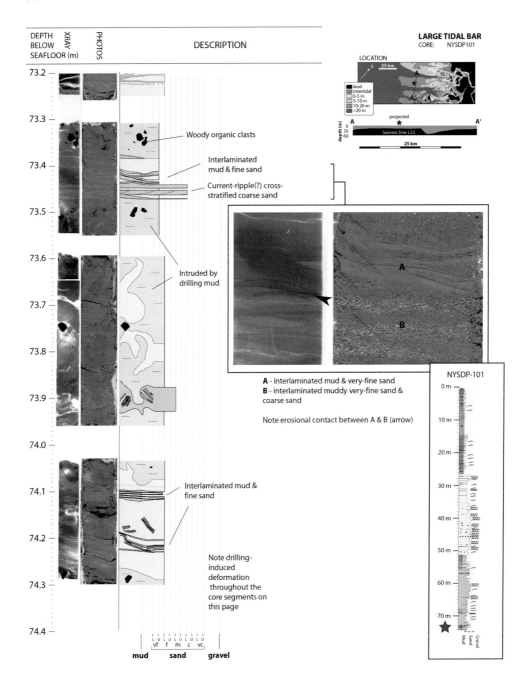

DEPTH BELOW SEAFLOOR (m)

XRAY

PHOTOS

DESCRIPTION

Woody organic clasts

Interlaminated mud & fine sand

Current-ripple(?) cross-stratified coarse sand

Intruded by drilling mud

Interlaminated mud & fine sand

Note drilling-induced deformation throughout the core segments on this page

mud   sand   gravel

LARGE TIDAL BAR
CORE: NYSDP101

LOCATION

25 km

land
intertidal
0-5 m
5-10 m
10-20 m
>20 m

projected

depth (m)   Seismic line: L22

25 km

A - interlaminated mud & very-fine sand
B - interlaminated muddy very-fine sand & coarse sand

Note erosional contact between A & B (arrow)

NYSDP-101

0 m

10 m

20 m

30 m

40 m

50 m

60 m

70 m

Mud Sand Gravel

APPENDIX 2: SEISMIC TRANSECTS

1993 Seismic Survey, Line A1

.25 km

Line A1

10 km

Two-way sound travel time
(milliseconds)

Depth
(meters)

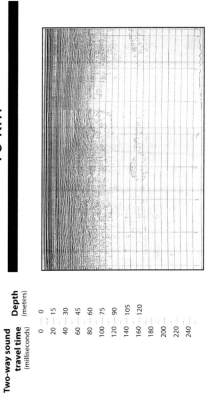

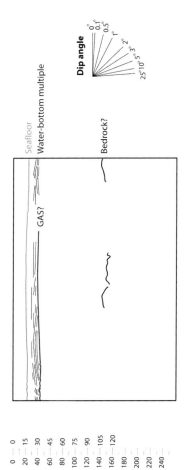

Seafloor
Water-bottom multiple
GAS?
Bedrock?

Dip angle

0°
0.1°
0.5°
1°
2°
3°
5°
10°
25°

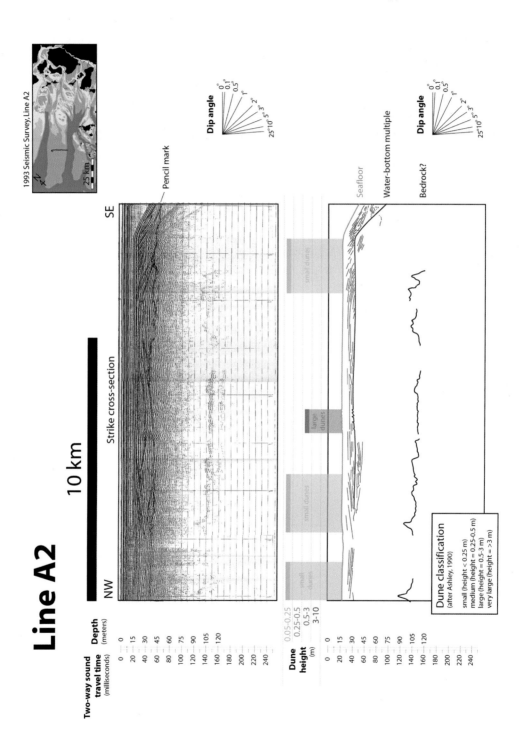

Line A2

10 km

Strike cross-section

NW   SE

1993 Seismic Survey, Line A2

.25 km

Pencil mark

Two-way sound travel time (milliseconds)

Depth (meters)

0	0
20	15
40	30
60	45
80	60
100	75
120	90
140	105
160	120
180	
200	
220	
240	

Dune height (m)

0.05–0.25
0.25–0.5
0.5–3
3–10

small dunes
small dunes
large dunes
small dunes

Dip angle
0° 0.1° 0.5° 1° 2° 3° 5° 10° 25°

Seafloor

Water-bottom multiple

Bedrock?

Dip angle
0° 0.1° 0.5° 1° 2° 3° 5° 10° 25°

Dune classification
(after Ashley, 1990)

small (height < 0.25 m)
medium (height = 0.25–0.5 m)
large (height = 0.5–3 m)
very large (height = >3 m)

Line A3

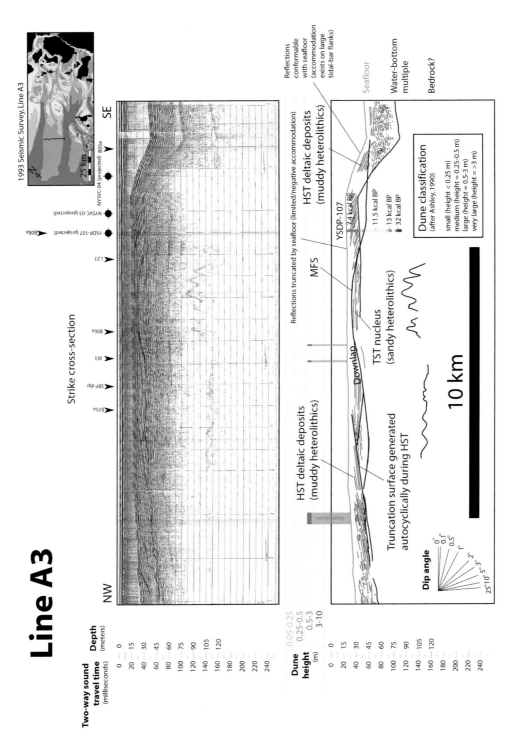

Line A4

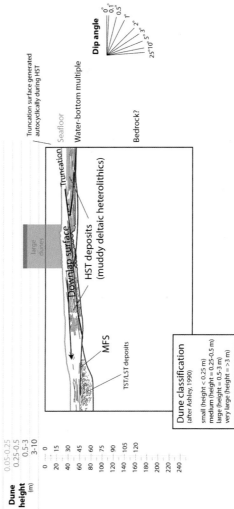

1993 Seismic Survey, Line A4

25 km

Two-way sound travel time (milliseconds)

Depth (meters)

10 km

NW   Strike cross-section   SE

Dune height (m)

0.05–0.25
0.25–0.5
0.5–3
3–10

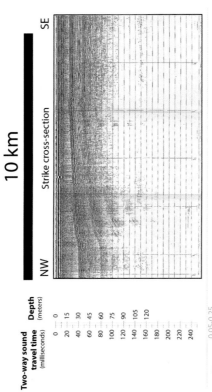

Truncation surface generated autocyclically during HST

Seafloor

Water-bottom multiple

Truncation

HST deposits (muddy deltaic heterolithics)

Downlap surface

large dunes

MFS

TST/LST deposits

Bedrock?

Dune classification (after Ashley, 1990)

small (height < 0.25 m)
medium (height = 0.25–0.5 m)
large (height = 0.5–3 m)
very large (height = >3 m)

Dip angle

0°
0.1°
0.5°
1°
2°
3°
5°
10°
25°

Line A4a, 1993 Seismic Survey

25 km

Line A4a

10 km

Two-way sound travel time (milliseconds)

Depth (meters)

0	0
20	15
40	30
60	45
80	60
100	75
120	90
140	105
160	120
180	
200	
220	
240	

NW   Strike cross-section   SE

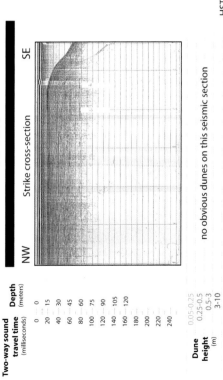

no obvious dunes on this seismic section

Dune height (m)

| 0.05-0.25 |
| 0.25-0.5 |
| 0.5-3 |
| 3-10 |

0	0
20	15
40	30
60	45
80	60
100	75
120	90
140	105
160	120
180	
200	
220	
240	

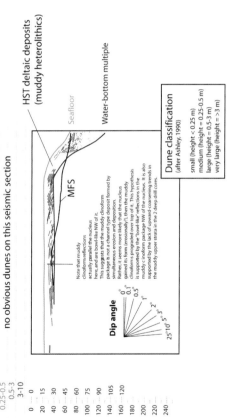

HST deltaic deposits (muddy heterolithics)

Seafloor

Water-bottom multiple

MFS

Note that muddy clinoform reflections actually parallel the nucleus here, and are bowl-like NW of it. This suggests that the muddy clinoform package is not a channel type deposit formed by simultaneous erosion and deposition. Rather, it seems more likely that the nucleus gained its form (erosionally?), then the muddy clinoforms prograded over top of it. This hypothesis is supported by the "bowl-like" reflections in the muddy c-inoform package NW of the nucleus. It is also supported by the lack of upward-coarsening trends in the muddy upper strata in the 2 deep drill cores.

Dip angle

0°
0.1°
0.5°
1°
2°
3°
5°
10°
25°

Dune classification (after Ashley, 1990)

small (height < 0.25 m)
medium (height = 0.25-0.5 m)
large (height = 0.5-3 m)
very large (height = >3 m)

Line A5

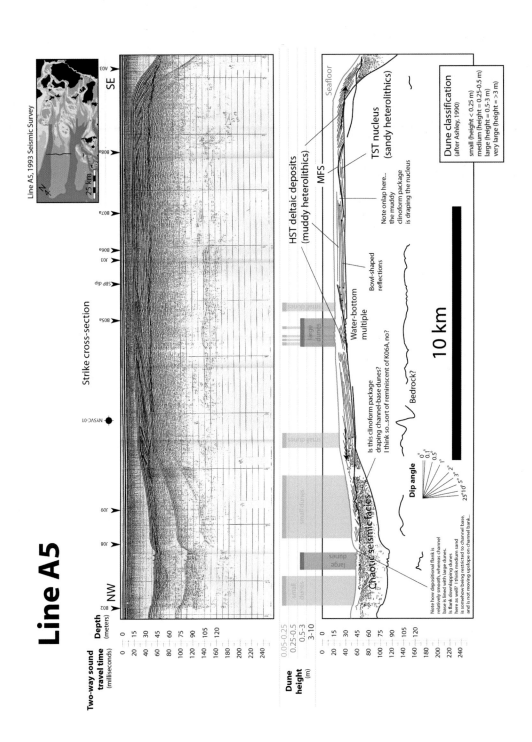

Line A5, 1993 Seismic Survey

Two-way sound travel time (milliseconds)

Depth (meters)

Strike cross-section

NW   SE

Dune height (m)

Dune classification
(after Ashley, 1990)

small (height < 0.25 m)
medium (height = 0.25-0.5 m)
large (height = 0.5-3 m)
very large (height > 3 m)

HST deltaic deposits (muddy heterolithics)

MFS

TST nucleus (sandy heterolithics)

Seafloor

Note onlap here... the muddy clinoform package is draping the nucleus

Bowl-shaped reflections

Water-bottom multiple

Is this clinoform package draping channel-base dunes? I think so...sort of reminiscent of K06A, no?

Bedrock?

Dip angle

10 km

Chaotic seismic facies

Note how depositional flank is relatively smooth, whereas channel base is lined with large dunes. Is flank downlapping dunes here as well? I think medium sand is somehow being restricted to channel base, and is not moving upslope on channel bank...

large dunes

small dunes

large dunes

small dunes

Line A7

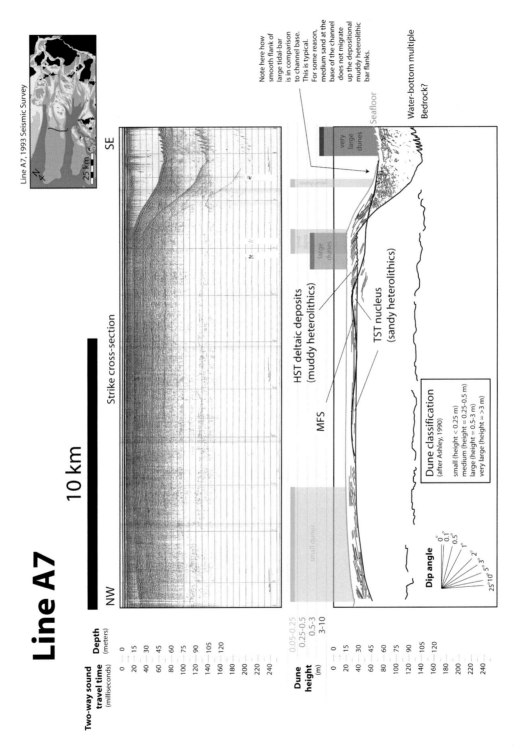

10 km

Strike cross-section

Line A7, 1993 Seismic Survey

25 km

NW

SE

Two-way sound travel time (milliseconds)

0
20
40
60
80
100
120
140
160
180
200
220
240

Depth (meters)

0
15
30
45
60
75
90
105
120

Dune height (m)

0.05-0.25
0.25-0.5
0.5-3
3-10

small dunes

large dunes

very large dunes

Seafloor

Note here how smooth flank of large tidal-bar is in comparison to channel base. This is typical. For some reason, medium sand at the base of the channel does not migrate up the depositional muddy heterolithic bar flanks.

Water-bottom multiple

Bedrock?

HST deltaic deposits (muddy heterolithics)

TST nucleus (sandy heterolithics)

MFS

Dune classification (after Ashley, 1990)

small (height < 0.25 m)
medium (height = 0.25-0.5 m)
large (height = 0.5-3 m)
very large (height = >3 m)

Dip angle

0°
0.1°
0.5°
1°
2°
3°
5°
10°
25°

0
20
40
60
80
100
120
140
160
180
200
220
240

Line A7a

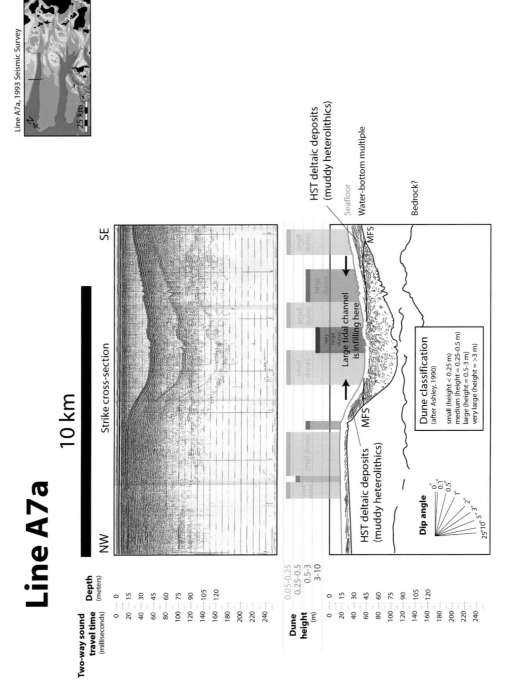

Line A7a, 1993 Seismic Survey

25 km

10 km

Two-way sound travel time (milliseconds)

Depth (meters)

NW   Strike cross-section   SE

Dune height (m)
0.05–0.25
0.25–0.5
0.5–3
3–10

HST deltaic deposits (muddy heterolithics)

small dunes

large dunes

very large dunes

small dunes

small dunes

Large tidal channel is infilling here

MFS

MFS

HST deltaic deposits (muddy heterolithics)

Seafloor

Water-bottom multiple

Bedrock?

Dune classification (after Ashley, 1990)

small (height < 0.25 m)
medium (height = 0.25–0.5 m)
large (height = 0.5–3 m)
very large (height = >3 m)

Dip angle
0°
0.1°
0.5°
1°
2°
3°
5°
10°
25°

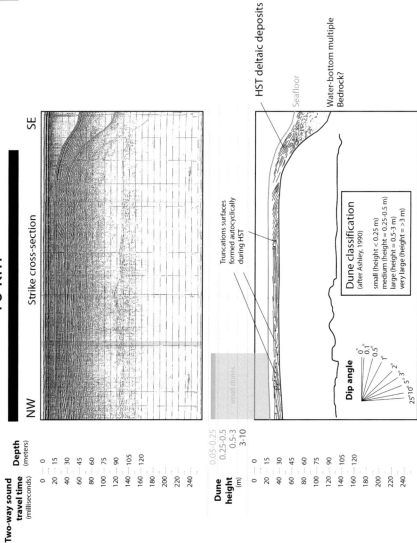

Line A8, 1993 Seismic Survey

25 km

Line A8

10 km

NW   Strike cross-section   SE

Two-way sound travel time (milliseconds)

0 — 0
20 — 15
40 — 30
60 — 45
80 — 60
100 — 75
120 — 90
140 — 105
160 — 120
180
200
220
240

Depth (meters)

Dune height (m)

0.05-0.25
0.25-0.5
0.5-3
3-10

small dunes

0 — 0
20 — 15
40 — 30
60 — 45
80 — 60
100 — 75
120 — 90
140 — 105
160 — 120
180
200
220
240

Seafloor

HST deltaic deposits

Water-bottom multiple
Bedrock?

Truncations surfaces
formed autocyclically
during HST

Dune classification
(after Ashley, 1990)

small (height < 0.25 m)
medium (height = 0.25-0.5 m)
large (height = 0.5-3 m)
very large (height = >3 m)

Dip angle

0°
0.1°
0.5°
1°
2°
3°
5°
10°
25°

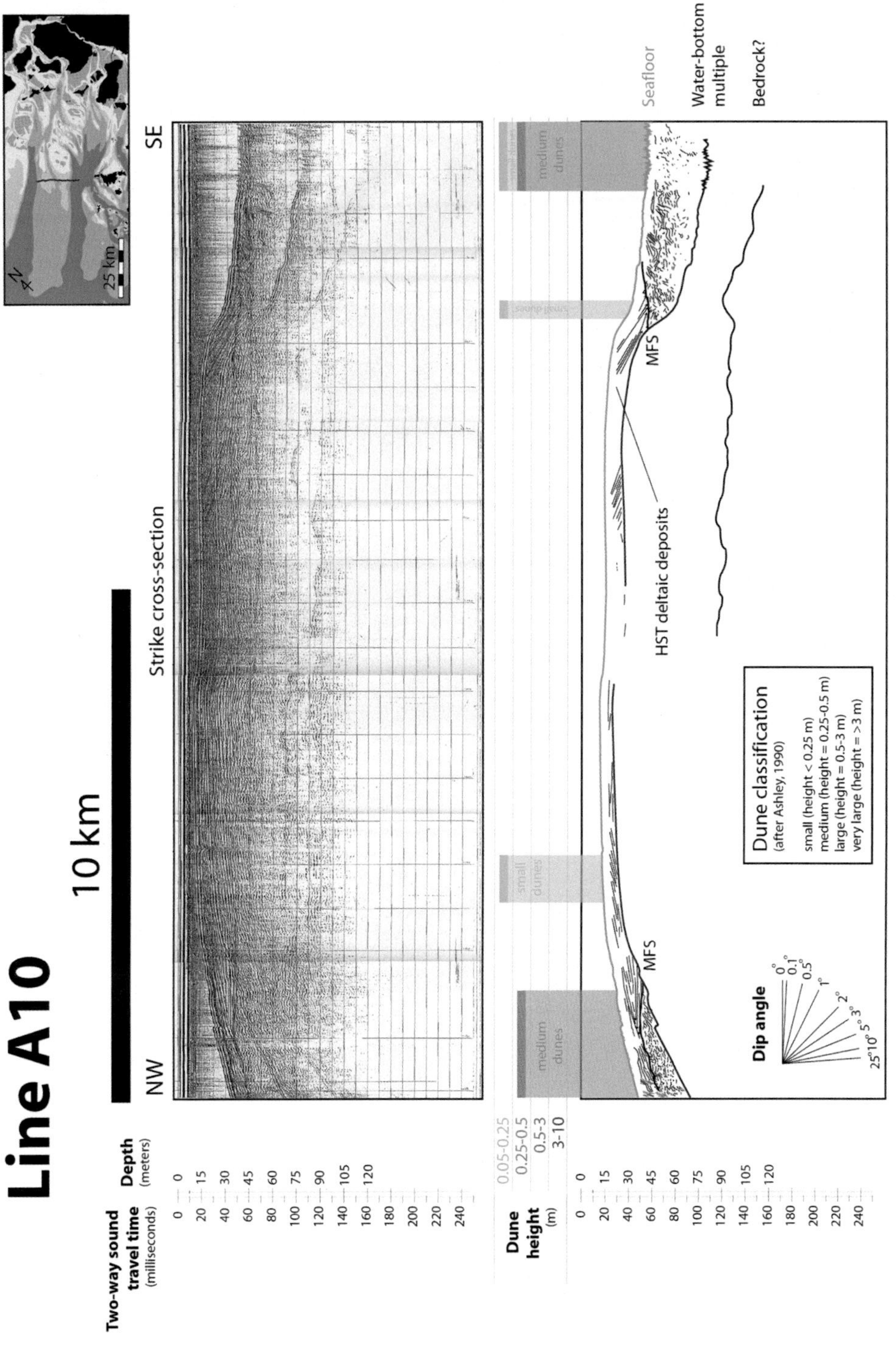

Line A10

Line A10, 1993 Seismic Survey

25 km

10 km

NW

SE

Strike cross-section

Two-way sound travel time (milliseconds)

Depth (meters)

Dune height (m)

0.05–0.25
0.25–0.5
0.5–3
3–10

medium dunes

small dunes

medium dunes

MFS

HST deltaic deposits

MFS

Seafloor

Water-bottom multiple

Bedrock?

Dune classification
(after Ashley, 1990)

small (height < 0.25 m)
medium (height = 0.25–0.5 m)
large (height = 0.5–3 m)
very large (height = >3 m)

Dip angle

0°
0.1°
0.5°
1°
2°
3°
5°
10°
25°

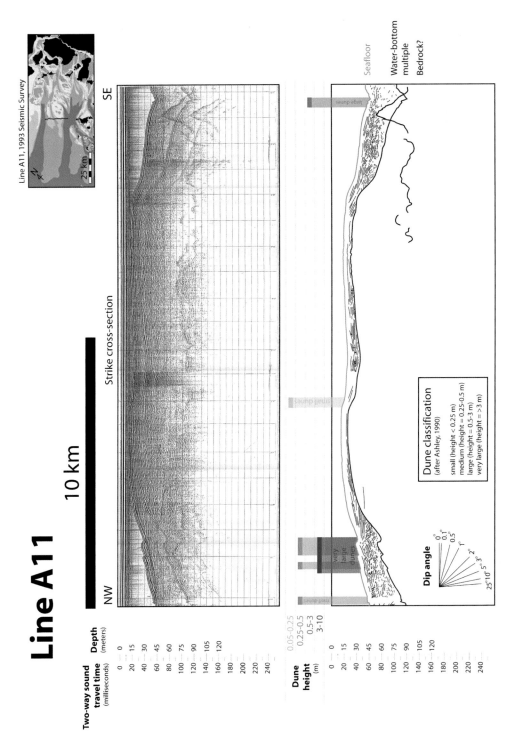

Line A11

Line A11, 1993 Seismic Survey

25 km

SE

Strike cross-section

NW

10 km

Seafloor

Water-bottom multiple

Bedrock?

Two-way sound travel time (milliseconds)

0
20
40
60
80
100
120
140
160
180
200
220
240

Depth (meters)

0
15
30
45
60
75
90
105
120

Dune height (m)

0.05–0.25
0.25–0.5
0.5–3
3–10

0
20
40
60
80
100
120
140
160
180
200
220
240

0
15
30
45
60
75
90
105
120

large dunes

small dunes

very large dunes

small dunes

Dune classification
(after Ashley, 1990)

small (height < 0.25 m)
medium (height = 0.25–0.5 m)
large (height = 0.5–3 m)
very large (height = >3 m)

Dip angle

0°
0.1°
0.5°
1°
2°
3°
5°
10°
25°

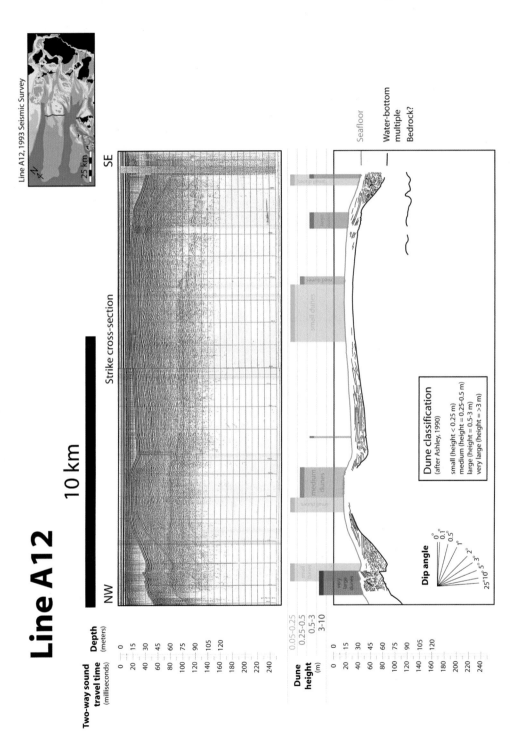

Line A12

Line A12, 1993 Seismic Survey

25 km

10 km

Two-way sound travel time (milliseconds)

Depth (meters)

NW

Strike cross-section

SE

Seafloor

Water-bottom multiple

Bedrock?

Dune height (m)

0.05–0.25
0.25–0.5
0.5–3
3–10

small dunes

medium dunes

very large dunes

Dune classification (after Ashley, 1990)

small (height < 0.25 m)
medium (height = 0.25–0.5 m)
large (height = 0.5–3 m)
very large (height = >3 m)

Dip angle

0°
0.1°
0.5°
1°
2°
3°
5°
10°
25°

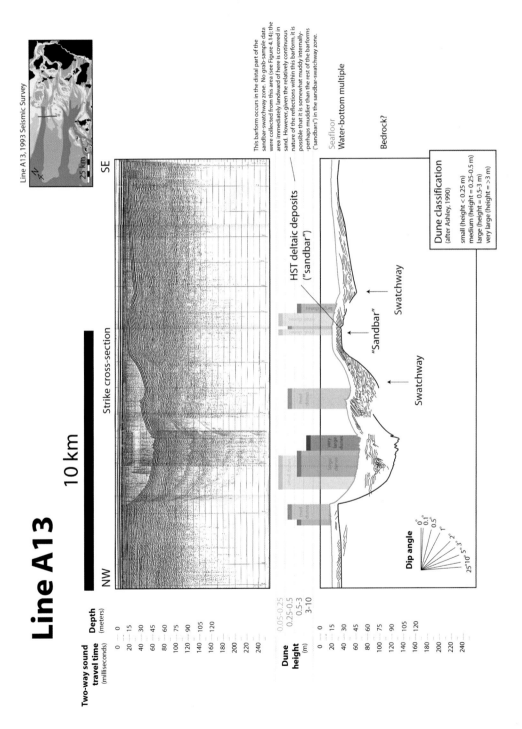

Line A13

Line A13, 1993 Seismic Survey

25 km

NW   Strike cross-section   SE

10 km

Two-way sound travel time (milliseconds)

Depth (meters)

0	0
20	15
40	30
60	45
80	60
100	75
120	90
140	105
160	120
180	
200	
220	
240	

Dune height (m)

0.05–0.25
0.25–0.5
0.5–3
3–10

This barform occurs in the distal part of the sandbar-swathway zone. No grab-sample data were collected from this area (see Figure 4.14); the area immediately landward of here is covered in sand. However, given the relatively continuous nature of the reflections within this barform, it is possible that it is somewhat muddier internally—perhaps muddier than the rest of the barforms ("sandbars") in the sandbar-swathway zone.

Seafloor
Water-bottom multiple

Bedrock?

HST deltaic deposits ("sandbar")

"Sandbar"

Swathway

Swathway

Swathway

Dune classification
(after Ashley, 1990)

small (height < 0.25 m)
medium (height = 0.25–0.5 m)
large (height = 0.5–3 m)
very large (height = >3 m)

Dip angle

0°
0.1°
0.5°
1°
2°
3°
5°
10°
25°

Line A13a

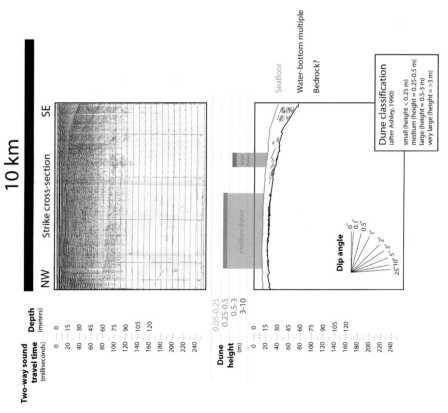

Line A13a, 1993 Seismic Survey

25 km

10 km

Two-way sound travel time (milliseconds)

Depth (meters)

NW   Strike cross-section   SE

Seafloor

Water-bottom multiple

Bedrock?

Dune classification
(after Ashley, 1990)

small (height < 0.25 m)
medium (height = 0.25–0.5 m)
large (height = 0.5–3 m)
very large (height = >3 m)

Dune height (m)

0.05–0.25
0.25–0.5
0.5–3
3–10

medium dunes

large dunes

Dip angle

0°
0.1°
0.5°
1°
2°
3°
5°
10°
25°

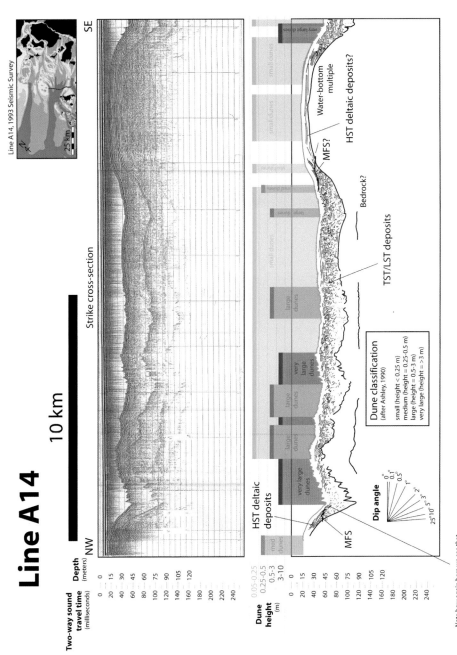

Line A14

10 km

Line A14, 1993 Seismic Survey

25 km

NW

SE

Two-way sound travel time (milliseconds)

Depth (meters)

Strike cross-section

Dune height (m)

0.05–0.25
0.25–0.5
0.5–3
3–10

HST deltaic deposits

MFS

med dunes

very large dunes

large dunes

very large dunes

large dunes

small dunes

large dunes

small dunes

small dunes

small dunes

very large dunes

Water-bottom multiple

HST deltaic deposits?

MFS?

Bedrock?

TST/LST deposits

Dune classification
(after Ashley, 1990)

small (height < 0.25 m)
medium (height = 0.25–0.5 m)
large (height = 0.5–3 m)
very large (height = >3 m)

Dip angle

0°
0.1°
0.5°
1°
2°
3°
5°
10°
25°

Note how again here we see that
the depositional large-tidal-bar flank
is smooth (dune-less), whereas the large-tidal-
channel base is covered in dunes.

Line A14a

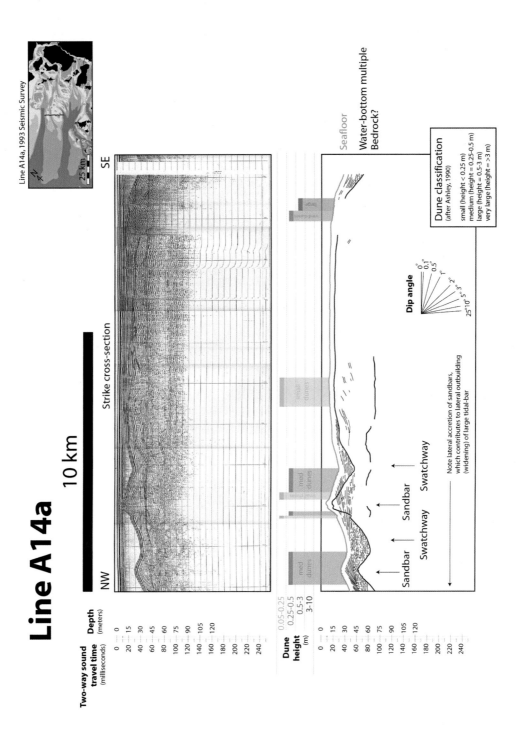

Line A14a, 1993 Seismic Survey

25 km

10 km

Strike cross-section

NW   SE

Two-way sound travel time (milliseconds)

Depth (meters)

Dune height (m)
0.05–0.25
0.25–0.5
0.5–3
3–10

Seafloor

Water-bottom multiple

Bedrock?

Sandbar   Swatchway   Sandbar   Swatchway

large dunes

small dunes

med dunes

med dunes

Note lateral accretion of sandbars, which contributes to lateral outbuilding (widening) of large tidal-bar

Dip angle
0°
0.1°
0.5°
1°
2°
3° 5°
25° 10°

Dune classification (after Ashley, 1990)

small (height < 0.25 m)
medium (height = 0.25–0.5 m)
large (height = 0.5–3 m)
very large (height = >3 m)

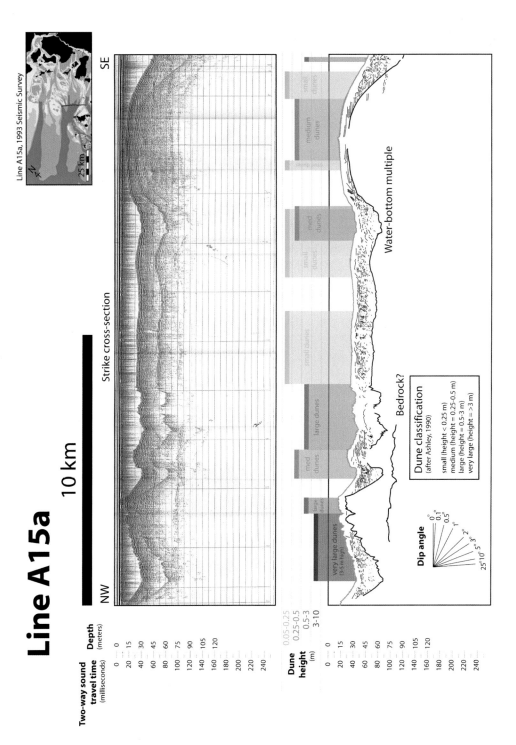

Line A15a

10 km

Line A15a, 1993 Seismic Survey

25 km

NW

SE

Strike cross-section

Two-way sound travel time (milliseconds)

Depth (meters)

0
20 — 15
40 — 30
60 — 45
80 — 60
100 — 75
120 — 90
140 — 105
160 — 120
180
200
220
240

very large dunes (3.5 m high)

large dunes

med dunes

small dunes

med dunes

small dunes

medium dunes

small dunes

large dunes

small dunes

Bedrock?

Water-bottom multiple

Dune classification
(after Ashley, 1990)

small (height < 0.25 m)
medium (height = 0.25–0.5 m)
large (height = 0.5–3 m)
very large (height = >3 m)

Dip angle

0°
0.1°
0.5°
1°
2°
3°
5°
10°
25°

Dune height (m)

0.05–0.25
0.25–0.5
0.5–3
3–10

0
20 — 15
40 — 30
60 — 45
80 — 60
100 — 75
120 — 90
140 — 105
160 — 120
180
200
220
240

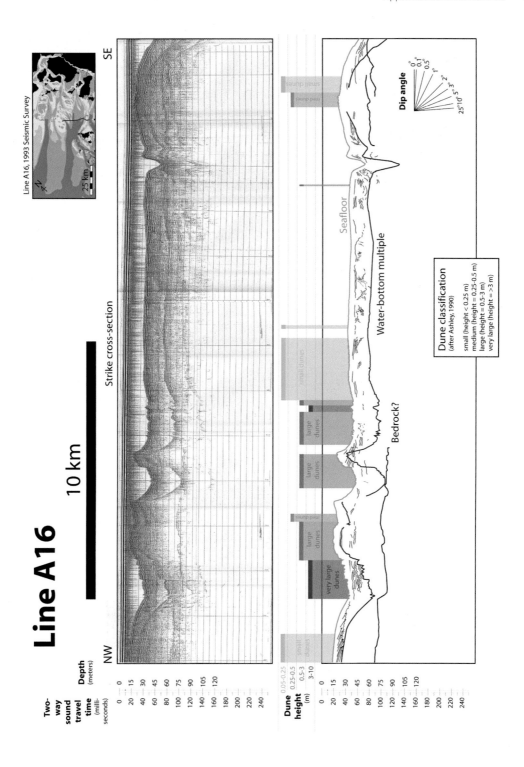

Line A16

Line A16, 1993 Seismic Survey

25 km

NW · SE

Strike cross-section

10 km

Two-way sound travel time (milli-seconds)

0, 20, 40, 60, 80, 100, 120, 140, 160, 180, 200, 220, 240

Depth (meters)

0, 15, 30, 45, 60, 75, 90, 105, 120

Dune height (m)

0.05–0.25
0.25–0.5
0.5–3
3–10

0, 15, 30, 45, 60, 75, 90, 105, 120, 140, 160, 180, 200, 220, 240

small dunes
med dunes
large dunes
very large dunes

Seafloor

Water-bottom multiple

Bedrock?

Dune classification
(after Ashley, 1990)

small (height < 0.25 m)
medium (height = 0.25–0.5 m)
large (height = 0.5–3 m)
very large (height = >3 m)

Dip angle

0°, 0.1°, 0.5°, 1°, 2°, 3°, 5°, 10°, 25°

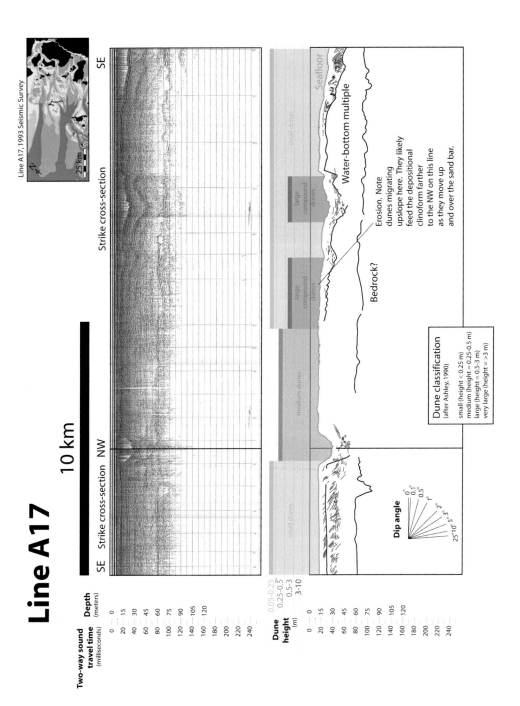

Line A17

Line A17, 1993 Seismic Survey

25 km

10 km

Strike cross-section

SE

NW   Strike cross-section   SE

SE

Two-way sound travel time (milliseconds)

0
20
40
60
80
100
120
140
160
180
200
220
240

Depth (meters)

0
15
30
45
60
75
90
105
120

Dune height (m)

0.05-0.25
0.25-0.5
0.5-3
3-10

0
20
40
60
80
100
120
140
160
180
200
220
240

Seafloor

small dunes

large compound dunes

large compound dunes

medium dunes

small dunes

small dunes

Water-bottom multiple

Bedrock?

Erosion. Note dunes migrating upslope here. They likely feed the depositional clinoform farther to the NW on this line as they move up and over the sand bar.

Dune classification
(after Ashley, 1990)

small (height < 0.25 m)
medium (height = 0.25-0.5 m)
large (height = 0.5-3 m)
very large (height = >3 m)

Dip angle

0°
0.1°
0.5°
1°
2°
3°
5°
10°
25°

Line A18

Line A18, 1993 Seismic Survey

25 km

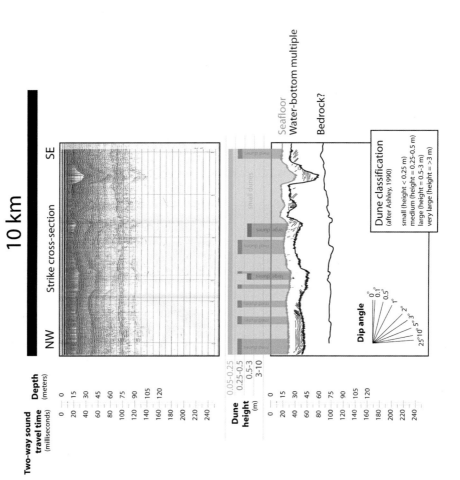

10 km

Two-way sound travel time (milliseconds)

Depth (meters)

NW   Strike cross-section   SE

Seafloor
Water-bottom multiple
Bedrock?

Dune classification
(after Ashley, 1990)

small (height < 0.25 m)
medium (height = 0.25-0.5 m)
large (height = 0.5-3 m)
very large (height = >3 m)

Dune height (m)
0.05-0.25
0.25-0.5
0.5-3
3-10

Dip angle
0°
0.1°
0.5°
1°
2°
3°
5°
10°
25°

small dunes
med dunes
large dunes

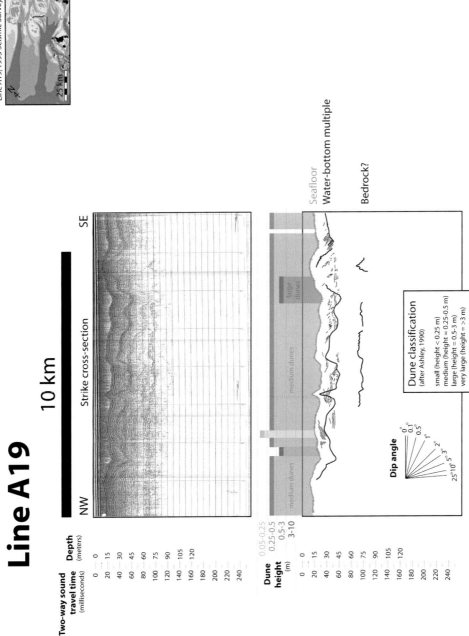

Line A19, 1993 Seismic Survey

25 km

Line A19

10 km

Strike cross-section

NW   SE

Two-way sound travel time
(milliseconds)

0
20
40
60
80
100
120
140
160
180
200
220
240

Depth
(meters)

0
15
30
45
60
75
90
105
120

Dune height
(m)

0.05-0.25
0.25-0.5
0.5-3
3-10

0
20
40
60
80
100
120
140
160
180
200
220
240

medium dunes

medium dunes

large dunes

Seafloor

Water-bottom multiple

Bedrock?

Dip angle

0°
0.1°
0.5°
1°
2°
5°
3°
10°
25°

Dune classification
(after Ashley, 1990)

small (height < 0.25 m)
medium (height = 0.25–0.5 m)
large (height = 0.5–3 m)
very large (height = >3 m)

Line A20, 1993 Seismic Survey

25 km

Line A20

10 km

Strike cross-section

NW   SE

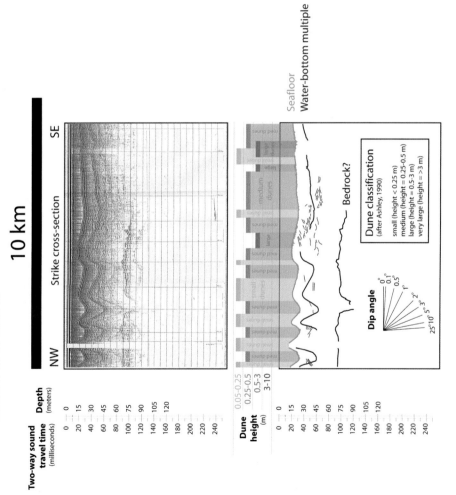

Two-way sound
travel time
(milliseconds)

Depth
(meters)

Dune
height
(m)

Seafloor

Water-bottom multiple

Bedrock?

Dune classification
(after Ashley, 1990)

small (height < 0.25 m)
medium (height = 0.25-0.5 m)
large (height = 0.5-3 m)
very large (height = >3 m)

Dip angle

0°
0.1°
0.5°
1°
2°
3°
5°
10°
25°

small dunes
medium dunes
large
very large

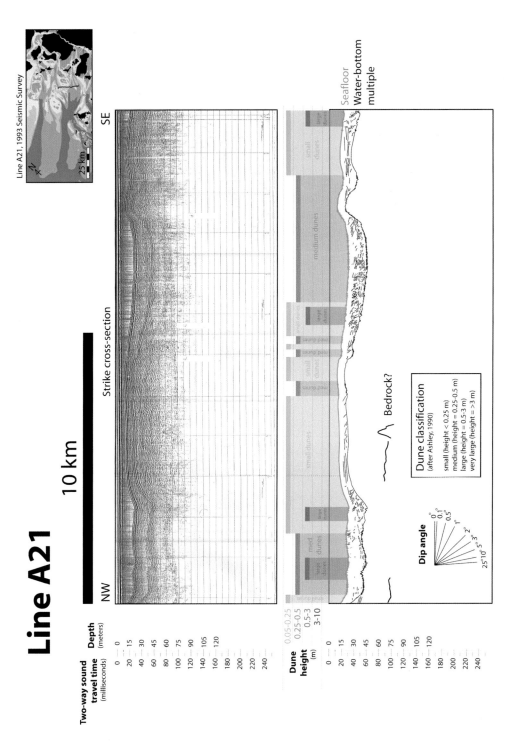

Line A21

10 km

Strike cross-section

NW   SE

Line A21, 1993 Seismic Survey

25 km

Two-way sound travel time (milliseconds)

Depth (meters)

Seafloor

Water-bottom multiple

Bedrock?

Dune classification
(after Ashley, 1990)

small (height < 0.25 m)
medium (height = 0.25-0.5 m)
large (height = 0.5-3 m)
very large (height = >3 m)

Dune height (m)

0.05-0.25
0.25-0.5
0.5-3
3-10

Dip angle

0.1°
0.5°
1°
2°
3°
5°
10°
25°

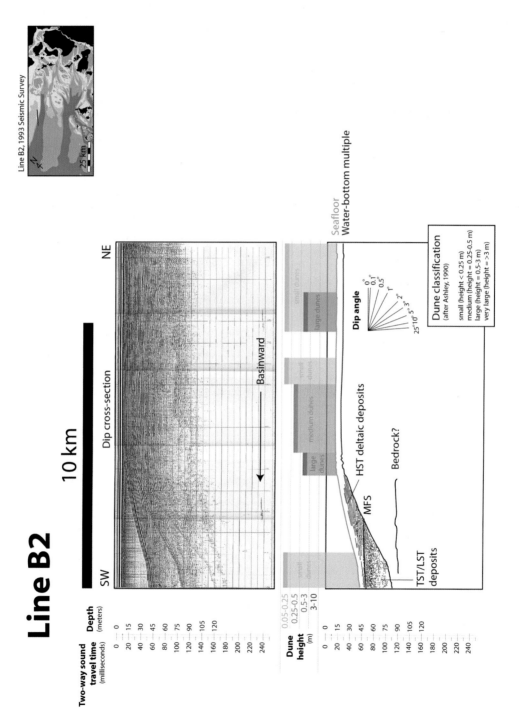

Line B2, 1993 Seismic Survey

Line B2

10 km

Two-way sound travel time (milliseconds)

Depth (meters)

SW   Dip cross-section   NE

Basinward

Seafloor
Water-bottom multiple

Dune height (m)
0.05–0.25
0.25–0.5
0.5–3
3–10

small dunes
large dunes
small dunes
medium dunes
large dunes
small dunes

HST deltaic deposits

MFS

Bedrock?

TST/LST deposits

Dip angle
0°
0.1°
0.5°
1°
2°
3°
5°
10°
25°

Dune classification
(after Ashley, 1990)

small (height < 0.25 m)
medium (height = 0.25–0.5 m)
large (height = 0.5–3 m)
very large (height = >3 m)

Line B5

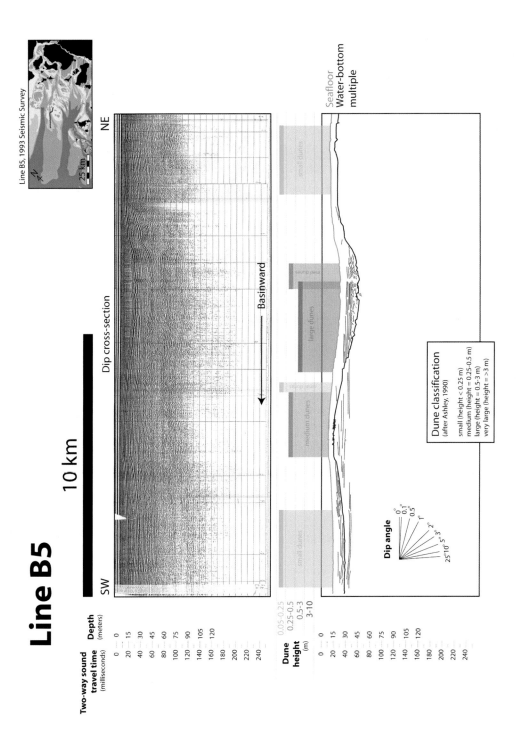

Line B5a

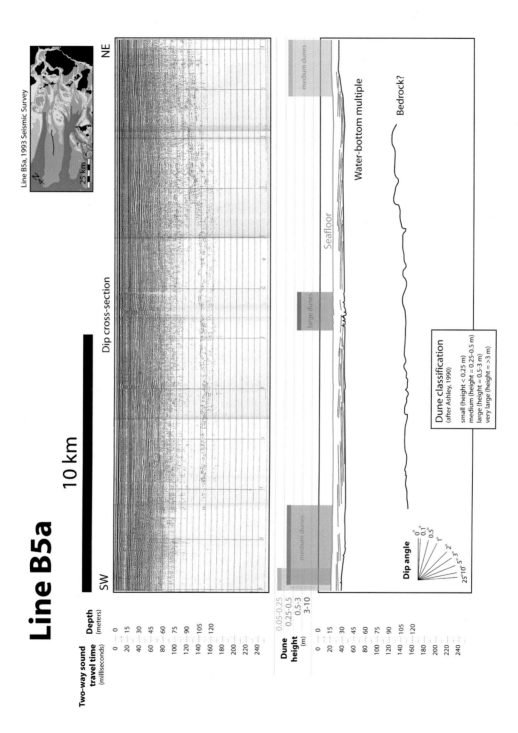

Line B5a, 1993 Seismic Survey

25 km

10 km

Dip cross-section

SW   NE

Two-way sound travel time (milliseconds)

Depth (meters)

0 — 0
20 — 15
40 — 30
60 — 45
80 — 60
100 — 75
120 — 90
140 — 105
160 — 120
180 —
200 —
220 —
240 —

Dune height (m)

0.05-0.25
0.25-0.5
0.5-3
3-10

0 — 0
20 — 15
40 — 30
60 — 45
80 — 60
100 — 75
120 — 90
140 — 105
160 — 120
180 —
200 —
220 —
240 —

medium dunes

large dunes

medium dunes

Seafloor

Water-bottom multiple

Bedrock?

Dune classification (after Ashley, 1990)

small (height < 0.25 m)
medium (height = 0.25-0.5 m)
large (height = 0.5-3 m)
very large (height = >3 m)

Dip angle

0°
0.1°
0.5°
1°
2°
3°
5°
10°
25°

Line B6

Line B6, 1993 Seismic Survey

25 km

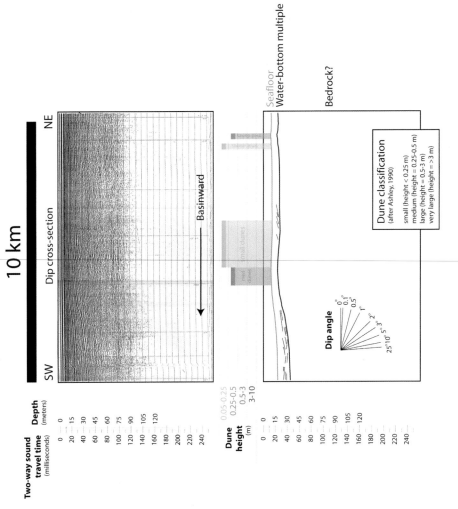

10 km

SW   Dip cross-section   NE

Basinward

Two-way sound travel time (milliseconds)

Depth (meters)

Seafloor

Water-bottom multiple

Bedrock?

Dune classification
(after Ashley, 1990)

small (height < 0.25 m)
medium (height = 0.25–0.5 m)
large (height = 0.5–3 m)
very large (height = >3 m)

Dune height (m)

0.05–0.25
0.25–0.5
0.5–3
3–10

Dip angle

0°
0.1°
0.5°
1°
2°
3°
5°
10°
25°

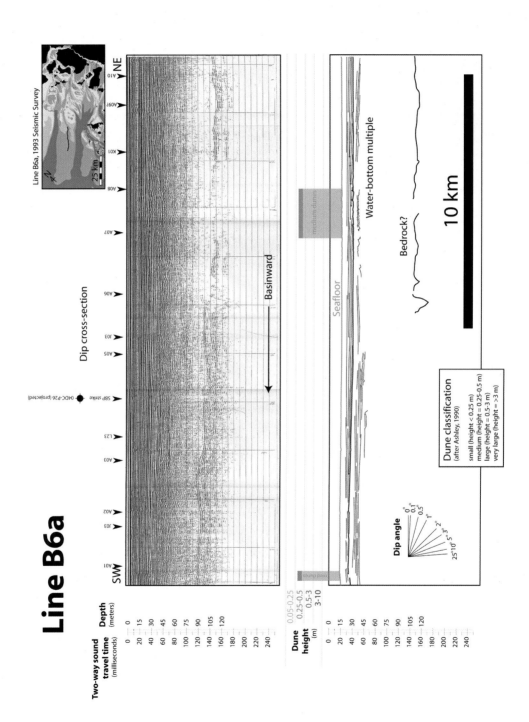

Line B6a

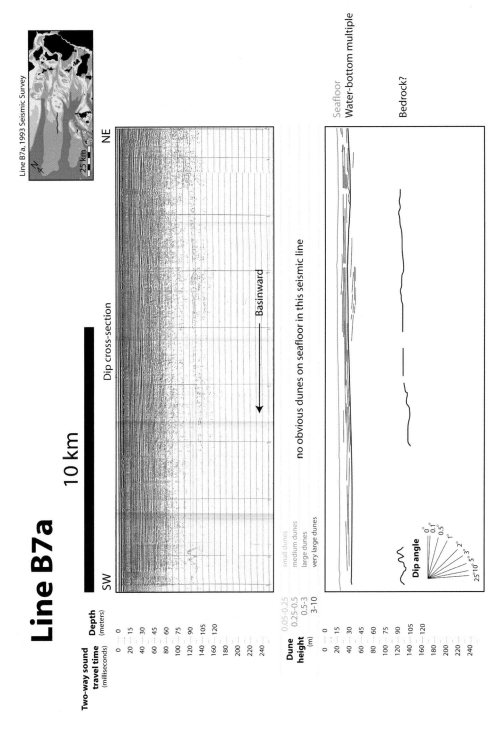

Line B7a

Line B7a, 1993 Seismic Survey

25 km

10 km

Dip cross-section

SW

NE

Two-way sound travel time (milliseconds)

Depth (meters)

Basinward

no obvious dunes on seafloor in this seismic line

Dune height (m)

small dunes   0.05–0.25
medium dunes   0.25–0.5
large dunes   0.5–3
very large dunes 3–10

Seafloor
Water-bottom multiple

Bedrock?

Dip angle

0°
0.1°
0.5°
1°
2°
3°
5°
10°
25°

Line B7b

Line B7b, 1993 Seismic Survey

25 km

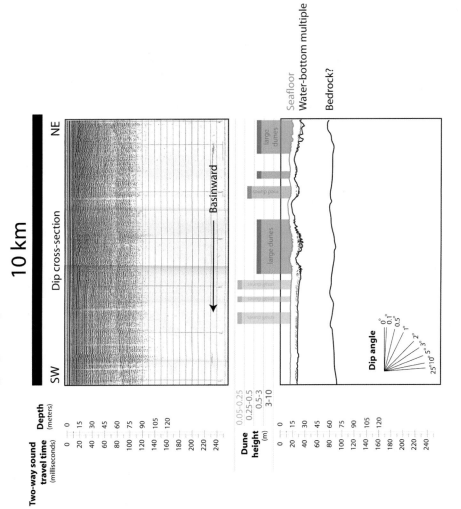

10 km

SW

Dip cross-section

NE

Two-way sound
travel time
(milliseconds)

Depth
(meters)

Basinward

Seafloor

Water-bottom multiple

Bedrock?

Dune
height
(m)

0.05–0.25
0.25–0.5
0.5–3
3–10

Dip angle

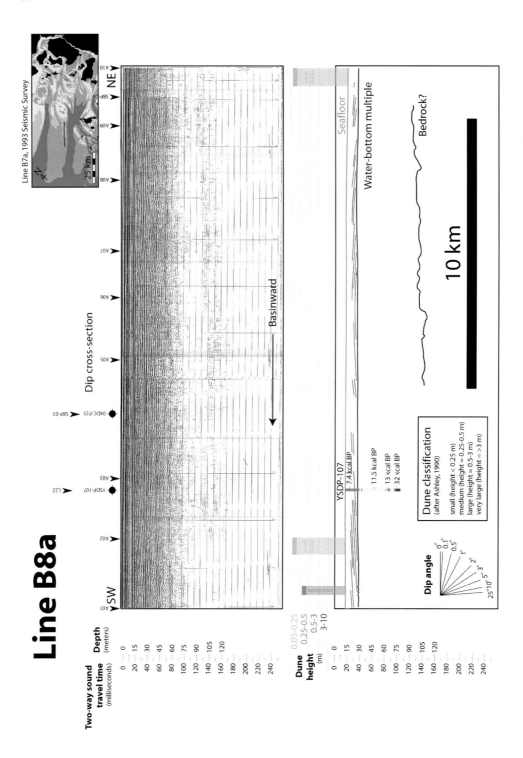

Line B8a

Line B7a, 1993 Seismic Survey

25 km

SW · NE

Dip cross-section

Basinward

Two-way sound travel time (milliseconds)

Depth (meters)

0 — 0
20 — 15
40 — 30
60 — 45
80 — 60
100 — 75
120 — 90
140 — 105
160 — 120
180 —
200 —
220 —
240 —

Dune height (m)
0.05–0.25
0.25–0.5
0.5–3
3–10

Seafloor

Water-bottom multiple

Bedrock?

10 km

YSDP-107

7.4 kcal BP
11.5 kcal BP
13 kcal BP
32 kcal BP

Dune classification
(after Ashley, 1990)

small (height < 0.25 m)
medium (height = 0.25–0.5 m)
large (height = 0.5–3 m)
very large (height = >3 m)

Dip angle
0°
0.1°
0.5°
1°
2°
3°
5°
10°
25°

Line B8b

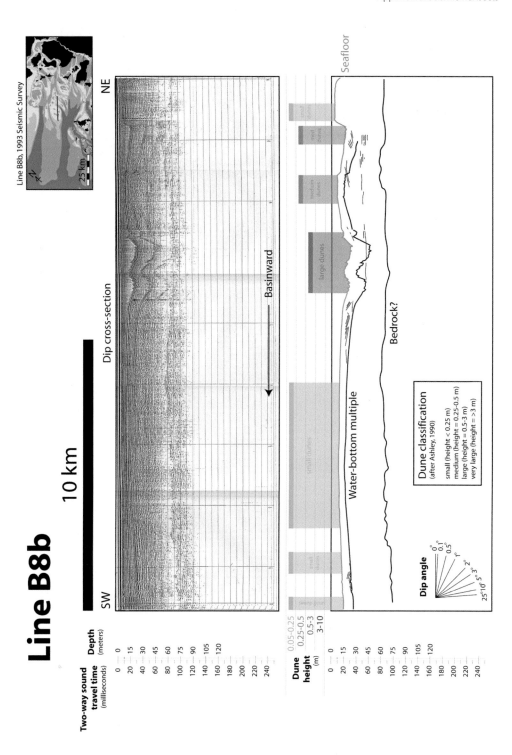

Line B8b, 1993 Seismic Survey

25 km

10 km

Dip cross-section

SW   NE

Basinward

Two-way sound travel time (milliseconds)

Depth (meters)

Seafloor

Water-bottom multiple

Bedrock?

large dunes

medium dunes

mid dunes

small dunes

small dunes

small dunes

Dune classification
(after Ashley, 1990)

small (height < 0.25 m)
medium (height = 0.25–0.5 m)
large (height = 0.5–3 m)
very large (height = >3 m)

Dune height (m)

0.05–0.25
0.25–0.5
0.5–3
3–10

Dip angle

0°
0.1°
0.5°
1°
2°
3°
5°
10°
25°

Line B9

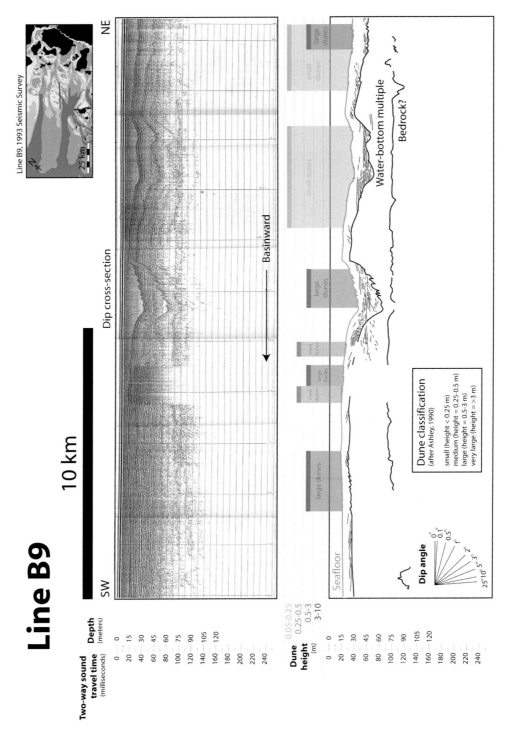

Line B9, 1993 Seismic Survey

25 km

SW

NE

Dip cross-section

10 km

Two-way sound travel time (milliseconds)

Depth (meters)

Basinward

large dunes

small dunes

small dunes

large dunes

med dunes

large dunes

med dunes

large dunes

Water-bottom multiple

Bedrock?

Seafloor

Dune classification
(after Ashley, 1990)

small (height < 0.25 m)
medium (height = 0.25–0.5 m)
large (height = 0.5–3 m)
very large (height = >3 m)

Dip angle

0°
0.1°
0.5°
1°
2°
3°
5°
10°
25°

Dune height (m)

0.05–0.25
0.25–0.5
0.5–3
3–10

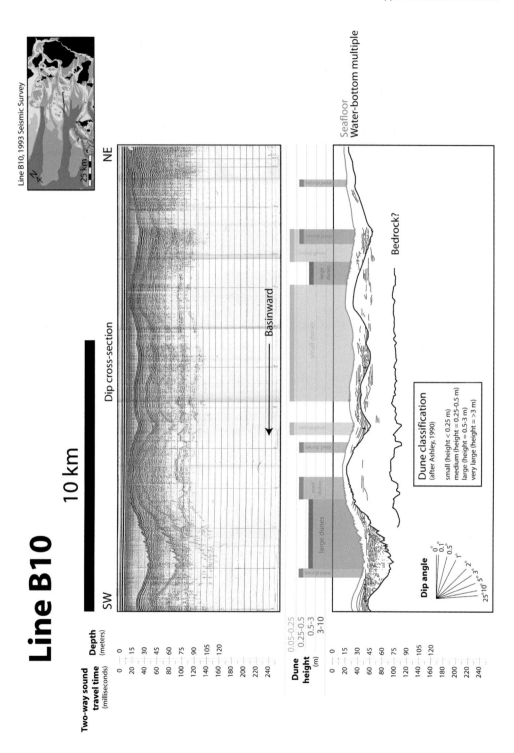

Line B10

Line B10, 1993 Seismic Survey

25 km

10 km

Dip cross-section

SW

NE

Basinward

Two-way sound travel time (milliseconds)

0
20
40
60
80
100
120
140
160
180
200
220
240

Depth (meters)

0
15
30
45
60
75
90
105
120

Dune height (m)

0.05–0.25
0.25–0.5
0.5–3
3–10

0
20
40
60
80
100
120
140
160
180
200
220
240

0
15
30
45
60
75
90
105
120

Seafloor

Water-bottom multiple

Bedrock?

Dune classification
(after Ashley, 1990)

small (height < 0.25 m)
medium (height = 0.25–0.5 m)
large (height = 0.5–3 m)
very large (height = >3 m)

Dip angle

0°
0.1°
0.5°
1°
2°
3°
5°
10°
25°

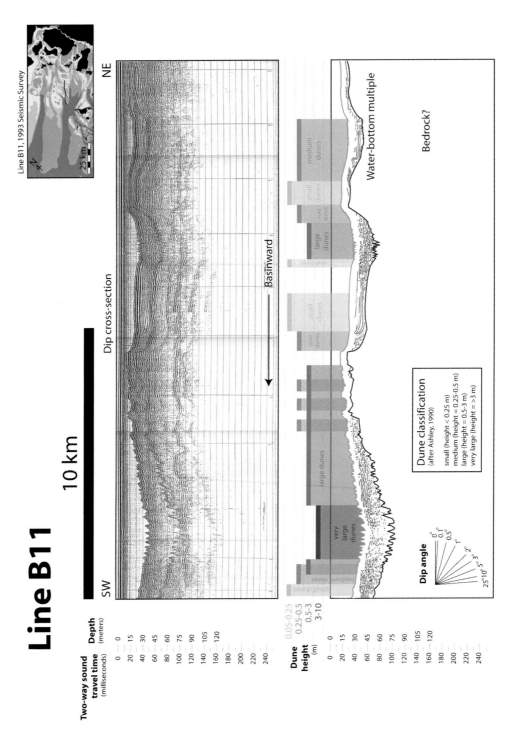

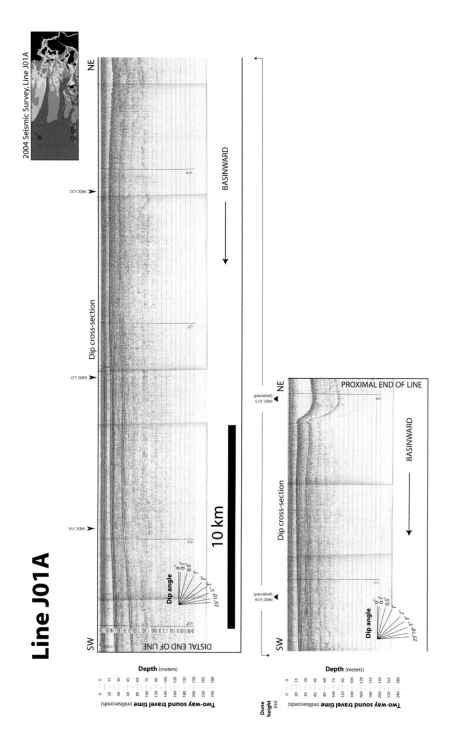

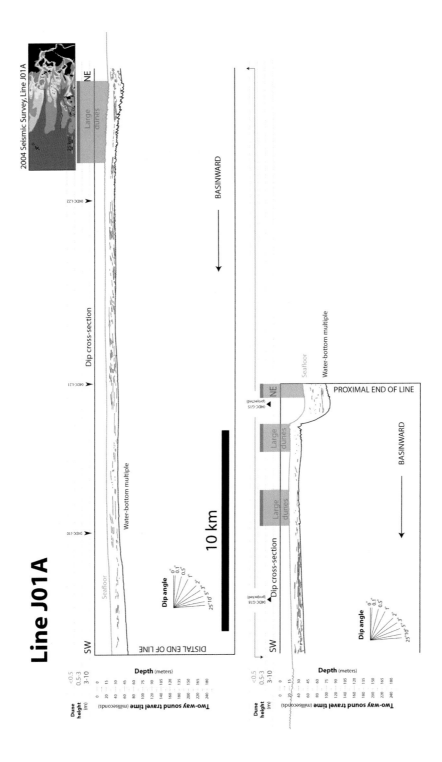

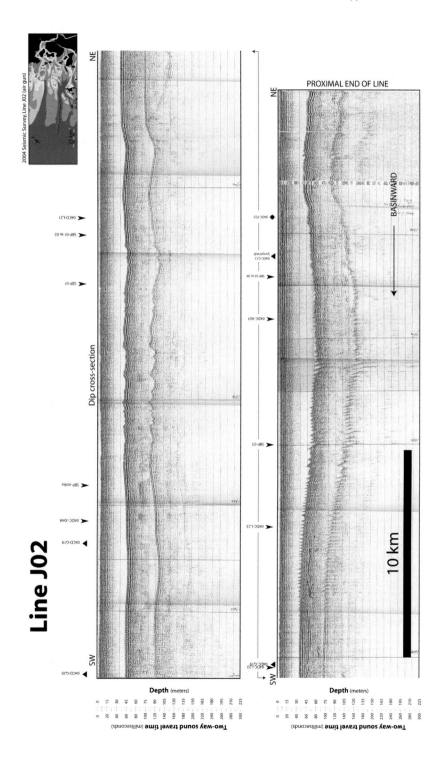

Line J02

2004 Seismic Survey, Line J02 (air gun)

Dip cross-section

PROXIMAL END OF LINE

BASINWARD

10 km

Depth (meters)

Two-way sound travel time (milliseconds)

Depth (meters)

Two-way sound travel time (milliseconds)

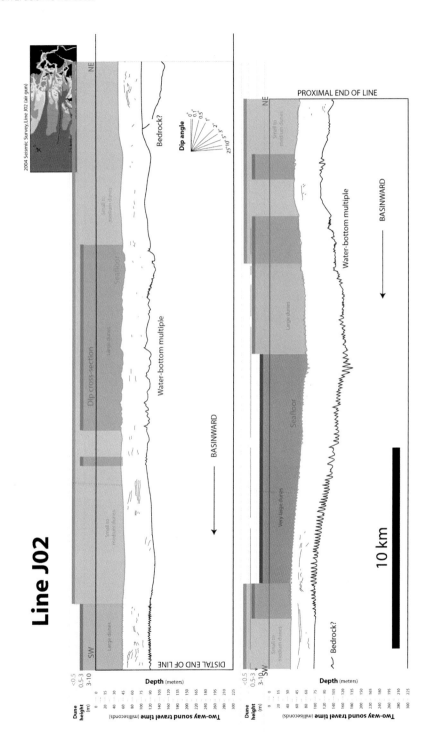

Line J02

2004 Seismic Survey, Line J02 (air gun)

SW — NE

Dip cross-section

Small to medium dunes

Seafloor

Large dunes

Small to medium dunes

Large dunes

Bedrock?

Water-bottom multiple

BASINWARD

DISTAL END OF LINE

Dip angle

0°
0.1°
0.5°
1°
2°
3°
5°
10°
25°

Dune height (m)
<0.5
0.5–3
3–10

Depth (meters)
0
20
40
60
80
100
120
140
160
180
200
220
240
260
280
300

Two-way sound travel time (milliseconds)
0
15
30
45
60
75
90
105
120
135
150
165
180
195
210
225

PROXIMAL END OF LINE

NE

Small to medium dunes

Large dunes

Seafloor

Very large dunes

Small to medium dunes

SW

Bedrock?

Water-bottom multiple

BASINWARD

10 km

Dune height (m)
<0.5
0.5–3
3–10

Depth (meters)
0
20
40
60
80
100
120
140
160
180
200
220
240
260
280
300

Two-way sound travel time (milliseconds)
0
15
30
45
60
75
90
105
120
135
150
165
180
195
210
225

2004 Seismic Survey, Line J09 (air gun)

25 km

Line J02A

SW

NE

Dip cross-section

BASINWARD

10 km

Depth (meters)

Two-way sound travel time (milliseconds)

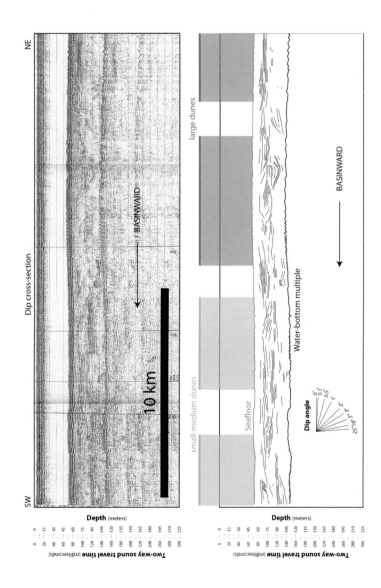

large dunes

small-medium dunes

Seafloor

BASINWARD

Water-bottom multiple

Dip angle

0°
0.1°
0.5°
1°
2°
3°
5°
10°
25°

Depth (meters)

Two-way sound travel time (milliseconds)

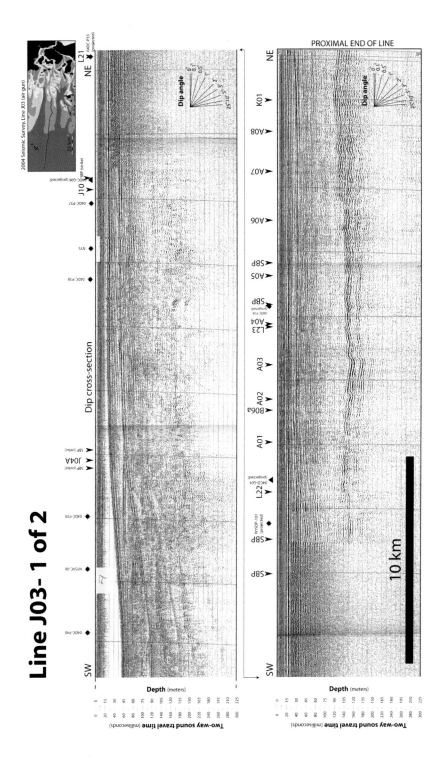

Line J03- 2 of 2

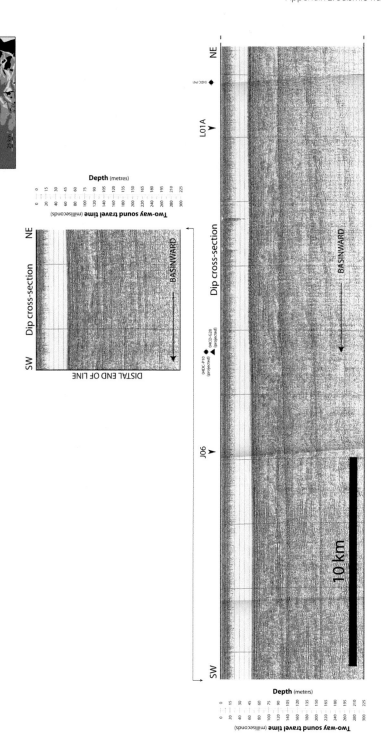

2004 Seismic Survey, Line J03 (air gun)

Depth (metres)

Two-way sound travel time (milliseconds)

NE

Dip cross-section

BASINWARD

SW

DISTAL END OF LINE

04DC-P10 (projected)
04DC-G28 (projected)

Dip cross-section

BASINWARD

NE

L01A

04DC-P41

J06

SW

10 km

Depth (meters)

Two-way sound travel time (milliseconds)

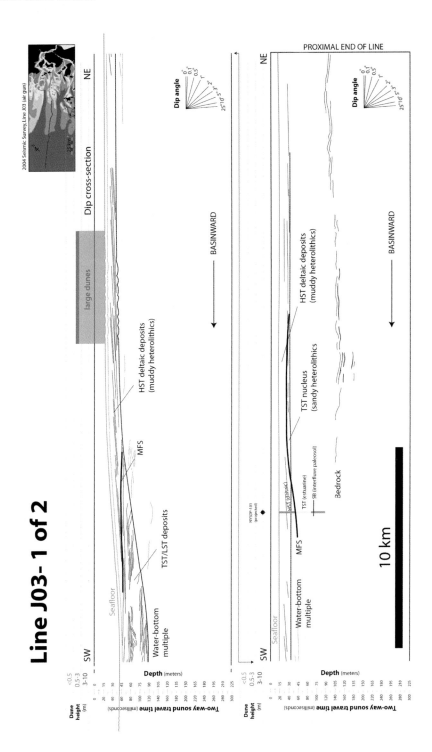

Line J03- 2 of 2

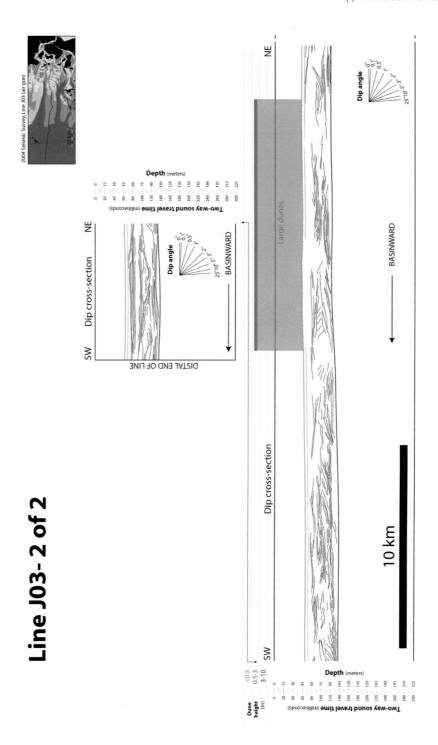

2004 Seismic Survey, Line J03 (air gun)

NE

Dip cross-section

SW

BASINWARD

Dip angle

Depth (meters)

Two-way sound travel time (milliseconds)

DISTAL END OF LINE

Large dunes

NE

BASINWARD

Dip angle

Dip cross-section

SW

10 km

Dune height (m)
<0.5
0.5-3
3-10

Depth (meters)

Two-way sound travel time (milliseconds)

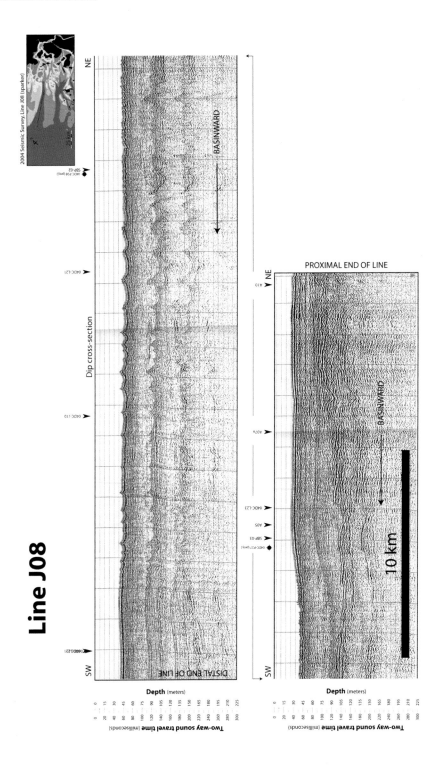

Line J08

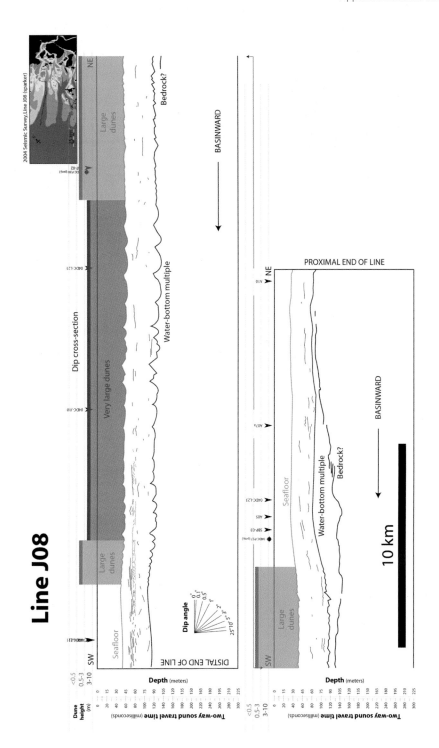

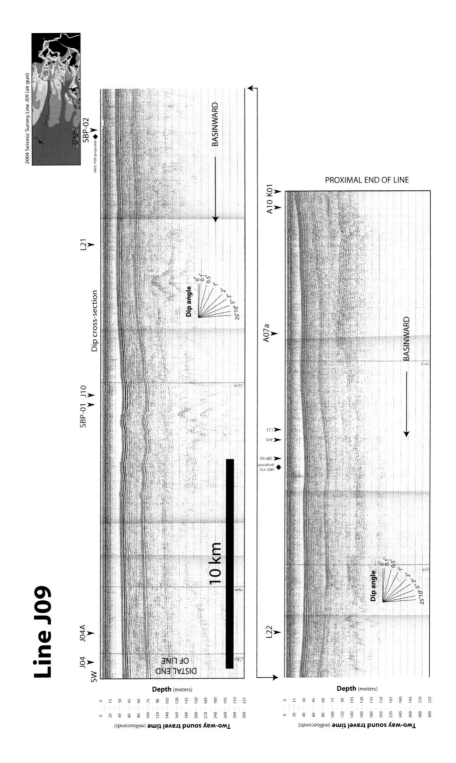

Line J09

Line J09

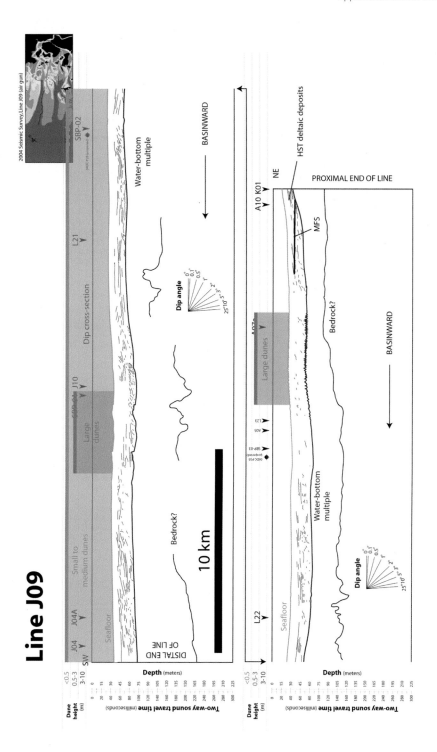

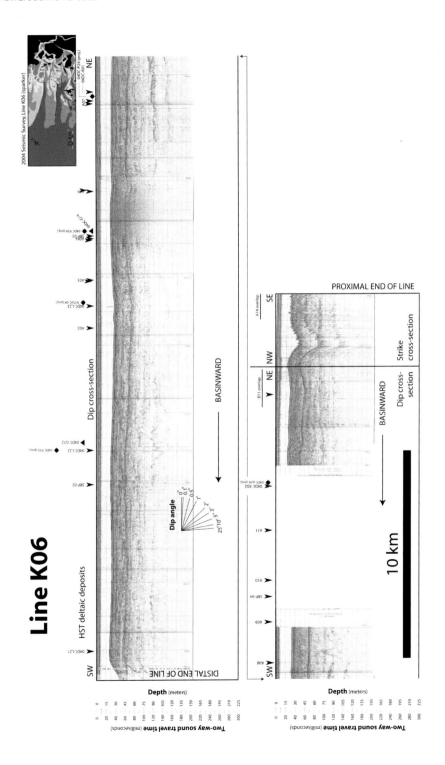

Line K06

Line K06

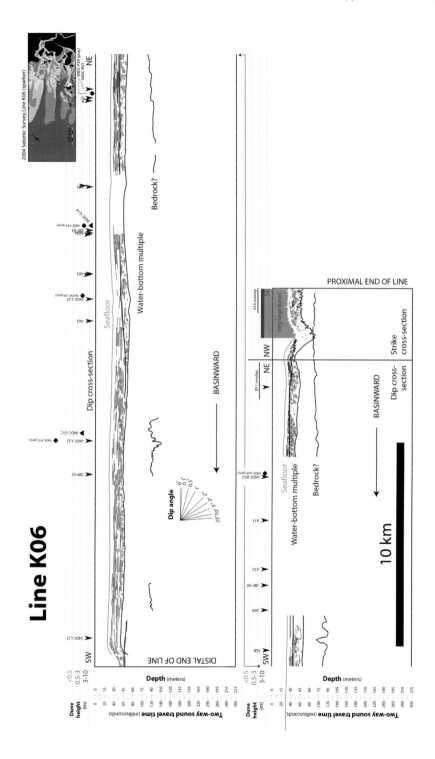

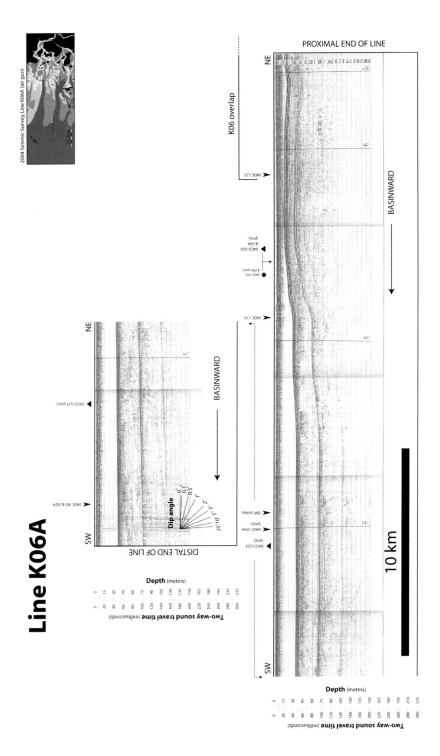

Line K06A

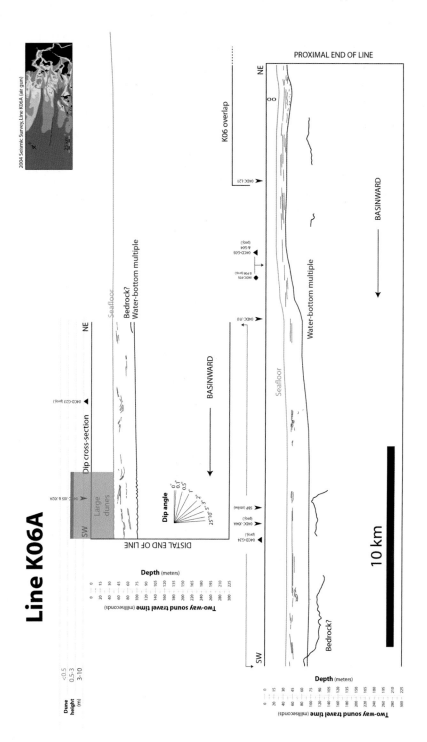

Line K06A

2004 Seismic Survey, Line K06A (air gun)

PROXIMAL END OF LINE

DISTAL END OF LINE

10 km

2004 Seismic Survey, Line J02 (air gun)

25 km

Line L01A

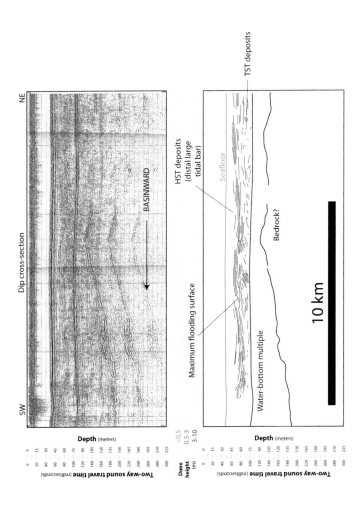

SW   Dip cross-section   NE

BASINWARD

Depth (meters)

Two-way sound travel time (milliseconds)

HST deposits
(distal large
tidal bar)

Seafloor

TST deposits

Maximum flooding surface

Bedrock?

Water-bottom multiple

10 km

Dune
height
(m)

<0.5
0.5-3
3-10

Depth (meters)

Two-way sound travel time (milliseconds)

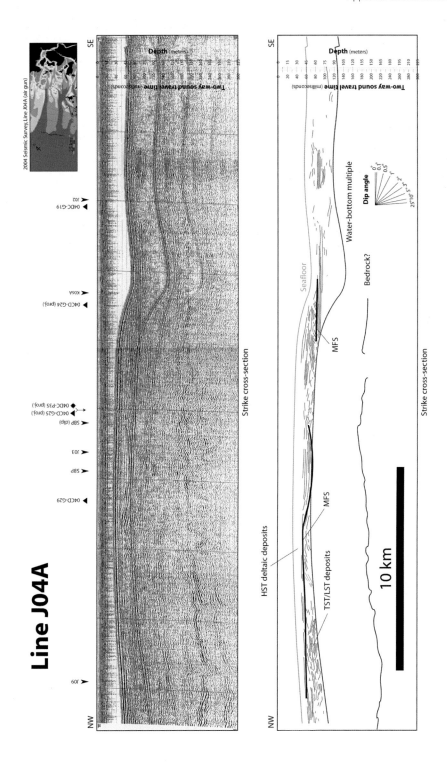

Line J04A

2004 Seismic Survey, Line J04A (air gun)

J02
04DC-G19
K06A
04CD-G24 (proj.)
04DC-P35 (proj.)
04CD-G25 (proj.)
SBP (dip)
J03
SBP
04CD-G29
J09

SE

Depth (meters)

Two-way sound travel time (milliseconds)

Strike cross-section

NW

SE

Depth (meters)

15
30
45
60
75
90
105
120
135
150
165
180
195
210
225

Two-way sound travel time (milliseconds)

Seafloor

Water-bottom multiple

Bedrock?

MFS

HST deltaic deposits

MFS

TST/LST deposits

Dip angle
0°
0.1°
0.5°
1°
2°
3°
5°
10°
25°

10 km

Strike cross-section

NW

2004 Seismic Survey, Line J05 (air gun)

Line J05

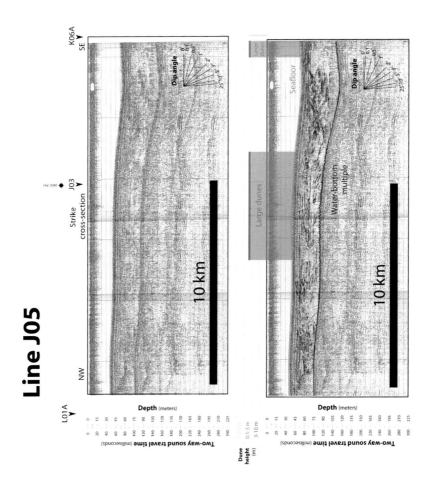

L01A

NW

Strike
cross-section

J03

SE

K06A

Dip angle

0°
0.5°
1°

2°
3°
5°
10°

25°

Depth (meters)

0
15
30
45
60
75
90
105
120
135
150
165
180
195
210
225

Two-way sound travel time (milliseconds)

0
20
40
60
80
100
120
140
160
180
200
220
240
260
280
300

10 km

Seafloor

Large dunes

Water-bottom
multiple

Dip angle

0°
0.5°
1°

2°
3°
5°
10°

25°

Depth (meters)

0
15
30
45
60
75
90
105
120
135
150
165
180
195
210
225

Two-way sound travel time (milliseconds)

0
20
40
60
80
100
120
140
160
180
200
220
240
260
280
300

10 km

Dune
height
(m)

0.5–3 m
3–10 m

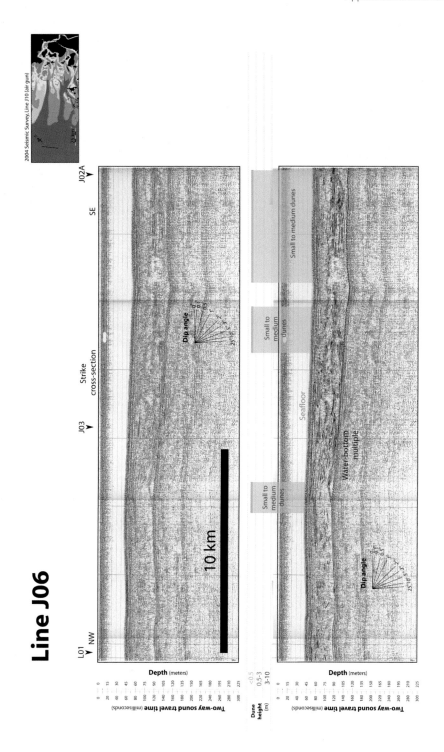

Line J06

2004 Seismic Survey, Line J10 (air gun)

NW — L01

JO3

Strike cross-section

SE

JO2A

10 km

Dip angle — 25°10°5°3°2°1°0.5°

Dip angle — 25°10°5°3°2°1°0.5°

Small to medium dunes

Small to medium dunes

Small to medium dunes

Small to medium dunes

Seafloor

Water-bottom multiple

Dune height (m) — <0.5 — 0.5–3 — 3–10

Depth (meters) — 0 15 30 45 60 75 90 105 120 135 150 165 180 195 210 225

Two-way sound travel time (milliseconds) — 0 20 40 60 80 100 120 140 160 180 200 220 240 260 280 300

Depth (meters) — 0 15 30 45 60 75 90 105 120 135 150 165 180 195 210 225

Two-way sound travel time (milliseconds) — 0 20 40 60 80 100 120 140 160 180 200 220 240 260 280 300

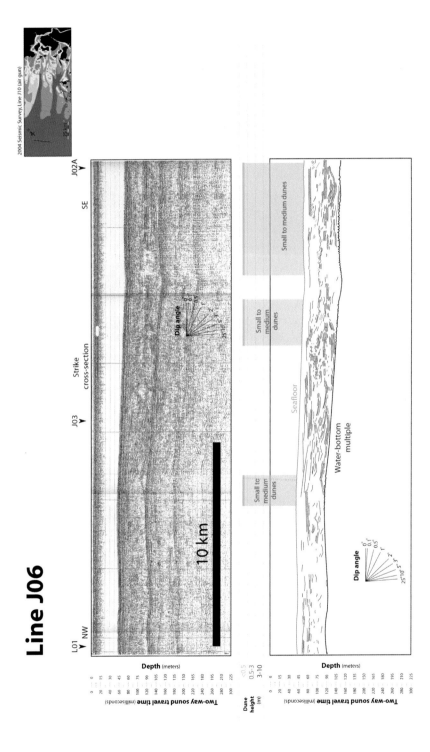

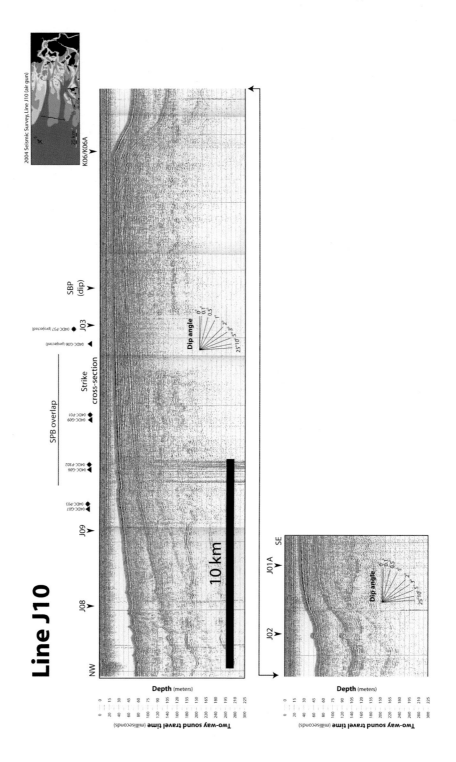

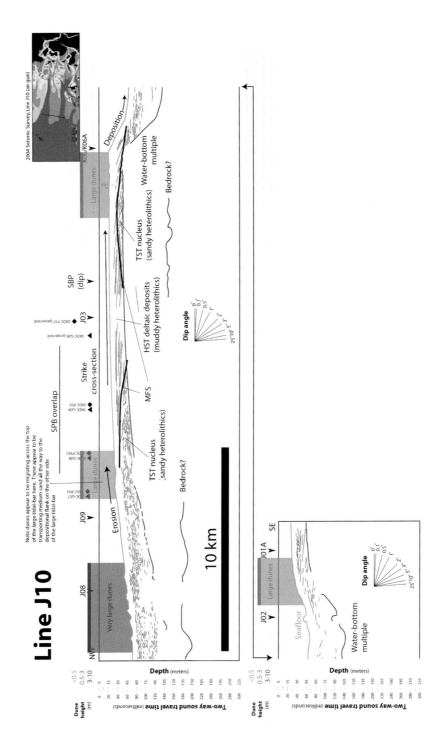

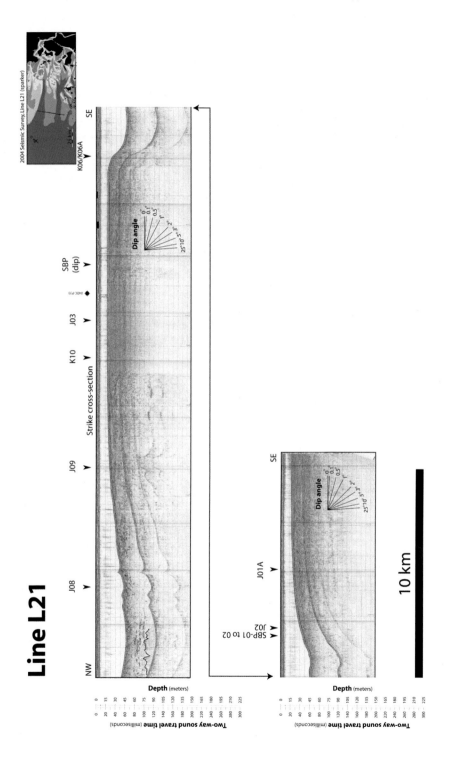

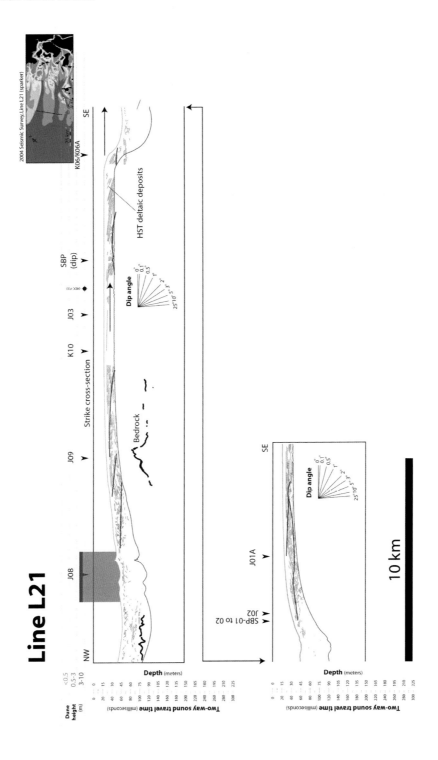

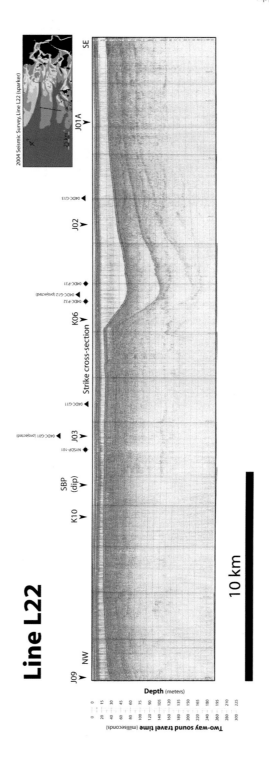

Line L22

10 km

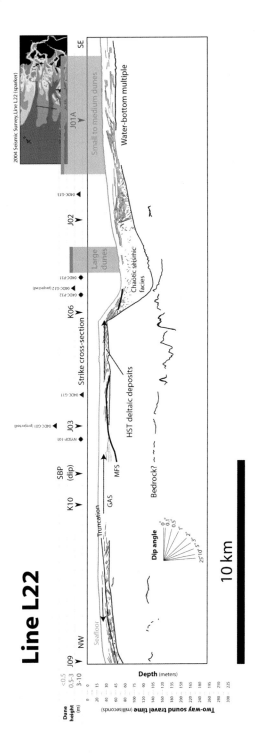

2004 Seismic Survey, Line L23 (sparker)

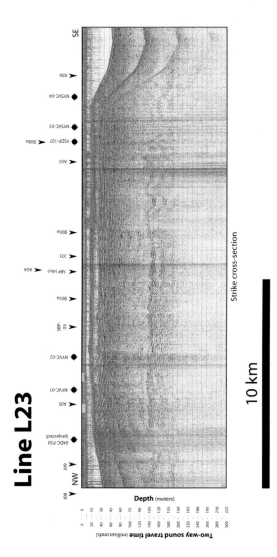

Line L23

NW

SE

Strike cross-section

10 km

Depth (meters)

Two-way sound travel time (milliseconds)

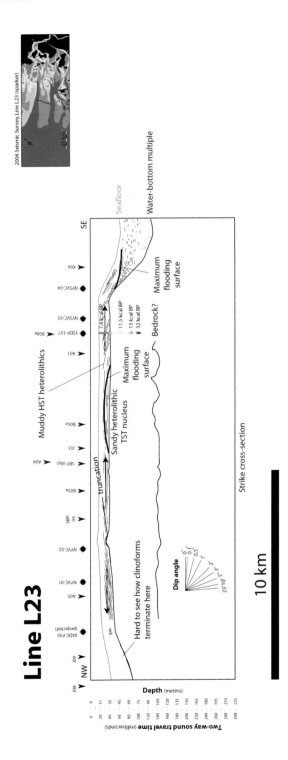

Line L23

REFERENCES

Alexander, C.R., Nittrouer, C.A., Demaster, D.J., Park, Y.-A., Park, S.-C., 1991. Macrotidal mudflats of the southwestern Korean coast; a model for interpretation of intertidal deposits. J. Sediment. Res. 61, 805–824.

Allen, G.P., Salomon, J.C., Bassoullet, P., Du Penhoat, Y., De Grandpre, C., 1980. Effects of tides on mixing and suspended sediment transport in macrotidal estuaries. Sediment. Geol. 26, 69–90.

Allen, J.R.L., 1980. Sandwaves: a model of origin and internal structure. Sediment. Geol. 26, 281–328.

Allen, P.A., 1997. Earth Surface Processes. Blackwell Science, Oxford, 404 p.

Allison, M.A., Nittrouer, C.A., Faria Jr., L.E.C., 1995. Rates and mechanisms of shoreface progradation and retreat downdrift of the Amazon River mouth. In: Nittrouer, C.A., Kuehl, S.A. (Eds.), Geological Significance of Sediment Transport and Accumulation on the Amazon Continental Shelf. Mar. Geol., 125. (Special issue), 373–392.

Allison, M.E., 1998. Historical changes in the Ganges–Brahmaputra delta front. J. Coast. Res. 14, 1269–1275.

Allison, M.E., Khna, S.R., Goodbred, S.L., Kuehl, S.A., 2003. Stratigraphic evolution of the late Holocene Ganges–Brahmaputra lower delta plain. Sediment. Geol. 155, 317–342.

Amos, C.L., 1995. Siliciclastic tidal flats. In: Perillo, G.M.E. (Ed.), Geomorphology and Sedimentology of Estuaries. Elsevier, Amsterdam, pp. 273–306.

Archer, A.W., 2005. Review of Amazonian depositional systems. In: Blum, M.D., Marriott, S.B., Leclair, S.F. (Eds.), Fluvial Sedimentology VII. Internat. Assoc. Sediment. Spec. Publ., vol. 35. pp. 17–39.

Archer, A.W., Hubbard, M.S., 2003. Highest tides of the world. In: Chan, A.M., Archer, A.W. (Eds.), Extreme Depositional Environments: Mega End Members in Geologic Time. Geol. Soc. Amer. Spec. Publ., vol. 370. pp. 151–173.

Arnott, R.W.C., 1993. Quasi-planar-laminated sandstone beds of the Lower Bootlegger Member, north-central Montana: evidence of combined-flow sedimentation. J. Sediment. Petrol. 63, 488–494.

Arnott, R.W.C., Southard, J.B., 1990. Exploratory flow-duct experiments on combined-flow bed configurations, and some implications for interpreting storm-event stratification. J. Sediment. Petrol. 60, 211–219.

Ashley, G.M., 1990. Classification of large-scale subaqueous bedforms: a new look at an old problem. SEPM Bedforms and Bedding Structures Research Symposium. J. Sediment. Petrol. 60, 160–172.

Baas, J.H., Best, J.L., 2002. Turbulence modulation in clay-rich sediment-laden flows and some implications for sediment deposition. J. Sediment. Res. 72, 336–340.

Baas, J.H., Best, J.L., 2008. The dynamics of turbulent, transitional and laminar clay-laden flow over a fixed current ripple. Sedimentology 55, 635–666.

Baas, J.H., Best, J.L., Peakall, J., 2011. Depositional processes, bedform development and hybrid bed formation in rapidly decelerated cohesive (mud–sand) sediment flows. Sedimentology 58, 1953–1987.

Baas, J.H., Best, J.L., Peakall, J., Wang, M., 2009. A phase diagram for turbulent, transitional, and laminar clay suspension flows. J. Sediment. Res. 79, 162–183.

Bagnold, R.A., 1966. An Approach to the Sediment Transport Problem from General Physics. US Geological Survey, Professional Paper 422-1, Washington, DC.

Bale, A.J., Uncles, R.J., Widdows, J., Brinsley, M.D., Barrett, C.D., 2002. Direct observation of the formation and break-up of aggregates in an annual flume using laser reflectance particle sizing. In: Winterwerp, J.C., Kranenburg, C. (Eds.), Fine Sediment Dyamics in the Marine Environment. Proc. Mar. Sci., vol. 5. pp. 189–201.

Bates, C.C., 1953. Rational theory of delta formation. Am. Assoc. Petrol. Geol. 37, 211–262.

Becker, M., Schrottke, K., Bartholomä, A., Ernstsen, V., Winter, C., Hebbeln, D., 2013. Formation and entrainment of fluid mud layers in trough of subtidal dunes in an estuarine turbidity zone. J. Geophys. Res. C. Oceans 118, 2175–2187.

Belderson, R.H., Johnson, M.A., Kenyon, N.H., 1982. Bedforms. In: Stride, A.H. (Ed.), Offshore Tidal Sands: Processes and Deposits. Chapman and Hall, New York, pp. 27–57.

Berné, S., Vagner, P., Guighard, F., Lericolais, G., Liu, Z., Trentesaux, A., Yin, P., Yi, H.I., 2002. Pleistocene forced regressions and tidal sand ridges in the East China Sea. Mar. Geol. 188, 293–315.

Best, J.L., 2005. The fluid dynamics of river dunes: a review and some future research directions. J. Geophys. Res. 110, F04S02. http://dx.doi.org/10.1029/2004JF000218.

Bhattacharya, J.P., 2010. Deltas. In: James, N.P., Dalrymple, R.W. (Eds.), Facies Models 4. Geological Association of Canada, pp. 233–264.

Bhattacharya, J.P., MacEachern, J.A., 2009. Hyperpycnal rivers and prodeltaic shelves in the Cretaceous seaway of North America. J. Sediment. Res. 79, 184–209.

Boersma, J.R., 1970. Distinguishing features of wave-ripple cross-stratification and morphology. Unpublished Ph.D. Thesis, University of Utrecht, Utrecht.

Boles, J.R., Franks, S.G., 1979. Clay diagenesis in Wilcox sandstones of Southwest Texas; implications of smectite diagenesis on sandstone cementation. J. Sediment. Petrol. 49, 55–70.

Boyd, R., 2010. Transgressive wave-dominated coasts. In: James, N.P., Dalrymple, R.W. (Eds.), Facies Models 4. Geol. Assoc. Canada, St. John's, pp. 263–292.

Byun, H.-R., Lee, D.-K., 2002. Defining three rainy seasons and the hydrological summer monsoon in Korea using available water resources index. J. Meteorol. Soc. Jpn. 80, 33–44.

Cartigny, M.J.B., Ventra, D., Postma, G., van den Berg, J.H., 2014. Morphodynamics and sedimentary structures of bedfroms under supercritical-flow conditions; new insights from flume experiments. Sedimentology 61, 712–748.

Castaing, P., Allen, G.P., 1981. Mechanisms controlling seaward escape of suspended sediment from the Gironde: a macrotidal estuary in France. Mar. Geol. 40, 101–118.

Cattaneo, A., Steel, R.J., 2003. Transgressive deposits; a review of their variability. Earth Sci. Rev. 62, 187–228.

Chang, P.H., Isobe, A., 2005. Interannual variation of freshwater in the Yellow and East China Seas: roles of the Changjiang discharge and wind forcing. J. Ocean. 61, 817–834.

Chang, T.S., Joerdel, O., Flemming, B.W., Bartholomä, A., 2006. The role of particle aggregation/disaggregation in muddy sediment dynamics and seasonal sediment turnover in a back-barrier tidal basin, East Frisian Wadden Sea, southern North Sea. Mar. Geol. 235, 49–61.

Chang, T.S., Kim, J.C., Yi, S., 2014. Discovery of Eemian marine deposits along the Baeksu tidal shore, southwest coast of Korea. In: Chen, M.-T., Liu, Z., Catto, N. (Eds.), Quaternary of East Asia and the Western Pacific: Part 2. Quatern. International, vol. 349. pp. 409–418.

Chatanantavet, P., Lamb, M.P., 2014. Sediment transport and topographic evolution of a coupled river and river plume system: An experimental and numerical study. J. Geophys. Res. Earth Surf. 119. http://dx.doi.org/10.1002/2013JF002910.

Cheel, R.J., Leckie, D.A., 1993. Hummocky cross-stratification. Sedimentol. Rev. 1, 103–122.

Chen, A., Saito, Y., Hori, K., Zhao, Y., Kitamura, A., 2003. Early Holocene mud-ridge formation in the Yangtze, offshore China: a tidal-controlled estuarine pattern and sea-level implications. Mar. Geol. 198, 245–257.

Chiu, J.-M., Kim, S.-G., 2004. Estimation of regional seismic hazard in the Korean peninsula using historical earthquake data between A.D. 2 and 1995. Bull. Seismol. Soc. Am. 94, 269–284.

Choi, J.H., 1991. Estimation of boundary shear velocities from tidal current in the Gyeonggi Bay, Korea. J. Oceanol. Soc. Kor. 26, 340–349.

Choi, J.H., Park, Y.A., 1992. Textural characteristics and transport mode of surface sediments of a tidal sand ridge in Gyeonggi Bay, Korea. J. Oceanol. Soc. Kor. 27, 145–153.

Choi, J.K., Eom, J.A., Ryu, J.H., 2011. Spatial relationships between surface sedimentary facies distribution and topography using remotely sensed data: example from the Ganghwa tidal flat, Korea. Mar. Geol. 280, 205–211.

Choi, J.Y., Kwon, Y.K., Chung, G.S., 2012. Late Quaternary stratigraphy and depositional environment of tidal sand ridge deposits in Gyeonggi Bay, west coast of Korea. J. Korean Earth Sci. Soc. 33, 1–10.

Choi, K.S., 2001. Late Quaternary Stratigraphy and Evolution of Tidal Deposits in Kyungi Bay, West Coast of Korea. PhD Thesis, Seoul National Univ., 223 p.

Choi, K.S., 2005. Pedogenesis of late Quaternary deposits, northern Kyonggi Bay, Korea: implications for relative sea-level change and regional stratigraphic correlation. Palaeogeogr. Palaeoclim. Palaeoecol. 220, 387–404.

Choi, K.S., 2011. External controls on the architecture of inclined heterolithic stratification (IHS) of macrotidal Sukmo Channel: wave versus rainfall. Mar. Geol. 285, 17–28.

Choi, K.S., 2014. Morphology, sedimentology and stratigraphy of Korean tidal flats–implications for future coastal managements. Ocean. Coast. Manage. 102, 437–448.

Choi, K.S., Dalrymple, R.W., 2004. Recurring tide-dominated sedimentation in Kyonggi Bay (west coast of Korea): similarity of tidal deposits in late Pleistocene and Holocene sequences. Mar. Geol. 21, 81–96.

Choi, K.S., Dalrymple, R.W., Chun, S.S., Kim, S.P., 2004a. Sedimentology of modern, inclined heterolithic stratification (IHS) in the macrotidal Han River delta, Korea. J. Sediment. Res. 74, 677–689.

Choi, K.S., Dalrymple, R.W., Jin, J.H., Chun, S.S., Min, G.H., 2004b. Morphodynamics of a giant tidal bar in Kyonggi Bay, west coast of Korea (poster). Tidalites 2004 Conference, August 2–5, Copenhagen, Denmark.

Choi, K.S., Hong, C.M., Kim, M.H., Oh, C.R., Jung, J.H., 2013. Morphologic evolution of macrotidal estuarine channels in Gomso Bay, west coast of Korea: implications for the architectural development of inclined heterolithic stratification. Mar. Geol. 346, 343–354.

Choi, K.S., Jo, J., 2015. Morphodynamics and stratigraphic architecture of compound dunes on the open-coast macrotidal flat in the northern Gyeonggi Bay, west coast of Korea. Mar. Geol. 366, 34–48.

Choi, K.S., Kim, B.O., Park, Y.A., 2001. Late Pleistocene tidal rhythmites in Kyunggi Bay, west coast of Korea: a comparison with simulated rhythmites based on modern tides and implications for intertidal positioning. J. Sed. Res. 71, 680–691.

Choi, K.S., Kim, S.P., 2006a. Identifying late Quaternary coastal deposits in Kyonggi Bay, Korea, by their geotechnical properties. Geo Mar. Lett. 26, 77–89.

Choi, K.S., Kim, S.P., 2006b. Late Quaternary evolution of macrotidal Kimpo tidal flat, Kyonggi Bay, west coast of Korea. Mar. Geol. 232, 17–34.

Choi, K.S., Park, Y.A., 2000. Late Pleistocene silty tidal rhythmites in the macrotidal flat between Youngjong and Yongyou islands, west coast of Korea. Mar. Geol. 167, 231–241.

Chough, S.K., Kim, D.C., 1981. Dispersal of fine-grained sediments in the southeastern Yellow Sea: a steady-state model. J. Sediment. Petrol. 51, 721–728.

Chough, S.K., Kim, J.W., Lee, S.H., Shinn, Y.J., Jin, J.H., Suh, M.C., Lee, J.S., 2002. High-resolution acoustic characteristics of epicontinental sea deposits, central–eastern Yellow Sea. Mar. Geol. 188, 317–331.

Chough, S.K., Kwon, S.T., Ree, J.H., Choi, D.K., 2000. Tectonic and sedimentary evolution of the Korean peninsula; a review and new view. Earth Sci Rev. 52, 175–235.

Chough, S.K., Lee, H.J., Chun, S.S., Shinn, Y.J., 2004. Depositional processes of late Quaternary sediments in the Yellow Sea: a review. Geosci. J. 8, 211–264.

Clifton, H.E., 1976. Wave-formed sedimentary structures–a conceptual model. In: Davis, R.A., Ethington, R.L. (Eds.), Beach and Nearshore Sedimentation. SEPM (Soc. Sediment. Geol.) Spec. Publ., vol. 24. pp. 126–148.

Clifton, H.E., 2006. A reexamination of clastic-shoreline facies models. In: Posamentier, H.W., Walker, R.G. (Eds.), Facies Models Revisited. SEPM (Soc. Sediment. Geol.) Spec. Publ., vol. 84. pp. 293–337.

Clifton, H.E., Dingler, J.R., 1984. Wave-formed structures and paleoenvironmental reconstruction. Mar. Geol. 60, 165–198.

Coleman, J.M., 1981. Deltas: Processes of Deposition and Models for Exploration. Springer.

Coleman, J.M., Wright, L.D., 1975. Modern river deltas: variability of processes and sand bodies. In: Broussard, M.L. (Ed.), Deltas, Models for Exploration. Houston Geological Society, Houston, TX, pp. 99–149.

Coughenour, C.L., Archer, A.W., Lacovara, K.J., 2009. Tides, tidalites, and secular changes in the Earth-Moon system Source. Earth Sci. Rev. 97, 96–116.

Crockett, J.S., 2003. Marine sedimentation of the Fly river, Papua New Guinea (abstract). Abstracts, Amer. Assoc. Petrol. Geol., Annual Meeting, Salt Lake City.

Cummings, D.I., Dumas, S., Dalrymple, R.W., 2009. Fine-grained versus coarse-grained wave ripples generated experimentally under large-scale oscillatory flow. J. Sediment. Res. 79, 83–93.

Dalrymple, R.W., 1984. Morphology and internal structure of sandwaves in the Bay of Fundy. Sedimentology 31, 365–382.

Dalrymple, R.W., 2010a. Introduction to siliciclastic facies models. In: James, N.P., Dalrymple, R.W. (Eds.), Facies Models 4. Geol. Assoc. Canada, St. John's, pp. 59–72.

Dalrymple, R.W., 2010b. Tidal depositional systems. In: James, N.P., Dalrymple, R.W. (Eds.), Facies Models 4. Geol. Assoc. Canada, St. John's, pp. 201–231.

Dalrymple, R.W., Baker, E.K., Harris, P.T., Hughes, M., 2003. Sedimentology and stratigraphy of a tide-dominated, foreland-basin delta (Fly River, Papua New Guinea). In: Sidi, F.H., Nummedal, D., Imbert, P., Darman, H., Posamentier, H.W. (Eds.), Tropical Deltas of Southeast Asia - Sedimentology, Stratigraphy, and Petroleum Geology. SEPM (Soc. Sediment. Geol.) Spec. Publ., vol. 76. pp. 147–173.

Dalrymple, R.W., Choi, K.S., 2003. Sediment transport by tides. In: Middleton, G.V. (Ed.), Encyclopedia of Sediments and Sedimentary Rocks. Springer, Dordrecht, pp. 606–609.

Dalrymple, R.W., Choi, K.S., 2007. Morphologic and facies trends through the fluvial-marine transition in tide-dominated depositional systems: a systematic framework for environmental and sequence-stratigraphic interpretation. Earth Sci. Rev. 81, 135–174.

Dalrymple, R.W., Jin, J.H., 2004. May 2004 cruise report: examination of tidal deposits offshore of the Han River, Kyonggi Bay, Korea. Unpublished Technical Report Submitted to the Tidal Signatures Industry Consortium, 31 p.

Dalrymple, R.W., Knight, R.J., Lambiase, J.J., 1978. Bedforms and their hydraulic stability relationships in a tidal environment, Bay of Fundy. Nature 275, 100–104.

Dalrymple, R.W., Mackay, D.A., Ichaso, A.A., Choi, K.S., 2012. Processes, morphodynamics and facies of tide-dominated estuaries. In: Davis Jr., R.A., Dalrymple, R.W. (Eds.), Principles of Tidal Sedimentology. Springer, Dordrecht, pp. 79–107.

Dalrymple, R.W., Makino, Y., Zaitlin, B.A., 1991. Temporal and spatial patterns of rhythmite deposition on mud flats in the macrotidal, Cobequid Bay-Salmon River estuary, Bay of Fundy, Canada. In: Smith, D.G., Reinson, G.E., Zaitlin, B.A., Rahmani, R.A. (Eds.), Clastic Tidal Sedimentology. Can. Soc. Petrol. Geol. Mem., vol. 16. pp. 137–160.

Dalrymple, R.W., Rhodes, R.N., 1995. Estuarine dunes and barforms. In: Perillo, G.M. (Ed.), Geomorphology and Sedimentology of Estuaries. Developments in Sedimentology, vol. 53. Elsevier, Amsterdam, pp. 359–422.

Dalrymple, R.W., Zaitlin, B.A., 1994. High-resolution sequence stratigraphy of a complex, incised valley succession, the Cobequid Bay- Salmon River estuary, Bay of Fundy, Canada. Sedimentology 41, 1069–1091.

Dalrymple, R.W., Zaitlin, B.A., Boyd, R., 1992. Estuarine facies models: conceptual basis and stratigraphic implications. J. Sediment. Petrol. 62, 1130–1146.

Dankers, P.J.T., Winterwerp, J.C., 2009. Hindered settling of mud flocs; theory and validation. Cont. Shelf Res. 27, 1893–1907.

Dashtgard, S.E., Venditti, J.G., Hill, P.R., Sisulak, C.F., Johnson, S.M., La Croix, A.D., 2012. Sedimentation across the tidal-fluvial transition in the Lower Fraser River, Canada. Sediment. Rec. 10, 4–9.

Davies, J.L., 1964. A morphogenetic approach to world shorelines. Z. Geomophol. 8, 127–142.

Davis Jr., R.A., 2012. Tidal signatures and their preservation potential in stratigraphic sequences. In: Davis Jr., R.A., Dalrymple, R.W. (Eds.), Principles of Tidal Sedimentology. Springer, Dordrecht, pp. 35–55.

de Raaf, J.F.M., Boersma, J.R., van Gelder, A., 1977. Wave-generated structures and sequences from a shallow marine succession, Lower Carboniferous, County Cork, Ireland. Sedimentology 24, 451–483.

Demarest, J.M., Kraft, J.C., 1987. Stratigraphic record of Quaternary sea levels: implications for more ancient strata. In: Nummedal, D., Pilkey, O., Howard, J.D. (Eds.), Sea-Level Fluctuations and Coastal Evolution, SEPM (Soc. Sediment. Geol.) Spec. Publ., vol. 41. pp. 223–239.

Doxaran, D., Froidefond, J.M., Castaing, P., Babin, M., 2009. Dynamics of the turbidity maximum zone in a macrotidal estuary (the Gironde, France): observations from field and MODIS satellite data. Estuar. Coast. Shelf Sci. 83, 321–332.

Dronkers, J., 1986. Tidal asymmetry and estuarine morphology. Neth. J. Sea Res. 20, 117–131.

Droppo, I., 2001. Rethinking what constitutes suspended sediment. Hydrol. Proc. 15, 1551–1564.

Dumas, S., Arnott, R.W.C., 2006. Origin of hummocky and swaley cross stratification—the controlling influence of unidirectional current strength and aggradation rate. Geology 34, 1073–1076.

Dumas, S., Arnott, R.W.C., Southard, J.B., 2005. Experiments on oscillatory-flow and combined-flow bed forms: implications for interpreting parts of the shallow-marine sedimentary record. J. Sediment. Res. 75, 501–513.

Dyer, K.R., 1995. Sediment transport processes in estuaries. In: Perillo, G.M.E. (Ed.), Geomorphology and Sedimentology of Estuaries. Develop. Sediment, vol. 53. Elsevier Science B.V, Amsterdam, pp. 423–449.

Dyer, K.R., Huntley, D.A., 1999. The origin, classification and modelling of sand banks and ridges. Cont. Shelf Res. 19, 1285–1330.

Emanuel, K., 2003. Tropical cyclones. Annu. Rev. Earth Planet. Sci. 31, 75–104.

Ettema, R., Daly, S., 2004. Sediment transport under ice. CCREL Tech. Rep. 54.

Faas, R.W., 1991. Rheological boundaries of mud. Where are the limits? Geo Marine Lett. 11, 143–146.

Fan, D.D., 2012. Open-coast tidal flats. In: Davis Jr., R.A., Dalrymple, R.W. (Eds.), Principles of Tidal Sedimentology. Springer, Dordrecht, pp. 187–229.

Fan, D.D., Guo, Y.X., Wang, P., 2006. Cross-shore variations in morphodynamic processes of an open-coast mudflat in the Changjiang Delta: with an emphasis on storm impacts. Cont. Shelf Res. 26, 517–538.

Fan, D.D., Li, C.X., Wang, D.J., Archer, A.W., Greb, S.F., 2004. Morphology and sedimentation on open-coast intertidal flats of the Changjiang Delta, China. J. Coast. Res. 43 (Special issue), 23–35.

Fenies, H., Tastet, J.-P., 1998. Facies and architecture of an estuarine tidal bar (the Trompeloup Bar, Gironde Estuary, SW France). Mar. Geol. 150, 149–169.

Ferguson, R.I., Church, M., 2004. A simple universal equation for grain settling velocity. J. Sediment. Res. 74, 933–937.

Fischer, A.G., 1961. Stratigraphic record of transgressing seas in light of sedimentation on the Atlantic coast of New Jersey. Am. Assoc. Petrol. Geol. Bull. 45, 1656–1666.

Fisk, H.N., 1944. Geological investigation of the Alluvial Valley of the Lower Mississippi River. Tulsa Geol. Soc. Digest. 15.

Fitzgerald, D., Buynevich, I., Hein, C., 2012. Morphodynamics and facies architecture of tidal inlets and tidal deltas. In: Davis Jr., R.A., Dalrymple, R.W. (Eds.), Principles of Tidal Sedimentology. Springer, Dordrecht, pp. 301–333.

Flemming, B.W., 1988. Zur klassifikation subaquatischer, strömungstransversaler transportkörper. Boch. Geol. U. Geotechn. Arb. 29, 44–47.

Flemming, B.W., 2012. Siliclastic back-barrier tidal flats. In: Davis Jr., R.A., Dalrymple, R.W. (Eds.), Principles of Tidal Sedimentology. Springer, Dordrecht, pp. 231–267.

Fox, J.M., Hill, P.S., Milligan, T.G., Boldrin, A., 2004. Flocculation and sedimentation on the Po River delta. Mar. Geol. 203, 95–107.

Frey, R.W., Howard, J.D., Han, S.J., Park, B.K., 1989. Sediments and sedimentary sequences on a modern macrotidal flat, Incheon, Korea. J. Sediment. Petrol. 59, 28–44.

Frey, R.W., Howard, J.D., Hong, J.S., 1987. Prevalent lebensspuren on a modern macrotidal flat, Incheon, Korea: ethological and environmental significance. Palaios 2, 571–593.

Friedrichs, C.T., Wright, L.D., 2004. Gravity-driven sediment transport on the continental shelf: implications for equilibrium profiles near river mouths. Coast. Eng. 51, 795–811.

Friedrichs, G.T., Aubrey, D.G., 1988. Non-linear tidal distortion in shallow well mixed estuaries: a synthesis. Estuar. Coast. Shelf Sci. 27, 521–546.

Galler, J.J., Allison, M.A., 2008. Estuarine controls on fine-grained sediment storage in the lower Mississippi and Atchafalaya Rivers. Geol. Soc. Am. Bull. 120, 386–398.

Galloway, W.E., 1975. Process framework for describing the morphologic and stratigraphic evolution of deltaic depositional systems. In: Broussard, M.L. (Ed.), Deltas, Models for Exploration. Houston Geological Society, Houston, TX, pp. 87–98.

Ganju, N.K., Schoellhamer, D.H., Warner, J.C., Barad, M.F., Schladow, S.G., 2004. Tidal oscillation of sediment between a river and a bay: a conceptual model. Estuar. Coast. Shelf Sci. 60, 81–90.

Garvine, R.W., 1999. Penetration of buoyant coastal discharge onto the continental shelf: a numerical model experiment. J. Phys. Ocean. 29, 1892–1909.

Geyer, W.R., 1993. The importance of suppression of turbulence by stratification on the estuarine turbidity maximum. Estuaries 16, 113–125.

Geyer, W.R., Hill, P.S., Kineke, G.C., 2004. The transport, transformation and dispersal of sediment by buoyant coastal flows. Cont. Shelf Res. 24, 927–949.

Gibling, M.R., Tandon, S.K., Sinha, R., Jain, M., 2005. Discontinuity-bounded alluvial sequences of the southern Gangetic Plains, India: aggradation and degradation in response to monsoonal strength. J. Sediment. Res. 75, 369–385.

Gilbert, G.K., 1885. The topographic features of lake shores. U.S. Geol. Surv. Ann. Rept. 5, 69–123.

Goda, Y., 2003. Revisiting Wilson's formulas for simplified wind-wave prediction. J. Waterway Port Coast. Ocean Eng. 129, 93–95.

Goodbred, S.L., Kuehl, S.A., 2000. Enormous Ganges–Brahmaputra sediment discharge during strengthened early Holocene monsoon. Geology 28, 1083–1086.

Goodbred, S.L., Saito, Y., 2012. Tide-dominated deltas. In: Davis Jr., R.A., Dalrymple, R.W. (Eds.), Principles of Tidal Sedimentology. Springer, Dordrecht, pp. 129–149.

Graber, H.C., Beardsley, R.C., Grant, W.D., 1989. Storm-generated surface waves and sediment resuspension in the East China and Yellow Seas. J. Phys. Ocean. 19, 1039–1059.

Granboulan, J., Feral, A., Villerot, M., Jouanneau, J.M., 1989. Study of the sedimentological and rheological properties of fluid mud in the fluvio-estuarine system of the Gironde Estuary. Ocean Shorel. Manage. 12, 23–46.

Grant, W.D., Madsen, O.S., 1979. Combined wave and current interaction with a rough bottom. J. Geophys. Res. 84, 1797–1808.

Grant, W.D., Madsen, O.S., 1986. The continental-shelf bottom boundary layer. Ann. Rev. Fluid Mech. 18, 265–305.

Green, M.O., Coco, G., 2014. Review of wave-driven sediment resuspension and transport in estuaries. Rev. Geophys. 52, 77–117.

Guo, X., Yanagi, T., 1998. Three-dimensional structure of tidal current in the East China Sea and the Yellow Sea. J. Ocean. 54, 651–668.

Harris, P.T., 1988. Large scale bedforms as indicators of mutually evasive sand transport and the sequential infilling of wide-mouthed estuaries. Sediment. Geol. 57, 273–298.

Harris, P.T., Hughes, M.G., Baker, E.K., Dalrymple, R.W., Keene, J.B., 2004. Sediment export from distributary channels to the pro-delta environment in a tidally-dominated delta: Fly River, Papua New Guinea. Cont. Shelf Res. 24, 2431–2454.

Hein, F.J., Dolby, G., Fairgrieve, B., 2013. A regional geologic framework for the Athabasca oil sands, northeastern Alberta, Canada. In: Hein, F.J., Leckie, D., Larter, S., Suter, J.R. (Eds.), Heavy-Oil and Oil-Sand Petroleum Systems in Alberta and Beyond. Amer. Assoc. Petrol. Geol. Studies in Geol., vol. 64. pp. 207–250.

Helland-Hansen, W., Hampson, G.J., 2009. Trajectory analysis: concepts and applications. Basin Research, vol. 21, 454–483.

Hill, P.S., Fox, J.M., Crockett, J.S., Curran, K.J., Friedrichs, C.T., Geyer, W.R., Milligan, T.G., Ogston, A.S., Puig, P., Scully, M.E., Traykovski, P.A., Wheatcroft, R.A., 2007. Sediment delivery to the seabed on continental margins. In: Nittrouer, C., Austin, J., Feld, M., Kravitz, J., Syvitski, J., Wiberg, P. (Eds.), Continental Margin Sedimentation: From Sediment Transport to Sequence Stratigraphy. International Association of Sedimentologists Special Publication, pp. 49–99.

Hong, G.H., Zhang, J., Kim, S.H., Chung, C.S., Yang, S.R., 2002. East Asian marginal seas: river-dominated ocean margin. In: Hong, G.H., Zhang, J., Chung, C.S. (Eds.), Imapct of Interface Exchange on the Biogeochemical Processes of the Yellow and East China Seas. Bumshin Press, Seoul, pp. 233–260.

Hong, R.J., Karim, M.F., Kennedy, J.F., 1984. Low temperature effects on flow in sand-bed streams. J. Hydraul. Eng. 110, 109–125.

Hori, K., Saito, Y., Zhao, Q., Cheng, X., Wang, P., Sato, Y., Li, C., 2001. Sedimentary facies and Holocene progradation rates of the Changjiang (Yangtze) delta, China. Geomorphology 41, 233–248.

Hori, K., Saito, Y., Zhao, Q., Wang, P., 2002. Architecture and evolution of the tide-dominated Changjiang (Yangtze) River delta, China. Sediment. Geol. 146, 249–264.

Hovikoski, J., Rasanen, M., Gingras, M.K., Ranzi, A., Melo, J., 2008. Tidal and seasonal controls in the formation of Late Miocene inclined heterolithic stratification deposits, western Amazonian foreland basin. Sedimentology 55, 499–530.

Hoyal, D.C.J.D., Van Wagoner, J.C., Adair, N.L., Deffenbaugh, M., Li, D., Sun, T., Huh, C., Giffin, D.E., 2003. Sedimentation from Jets: A Depositional Model for Clastic Deposits of All Scales and Environments. Am. Assoc. Petrol. Geol. Search and Discovery Article #40082.

Hübscher, C., Figueredo, A.G., Kruse, L., Spiess, V., 2002. High-resolution analysis of the deposition pattern on the Amazon sub-aquatic delta and outer continental shelf. Mar. Geophys. Res. 23, 209–222.

Huthnance, J.M., 1982a. On one mechanism forming linear sand banks. Estuar. Coast. Shelf Sci. 14, 79–99.

Huthnance, J.M., 1982b. On the formation of sand banks of finite extent. Estuar. Coast. Shelf Sci. 15, 277–299.

Ichaso, A.A., Dalrymple, R.W., 2009. Tide- and wave-generated fluid mud deposits in the Tilje Formation (Jurassic), offshore Norway. Geology 37, 539–542.

Ichaso, A.A., Dalrymple, R.W., 2014. Eustatic, tectonic and climatic controls on an early syn-rift mixed-energy delta, Tilje Formation, (Early Jurassic, Smørbukk Field, offshore mid-Norway. In: Martinius, A.W., Ravnås, R., Howell, J.A., Steel, R.J., Wonham, J.P. (Eds.), From Depositional Systems to Sedimentary Successions on the Norwegian Continental Shelf' Internat. Assoc. Sediment. Spec. Publ., vol. 46. pp. 339–388.

Ichikawa, H., Beardsley, R.C., 2002. The current system in the Yellow and East China Seas. J. Ocean. 58, 77–92.

Inglis, C.C., Allen, F.H., 1957. The Regimen of the Thames Estuary as Affected by Currents, Salinities, and River Flow, vol. 7. Proceedings of the Institute of Civil Engineering, London, pp. 827–868.

Jablonski, B.V.J., 2012. Process Sedimentology and Three-Dimensional Facies Architecture of a Fluvially Dominated, Tidally Influenced Point Bar: Middle McMurray Formation, Lower Steepbank River Area, Northeastern Alberta, Canada. Unpublished M.Sc. Thesis, Queen's University, Kingston, 356 p.

Jaeger, J.M., Nittrouer, C.A., 1995. Tidal controls on the formation of finescale sedimentary strata near the Amazon River mouth. Mar. Geol. 125, 259–281.

Jeong, J.H., Lee, D.I., Wang, C.C., Jang, S.M., You, C.H., Jang, M., 2012. Environment and morphology of mesoscale convective systems associated with the Changma front during 9–10 July 2007. Ann. Geophys. 30, 1235–1248.

Jiang, J., Mehta, A.J., 2000. Lutocline behavior in high-concentration estuary. J. Waterway Port Coast. Ocean Eng. 126, 324–328.

Jin, J.H., 2001. A Sedimentological Study of Long Sediment Cores in the Eastern Yellow Sea. Unpublished Ph.D. Thesis, Seoul National University, Seoul, 176 p.

Jin, J.H., Chough, S.K., 1998. Partitioning of transgressive deposits in the southeastern Yellow Sea: a sequence stratigraphic interpretation. Mar. Geol. 149, 79–92.

Jin, J.H., Chough, S.K., 2002. Erosional shelf ridges in the mid-eastern Yellow Sea. Geomarine Lett. 21, 219–225.

Jin, Y.H., Kawamura, A., Jinno, K., Berndtsson, R., 2005. Detection of the ENSO influence on the monthly precipitation in South Korea. Hydrol. Proc. 19, 4081–4092.

Johnson, S.M., Dashtgard, S.E., 2014. Inclined heterolithic stratification in a mixed tidal-fluvial channel: differentiating tidal versus fluvial controls on sedimentation. Sediment. Geol. 301, 41–53.

Jolivet, L., Tamaki, K., Fournier, M., 1994. Japan Sea, opening history and mechanism: a synthesis. J. Geophys. Res. 99, 22,237–22,259.

Jung, K.T., Kang, H.W., So, J.K., Lee, H.J., 2001. A model-generated circulation in the Yellow Sea and the East China Sea: 1. Depth-mean flow fields. Ocean. Polar Res. 23, 232–242.

Jung, W.Y., Suk, B.C., Min, G.H., Lee, Y.K., 1998. Sedimentary structure and origin of a mud-cored pseudo-tidal sand ridge, eastern Yellow Sea, Korea. Mar. Geol. 151, 73–88.

KACA, 1996. Geotechnical Investigation Report on Incheon International Airport Transportation Center. Korea Airport Construction Authority, Incheon, 342 p.

Kang, S.K., Foreman, M.G.G., Lie, H.J., Lee, J.H., Cherniawsky, J., Yum, K.D., 2002. Two-layer tidal modeling of the Yellow and East China Seas with application to seasonal variability of the M_2 tide. J. Geophys. Res. 107, 6-1–6-19.

KIGAM, 1997. Yellow Sea drilling program for studies of Quaternary geology: analyses of YSDP-106, YSDP-107 cores (in Korean, with English summary). KIGAM Rep. KR-97 ((B)-23), 204.

Kim, C.S., Lim, H.S., 2009. Sediment dispersal and deposition due to sand mining in the coastal waters of Korea. Cont. Shelf Res. 29, 194–204.

Kim, C.S., Lim, H.S., Kim, J.A., Kim, S.J., 2009. Residual flow and its implication to macro-tidal flats in Kyunggi Bay. J. Coast. Res. SI56, 976–980.

Kim, J.C., Chang, T.S., Yi, S., Hong, S.S., Nahm, W.H., 2014. OSL dating of coastal sediments from the southwestern Korean Peninsula: a comparison of different size fractions of quartz. Quat. Int. 1–9.

Kim, J.G., 2006. Characteristics of Surface Sediment Distribution and Sedimentary Facies and Processes on the Southern Intertidal Flat of Gwanghwa Island, Western Coast of Korean Peninsula. Unpublished M.Sc. Thesis, Chonnam National University, Chonnam, South Korea, 58 p.

Kim, J-H., 2009. Morphometric Analysis of Tidal Sand Dunes in Kyunggi Bay, Korea, Using Multibeam and Side-scan Sonar Images (In Korean, with English Summary). Unpublished MSc Thesis, Gyeongsang National University, Jinju, South Korea, 50 p.

Kim, J.S., Jain, S., Yoon, S.K., 2010. Warm season streamflow variability in the Korean Han River basin: links with atmospheric teleconnections. Int. J. Climatol. 32, 635–640.

Kim, K.W., Yoo, C.S., Park, M.K., Kim, H.J., 2007. Evaluation for usefulness of Chukwookee data in rainfall frequency analysis. J. Korea Water Resour. Assoc. 40, 851–859.

Kim, N.W., Lee, J.E., Kim, J.T., 2012. Assessment of flow regulation effects by dams in the Han River, Korea, on the downstream flow regimes using SWAT. J. Water Resour. Plan. Manag. 138, 24–35.

Kim, W., Yoon, K.S., Woo, H., 2002. Analysis of hydraulic characteristics in the river mouth with large tidal difference—ADCP application. In: Wahl, T.L., Pugh, C.A., Oberg, K.A., Vermeyen, V.B. (Eds.), Hydraulic Measurements and Experimental Methods Conference 2002, July 28-Aug 1, Estes Park, California.

Kim, W.C., Woo, H., Kim, W., Lee, D.H., Yeon, K.S., 2004. Re-channelization of stream channels affected by an extreme flood due to the 2002 Typhoon Rusa in Korea. Proceedings. Amer. Soc. Civil Eng., World Water and Environmental Resources 2004 Congress, Salt Lake City.

Kineke, G.C., Sternberg, R.W., 1995. Distribution of fluid muds on the Amazon continental shelf. Mar. Geol. 125, 193–233.

Kineke, G.C., Sternberg, R.W., Trowbridge, J.H., Geyer, W.R., 1996. Fluid-mud processes on the Amazon continental shelf. Cont. Shelf Res. 16, 667–696.

Kirby, R., 1991. Distinguishing features of layered muds deposited from shallow water high concentration suspensions. In: Bennett, R.H., Bryant, W.R., Hulbert, M.H. (Eds.), Microstructure of Fine Grained Sediments; from Mud to Shale. Frontiers in Sedimentary Geology, pp. 167–173.

Kirby, R., 2010. Distribution, transport and exchanges of fine sediment, with tidal power implications: Severn Estuary, UK. Mar. Poll. Bull. 6, 21–36.

Kirby, R., Parker, W.R., 1983. Distribution and behavior of fine sediment in the Severn Estuary and Inner Bristol Channel, U.K. Can. J. Fish. Aquat. Sci. 40, 83–95.

Koh, C.H., 2001. The Korean Getbol. Seoul National University Press, 1073 p. (in Korean).

Koh, C.H., Khim, J.S., 2014. The Korean tidal flat of the Yellow Sea: physical setting, ecosystem and management. Ocean Coast. Manage. 102, 398–414.

Komar, P.D., 1974. Oscillatory ripple marks and the evaluation of ancient wave conditions and environments. J. Sediment. Petrol. 44, 169–180.

Komar, P.D., Miller, M.C., 1973. The threshold of sediment movement under oscillatory water waves. J. Sediment. Petrol. 43, 1101–1110.

Komar, P.D., Miller, M.C., 1975. The initiation of oscillatory ripple marks and the development of plane-bed at high shear stresses under waves. J. Sediment. Petrol. 45, 697–703.

Korea Hydrographical and Oceanographic Administration, 2011. Asia, Korea, West coast, Gyeongnyeolbi Yeoldo to Daechung Gundo. Bathymetr. Chart 323 Mercator Projection, 1:250,000 scale.

Korea Meteorological Administration website. See http://www.kma.go.kr/.

Korea Meteorological Administration website (in Korean). See www.kma.go.kr/kor/weather/climate/climate_08_02.jsp.

Korea Water Resources Corporation (KOWACO) website. See http://english.kwater.or.kr/.

Kranck, K., 1980. Experiments on the significance of flocculation in the settling of fine-grained sediment in still water. Can. J. Earth Sci. 17, 1517–1526.

Kuehl, S.A., Allison, M.A., Goodbred, S.L., Kudrass, H., 2005. The Ganges–Brahmaputra delta. In: Giovsan, L., Bhattacharya, J.P. (Eds.), River Deltas—Concepts, Models, and Examples, vol. 83. SEPM Special Publication, pp. 413–434.

Kum, B.C., Shin, D.H., 2013. Dune migration on an offshore sand ridge in the southern Gyeonggi Bay, Korea. Ocean Polar Res. 35, 51–61.

Kum, B.C., Shin, D.H., Jung, S.K., Jang, S., Jang, N.D., Oh, J.K., 2010. Morphological features of bedforms and their changes due to marine sand mining in southern Gyeonggi Bay. Ocean Polar Res. 32, 337–350.

Kumar, B.P., Pang, I.C., Rao, A.D., Kim, T.H., Nam, J.C., Hon, C.S., 2003. Sea state hindcast for the Korean seas with a spectral wave model and validation with buoy observation during January 1997. J. Korean Earth Sci. Soc. 24, 7–21.

Kvale, E.P., 2006. The origin of neap—spring cycles. Mar. Geol. 235, 5–18.

Kvale, E.P., 2012. Tidal constituents of modern and ancient tidal rhythmites: criteria for recognition and analyses. In: Davis Jr., R.A., Dalrymple, R.W. (Eds.), Principles of Tidal Sedimentology, pp. 1–17.

Kwon, Y.K., 2012. Late Quaternary sequence stratigraphy in Kyeonggi Bay, mid-eastern Yellow Sea. J. Korean Earth Sci. Soc. 33, 242–258.

La Croix, A.D., Dashtgard, S.E., 2014. Of sand and mud: sedimentological criteria for identifying the turbidity maximum zone in a tidally influenced river. Sedimentology 61, 1961–1981.

Lane, E.W., 1957. A Study of the Shape of Channels Formed by Natural Streams Flowing in Erodible Material. US Army Corps of Engineers, Missouri River Division. Nebraska, Omaha, 121 p.

Lankov, A., October 4, 2005. The dawn of modern Korea: the Han River. Korean Times.

Le Bot, S., Trentesaux, A., 2004. Types of internal structure and external morphology of submarine dunes under the influence of tide- and wind-driven processes (Dover Strait, northern France). Mar. Geol. 211, 143–168.

Leckie, D.A., 1988. Wave-formed, coarse-grained ripples and their relationship to hummocky cross stratification. J. Sediment. Petrol. 58, 607–622.

Leclair, S., Bridge, J.S., 2001. Qualitative interpretation of sedimentary structures formed by river dunes. J. Sediment. Res. 71, 713–716.

Lee, B.S., 1976. Weather-climatological study of Changma and Kaul Changma in Korea in relation to two rainy seasons of East Asia. Sadae Nonchong 14, 185–218.

Lee, D.K., Niler, P., 2003. Ocean response to Typhoon Rusa in the South sea of Korea and in the East China sea. J. Korean Soc. Ocean. 38, 60–67.

Lee, H.J., Lee, S.L., Cho, C.H., Kim, C.H., 2002a. Tidal front in the main tidal channel of Kyunggi Bay, Eastern Yellow Sea. J. Korean Soc. Ocean 37, 10–19.

Lee, H.J., Park, J.Y., Lee, S.H., Lee, J.M., Kim, T.K., 2013. Suspended sediment transport in a rock-bound, macrotidal estuary: Han Estuary, Eastern Yellow Sea. J. Coast. Res. 29, 358–371.

Lee, J.C., Kim, C.S., Jung, K.T., 2001. Comparison of bottom friction formulations for single-constituent tidal simulations in Gyeonggi Bay. Estuarine. Coast. Shelf Sci. 53, 701–715.

Lee, S., Chwae, U., Min, K., 2002b. Landslide susceptibility mapping by correlation between topography and geological structure: the Janghung area, Korea. Geomorphology 46, 149–162.

Lee, Y.K., Ryu, J.H., Choi, J.L., Soh, J.G., Eom, J.A., Won, J.S., 2011. A study of decadal sedimentation trend changes by waterline comparisons within the Ganghwa tidal flats initiated by human activities. J. Coast. Res. 27, 857–869.

Legler, B., Johnson, H.D., Hampson, G.J., Massart, B.Y.G., Jackson, C.A.-L., Jackson, M.D., El-Barkooky, A., Ravnås, R., 2013. Facies model of a fine-grained, tide-dominated delta: Lower Dir Abu Lifa member (Eocene), Western Desert, Egypt. Sedimentology 60, 1313–1356.

Lim, D.I., Jung, H.S., Kim, B.O., Choi, J.Y., Kim, H.N., 2004. A buried palaeosol and late Pleistocene unconformity in coastal deposits of the eastern Yellow Sea, East Asia. Quat. Int. 121, 109–118.

Lim, D.I., Jung, H.S., Yoo, H.S., Seo, J.M., Paeng, W.H., 2003. Late Pleistocene unconformity in tidal-flat deposit of Gyeonggi Bay, western coast of Korea. J. Korean Earth Sci. Soc. 24, 657–667.

Lim, D.I., Park, Y.A., 2003. Late Quaternary stratigraphy and evolution of a Korean tidal flat, Haenam Bay, southeastern Yellow Sea, Korea. Mar. Geol. 193, 177–194.

Liu, J.P., Milliman, J.D., Gao, S., Cheng, P., 2004. Holocene development of the Yellow River's subaqueous delta, North Yellow Sea. Mar. Geol. 209, 45–67.

Liu, Z.X., Xia, D.X., Berne, S., Wang, K.Y., Marsset, T., Tang, Y.X., Bourillet, J.F., 1998. Tidal deposition systems of China's continental shelf, with special reference to the eastern Bohai Sea. Mar. Geol. 145, 225–253.

Livingstone, I., Warren, A., 1996. Aeolian Geomorphology: An Introduction. Longman, Harlow, 209 p.

Lofi, J., Vagner, P., Berné, S., Weber, O., 2005. The tidal sand ridges of the East China Sea outer shelf: evidences for recent activity under modern sea-level conditions. Geophys. Res. Abstr. 7, 00440.

MacEachern, J.A., Bann, K.L., Bhattacharya, J.P., Howell, C.D., 2005. Ichnology of deltas. In: Giosan, L., Bhattacharya, J.P. (Eds.), River Deltas: Concepts, Models, and Examples, vol. 83. SEPM (Soc. Sediment. Geol.) Spec. Publ., pp. 49–85.

Mackay, D.A., Dalrymple, R.W., 2011. Dynamic mud deposition in a tidal environment: the record of fluid-mud deposition in the Cretaceous Bluesky formation, Alberta, Canada. J. Sediment. Res. 81, 901–920.

Manning, A.J., Langston, W.J., Jonas, P.J.C., 2010. A review of sediment dynamics in the Severn Estuary: influence of flocculation. Mar. Poll. Bull. 61, 37–51.

Martin, D.P., Nittrouer, C.A., Ogston, A.S., Crockett, J.S., 2008. Tidal and seasonal dynamics of a muddy inner shelf environment, Gulf of Papua. J. Geophys. Res. Part F. Earth Surf. 113, F01S07.

Martinius, A.W., Hegner, J., Kaas, I., Mjos, R., Berjarano, C., Mathieu, X., 2013. Geologic reservoir characterization and evaluation of the Petrocedeno Field, early Miocene Oficina Formation, Orinoco heavy oil belt, Venezuela. In: Hein, F.J., Leckie, D., Larter, S., Suter, J.R. (Eds.), Heavy-oil and Oil-sand Petroleum Systems in Alberta and Beyond. Amer. Assoc. Petrol. Geol. Studies in Geology, vol. 64. pp. 103–131.

Martinius, A.W., Kaas, I., Næss, A., Helgesen, G., Kjærefjord, J.M., Leith, D.A., 2001. Sedimentology of the heterolithic and tide-dominated Tilje Formation (Early Jurassic, Halten Terrace, offshore mid-Norway). In: Martinsen, O.J., Dreyer, T. (Eds.), Sedimentary Environments Offshore Norway – Paleozoic to Recent, vol. 10. NPF Spec. Publ., pp. 103–144.

McAnally, W.H., Friedrichs, C., Hamilton, D., Hayter, E., Shrestha, P., Rodriguez, H., Sheremet, A., Teeter, A., 2007. Management of fluid mud in estuaries, bays, and lakes. 1: present state of understanding on character and behavior. J. Hydraul. Eng. 133, 9–22.

Meade, R.H., 1996. River-sediment inputs to major deltas. In: Milliman, J.D., Haq, B.U. (Eds.), Sea-Level Rise and Coastal Subsidence—Causes, Consequences, and Strategies. Kluwer Academic Publishers. city, pp. 63–85.

Mehta, A.J., 1989. On estuarine cohesive sediment suspension behavior. J. Geophys. Res. 94, 14,303–14,314.

Mehta, A.J., 1991. Understanding fluid mud in a dynamic environment. Geo. Mar. Lett. 11, 113–118.

Meyer-Peter, E., Mueller, R., 1948. Formulas for bed-load transport. Proc. Int. Ass. Hydraul. Res., 2nd Congress, Stockholm, Sweden, pp. 39–64.

Middleton, G.V., 1991. A short historical review of clastic tidal sedimentology. In: Smith, D.G., Zaitlin, B.A., Reinson, G.E., Rahmani, R.A. (Eds.), Clastic Tidal Sedimentology. Can. Soc. Petrol. Geol. Mem., vol. 16. pp. ix–xv.

Milliman, J.D., Meade, R.H., 1983. World-wide delivery of river sediment to the oceans. J. Geol. 91, 1–21.

Milliman, J.D., Syvitski, J.M., 1992. Geomorphic/tectonic control of sediment discharge to the ocean: the importance of small mountainous rivers. J. Geol. 100, 525–544.

Molnar, P., Tapponier, P., 1975. Cenozoic tectonics of Asia: effects of a continental collision. Science 189, 419–426.

Moon, I.J., Oh, I.S., Murty, T., Youn, Y.H., 2003. Causes of the unusual coastal flooding generated by Typhoon Winnie on the west coast of Korea. Nat. Hazards 29, 485–500.

Musial, G., Reynaud, J.-Y., Gingras, M.K., Fenies, H., 2011. Subsurface and outcrop characterization of large tidally influenced point bars of the Cetaceous McMurray Formation (Alberta, Canada). Sediment. Geol. 17. http://dx.doi.org/10.1016/j.sedgeo.2011.04.020.

Nardin, T.R., Feldman, H.R., Carter, B.J., 2013. Stratigraphic architecture of a large-scale point-bar complex in the McMurray formation; Syncrude's Mildred Lake Mine, Alberta, Canada. In: Hein, F.J., Leckie, D., Larter, S., Suter, J.R. (Eds.), Heavy-Oil and Oil-Sand Petroleum Systems in Alberta and Beyond. Amer. Assoc. Petrol. Geol. Studies in Geol., vol. 64. pp. 273–311.

Nestor, M.J.R., 1977. The environment of South Korea and adjacent sea areas. Naval environment prediction research facility. Tech. Rep. Tr. 77, 274.

Neuman, C., 1993. Global overview. In: Holland, G.J. (Ed.), Global Fuide to Tropical Cyclone Forecasting WMO/TD-560. World Meteorological Organization, Geneva, pp. 1–46.

Nichols, M.M., 1985. Fluid mud accumulation processes in an estuary. Geo Mar. Lett. 4, 171–176.

Nio, S., Yang, C., 1991. Diagnostic attributes of clastic tidal deposits: a review. In: Smith, D.G., Reinson, G.E., Zaitlin, B.A., Rahmani, R.A. (Eds.), Clastic Tidal Sedimentology. Can. Soc. Pet. Geol., Mem., vol. 16. pp. 3–28.

Nittrouer, C.A., Kuehl, S.A., DeMaster, D.J., Kowsmann, R.O., 1986. The deltaic nature of Amazon shelf sedimentation. Geol. Soc. Am. Bull., vol. 97, 444–458.

Nittrouer, C.A., Kuehl, S.A., Figueiredo, G., Allison, M.A., Sommerfield, C.K., Rine, J.M., Faria, E.C., Silveira, O.M., 1996. The geological record preserved by Amazon shelf sedimentation. Cont. Shelf Res. 16, 817–841.

Nittrouer, J.A., 2013. Backwater hydrodynamics and sediment transport in the lowermost Mississippi River delta; implications for the development of fluvial-deltaic landforms in a large lowland river. In: Young, G., Perillo, G.M.E., Aksov, H., Bogen, J., Gelfan, A., Mahe, G., Marsh, P., Savenije, H.H.G. (Eds.), Deltas; Landforms, Ecosystems and Human Activities. IAHS-AISH Publication, vol. 358. pp. 48–61.

Nittrouer, J.A., Shaw, J., Lamb, M.P., Mohrig, D., 2012. Spatial and temporal trends for water-flow velocity and bed-material sediment transport in the lower Mississippi River. Geol. Soc. Am. Bull. 124, 400–414.

Noffke, N., Krumbein, W.E., 1999. A quantitative approach to sedimentary surface structures contoured by the interplay of microbial colonization and physical dynamics. Sedimentology 46, 417–426.

Off, T., 1963. Rhythmic linear sand bodies caused by tidal currents. Am. Assoc. Petrol. Geol. Bull. 47, 324–341.

Oh, I.S., Lee, D.E., 1998. Tides and tidal currents of the Yellow and East China Seas during the last 13,000 years. J. Korean Soc. Oceanogr. 33, 137–145.

Oh, J.K., 1995. Sedimentation processes of the suspended sediments in Yumha Channel of the Han River estuary. Korea. J. Korean Earth Sci. Soc. 16, 20–29.

Olariu, C., Bhattacharya, J.P., 2006. Terminal distributary channels and delta front architecture of river-dominated delta systems. J. Sediment. Res. 76, 212–233.

Paola, C., Borgman, L.E., 1991. Reconstructing random topography from preserved stratification. Sedimentology 38, 553–565.

Paola, C., Mohrig, D., 1996. Palaeohydraulics revisited: palaeoslope estimation in coarse-grained braided rivers. Basin Res. 8, 243–254.

Papenmeier, S., Schrottke, K., Bartholomä, A., Flemming, B.W., 2013. Sedimentological and rheological properties of the water-solid bed interface in the Weser and Ems Estuaries, North Sea, Germany: implications for fluid mud classification. J. Coast. Res. 29, 797–808.

Park, K., Oh, J.H., Kim, H.S., Im, H.H., 2002. Case study: mass transport mechanism in Kyunggi Bay around Han river mouth, Korea. J. Hydraul. Eng. 128, 257–267.

Park, S.-C., Lee, B.-H., Han, H.-S., Yoo, D.-G., Lee, C.-W., 2006. Late Quaternary stratigraphy and development of tidal sand ridges in the eastern Yellow Sea. J. Sediment. Res. 76, 1093–1105.

Park, Y.A., Choi, K.S., 2002. Late Quaternary stratigraphy of the muddy tidal deposits, west coast of Korea. In: Healy, T., Wang, Y., Healy, J.-A. (Eds.), Muddy Coasts of the World; Processes, Deposits and Function. Proc. Mar. Sci., vol. 4. pp. 391–409.

Pirmez, C., Pratson, L.F., Steckler, M.S., 1998. Clinoform development by advection-diffusion of suspended sediment: modeling and comparison to natural systems. J. Geophys. Res. 103, 24141–24157.

Posamentier, H.W., Allen, G.P., 1999. Siliciclastic sequence stratigraphy–concepts and applications. Concepts in Sedimentology and Paleontology, vol. 7. SEPM (Soc. Sediment. Geol.).

Postma, H., 1967. Sediment transport and sedimentation in the marine environment. In: Lauff, G.H. (Ed.), Estuaries. American Association for the Advancement of Science, Publication, vol. 83. American Association for the Advancement of Science, Washington, DC, pp. 158–186.

Prins, M.A., Postma, G., 2000. Effects of climate, sea level and tectonics unraveled for last deglaciation turbidite records of the Arabian Sea. Geology 28, 375–378.

Pritchard, D.W., 1967. What is an estuary? Physical viewpoint. In: Lauff, G.H. (Ed.), Estuaries. Amer. Assoc. Advance. Sci., Publ, vol. 83. American Association for the Advancement of Science, Washington, pp. 3–5.

Reineck, H.-E., Wunderlich, F., 1968. Classification and origin of flaser and lenticular bedding. Sedimentology 11, 99–104.

Reith, G., 2013. Environmental Reconstructions from Structures and Fabrics within Thick Mudstone Layers (Fluid Muds), Tilje Formation (Jurassic), Norwegian Continental Shelf. Unpublished M.Sc. Thesis, Queen's University, Kingston, Ontario, Canada, 122 p.

Ren, J., Tamaki, K., Sitian, L., Junxia, Z., 2002. Late Mesozoic and Cenozoic rifting and its dynamic setting in Eastern China and adjacent areas. Tectonophysics 344, 175–205.

Reynaud, J.-Y., Dalrymple, R.W., 2012. Shallow-marine tidal deposits. In: Davis Jr., R.A., Dalrymple, R.W. (Eds.), Principles of Tidal Sedimentology. Springer, Dordrecht, pp. 335–369.

Rich, J.L., 1951. Three critical environments of deposition, and criteria for recognition of rocks deposited in each of them. Geol. Soc. Am. Bull. 61, 1–19.

Richardson, L.F., 1920. The supply of energy from and to atmospheric eddies. Proc. R. Soc. Lond. Ser. A 27, 354–373.

Rieu, R., van Heteren, S., van der Spek, A.J.F., De Boer, P.L., 2005. Development and preservation of a mid-Holocene tidal-channel network offshore the Western Netherlands. J. Sediment. Res. 75, 409–419.

Rine, J.M., Ginsburg, R.N., 1985. Depositional facies of a mud shoreface in Suriname, South America – a mud analogue to study, shallow-marine deposits. J. Sediment. Petrol. 55, 633–652.

Roberts, H.H., Sydow, J., 2003. Late Quaternary stratigraphy and sedimentology of the offshore Mahakam Delta, East Kalimantan (Indonesia). In: Sidi, F.H., Nummedal, D., Imbert, P., Darman, H., Posamentier, H.W. (Eds.), Tropical Deltas of Southeast Asia; Sedimentology, Stratigraphy, and Petroleum Geology. SEPM (Soc. Sediment. Geol.) Spec. Publ., vol. 76. pp. 125–145.

Robinson, A.H.W., 1960. Ebb-flood channel systems in sandy bays and estuaries. Geography 45, 183–199.

Ross, M.A., Mehta, A.J., 1989. On the mechanics of lutoclines and fluid mud. J. Coast. Res. 5 (Special issue), 51–61.

Rotondo, K., 2004. Transport and Deposition of Fluid Mud Event Layers along the Western Louisiana Inner Shelf. Unpublished M.Sc. Thesis, Louisiana State University, Baton Rouge, 118 p.

Salomon, J.C., Allen, G.P., 1983. Role sedimentologique de la mare dans les estuaires a fort marnage. Comp. Francais Pétroles, Notes et Mem. 18, 35–44.

Sampe, T., Xie, S.-P., 2010. Large-scale dynamics of the Meiyu-Baiu rainband: environmental forcing by the westerly jet. J. Clim. 23, 113–134.

Schellart, W.P., Lister, G.S., 2005. The role of the East Asian active margin in widespread extensional and strike-slip deformation in East Asia. J. Geol. Soc. 162, 959–972.

Schrottke, K., Becker, M., Bartholomä, A., Flemming, B.W., Hebbeln, D., 2006. Fluid mud dynamics in the Weser estuary turbidity zone tracked by high resolution side-scan sonar and parametric sub-bottom profiler. Geo Mar. Lett. 26, 185–198.

Shackleton, N.J., 1987. Oxygen isotopes, ice volume and sea level. Quatern. Sci. Rev. 6, 183–190.

Shanley, K.W., McCabe, P.J., Hettinger, R.D., 1992. Tidal influence in Cretaceous fluvial strata from Utah, USA: a key to sequence stratigraphic interpretation. Sedimentology 39, 905–930.

Sheriff, R.E., Geldart, L.P., 1995. Exploration Seismology. Cambridge University Press, 624 p.

Shinn, Y.J., Chough, S.K., Kim, J.W., Lee, S.H., Woo, J., Jin, J.H., Hwang, S.Y., Choi, S.H., Suh, M.C., 2004. High-resolution seismic reflection studies of late Quaternary sediments in the eastern Yellow Sea. In: Clift, P., Kuhnt, W., Wang, P., Hayes, D. (Eds.), Continent–Ocean Interactions within East Asian Marginal Seas. American Geophys. Union, Geophysical Monograph Series, vol. 149. pp. 175–191.

Song, C., Sun, X., Wang, J., Li, M., Zheng, L., 2015. Spatio-temporal characteristics and causes of changes in erosion accretion in the Yangtze (Changjiang) submerged delta from 1982 to 2010. J. Geogr. Sci. 25, 899–916.

Southard, J.B., 1991. Experimental determination of bed-form stability. Annu. Rev. Earth Planet. Sci. 19, 423–455.

Southard, J.B., Boguchwal, L.A., 1990. Bed configurations in steady unidirectional water flows. Part 2. Synthesis of flume data. J. Sediment. Petrol. 60, 658–679.

Southard, J.B., Lambie, J.M., Federico, D.C., Pile, H.T., Weidman, C.R., 1990. Experiments on bed configurations in fine sands under bidirectional purely oscillatory flow, and the origin of hummocky cross-stratification. J. Sediment. Res. 60, 1–17.

Stuiver, M., Polach, H.A., 1977. Discussion: reporting of ^{14}C data. Radiocarbon 19, 355–363.

Stuiver, M., Reimer, P.J., Bard, E., Beck, J.W., Burr, G.S., Hughen, K.A., Kromer, B., McCormac, G., van der Plicht, J., Spurk, M., 1998. INTCAL98 radiocarbon age calibration, 24,000–0 cal BP. Radiocarbon 40, 1041–1083.

Swenson, J.B., Paola, C., Pratson, L., Voller, V.R., Murray, A.B., 2005. Fluvial and marine controls on combined subaerial and subaqueous delta progradation: morphodynamic modeling of compound-clinoform development. J. Geophys. Res. 110, 1–16.

Swift, D.J.P., 1968. Coastal erosion and transgressive stratigraphy. J. Geol. 76, 444–456.

Swift, D.J.P., Niederoda, A.W., 1985. Fluid and sediment dynamics on continental shelves. In: Tillman, R.W., Swift, D.J.P., Walker, R.G. (Eds.), Shelf Sands and Sandstone Reservoirs. SEPM (Soc. Sediment. Geol.) Short Course Notes, vol. 13. pp. 47–133.

Taylor, A.M., Goldring, R., 1993. Description and analysis of bioturbation and ichnofabric. J. Geol. Soc. Lond. 150, 141–148.

Tessier, B., 2012. Stratigraphy of tide-dominated estuaries. In: Davis Jr., R.A., Dalrymple, R.W. (Eds.), Principles of Tidal Sedimentology. Springer, Dordrecht, pp. 109–128.

Thomas, R.G., Smith, D.G., Wood, J.M., Visser, J., Calverley-Range, E.A., Koster, E.H., 1987. Inclined heterolithic stratification—terminology, description, interpretation and significance. Sediment. Geol. 53, 123–179.

Traykovski, P., Geyer, W.R., Irish, J.D., Lynch, J.F., 2000. The role of wave-induced density-driven fluid mud flows for cross-shelf transport on the Eel River continental shelf. Cont. Shelf Res. 20, 2113–2140.

Uehara, K., Saito, Y., 2003. Late Quaternary evolution of the Yellow/East China Sea tidal regime and its impacts on sediments dispersal and seafloor morphology. Sediment. Geol. 162, 25–38.

Uncles, R.J., Stephens, J.A., Harris, C., 2006a. Runoff and tidal influences on the estuarine turbidity maximum of a highly turbid system: the upper Humber and Ouse Estuary, UK. Mar. Geol. 235, 213–228.

Uncles, R.J., Stephens, J.A., Harris, C., 2006b. Properties of suspended sediment in the estuarine turbidity maximum of the highly turbid Humber Estuary system, UK. Ocean Dynam. 56, 235–247.

Uncles, R.J., Stephens, J.A., Law, D.J., 2006c. Turbidity maximum in the macrotidal, highly turbid Humber Estuary, UK: flocs, fluid mud, stationary suspensions and tidal bores. Estuar. Coast. Shelf Sci. 67, 30–52.

US Army Corps of Engineers, 1984. Shore Protection Manual. Coastal Engineering Research Center, Waterways Experiment Station, Vicksburg, Mississippi.

US Army Corps of Engineers, 1985. Direct Methods for Calculating Wavelength. Coastal Engineering Research Center, Coastal Engineering Technical Note CETN-1–17. Vicksburg, Mississippi.

US military website. www.nrlmry.navy.mil/port_studies/thh-nc/korea/incheon/text/sect5.htm.

US Navy unclassified publication. Typhoon Havens Handbook for the Western Pacific and Indian Oceans. See http://pao.cnmoc.navy.mil/pao/Heavyweather/thh_nc/korea/incheon/text/sect5.htm.

Van den Berg, J.H., 1981. Rhythmic seasonal layering in a mesotidal channel fill sequence, Oosterchelde Mouth, the Netherland. In: Nio, S.-D., Shuttenhelm, R.T.E., van Weering, Tj.C.E. (Eds.), Holocene Marine Sedimentation in the North Sea Basin. Internat. Assoc. Sediment. Spec. Publ., vol. 5. pp. 147–159.

van Olphen, H., 1963. An Introduction to Clay Colloid Chemistry. John Wiley and Sons, New York, 301 p.

van Straaten, L.M.J.U., 1961. Sedimentation in tidal flat areas. Alta. Soc. Petrol. Geol. J. 9, 203–226.

van Straaten, L.M.J.U., Kuenen, Ph.H., 1958. Tidal action as a cause of clay accumulation. J. Sediment. Petrol. 28, 406–413.

van Veen, J., 1950. Eb- en vloedschaarsystemen in de Nederlandse getijwateren [in Dutch]. J. R. Dutch Geogr. Soc. 67, 303–325.

van Veen, J., van der Spek, A.J.F., Stive, M.J.F., Zitman, T., 2005. Ebb and flood channel systems in the Netherlands tidal waters. J. Coast. Res. 21, 1107–1120.

Van Wagoner, J.C., Mitchum, R.M., Campion, K.M., Rahmanian, V.D., 1990. Siliciclastic Sequence Stratigraphy in Well Logs, Cores, and Outcrops; Concepts for High-resolution Correlation of Time and Facies. Methods in Explorer. Series, vol. 7. 55 pp.

Walker, R.G., 1992. Facies, facies models and modern stratigraphic concepts. In: Walker, R.G., James, N.P. (Eds.), Facies Models; Response to Sea Level Change. Geological Association of Canada, St. John's, Newfoundland, pp. 1–14.

Walker, R.G., Eyles, C.H., 1991. Topography and significance of a basinwide sequence-bounding erosion surface in the Cretaceous Cardium Formation, Alberta, Canada. J. Sediment. Petrol. 61, 473–496.

Walsh, J.P., Nittrouer, C.A., 2009. Understanding fine-grained river-sediment dispersal on continental margins. Mar. Geol. 263, 34–45.

Walsh, J.P., Nittrouer, C.A., Palinkas, C.M., Ogston, A.S., Sternberg, R.W., Brunskill, G.J., 2004. Clinoform mechanics in the Gulf of Papua, New Guinea. Cont. Shelf Res. 24, 2487–2510.

Wang, P., 2012. Principles of sediment transport applicable in tidal environments. In: Davis, R.A., Dalrymple, R.W. (Eds.), Principles of Tidal Sedimentology. Springer, pp. 19–34.

Wellner, R.W., Bartek, L., 2003. The effect of sea level, climate, and shelf physiography of the development of incised valley complexes: a modern example from the East China Sea. J. Sediment. Res. 73, 926–940.

Wells, J.T., 1988. Distribution of suspended sediment in the Korea Strait and southeastern Yellow Sea: onset of winter monsoons. Mar. Geol. 83, 273–284.

Wells, J.T., Adams, C.E., Park, Y.A., Frankenberg, E.W., 1990. Morphology, sedimentology and tidal channel processes on a high-tide-range mudflat, west coast of South Korea. Mar. Geol. 95, 111–130.

Wiberg, P.L., Sherwood, C.R., 2008. Calculating wave-generated bottom orbital velocities from surface-wave parameters. Comput. Geosci. 34, 1243–1262.

Willis, B.J., 2005. Deposits of tide-influenced river deltas. In: Giosan, L., Bhattacharya, J.P. (Eds.), River Deltas–Concepts, Models, and Examples. SEPM (Soc. Sediment. Geol.) Spec. Publ., vol. 83. pp. 87–129.

Wolanski, E., King, B., Galloway, D., 1995. Dynamics of the turbidity maximum in the Fly River estuary, Papua New Guinea. Estuar. Coast. Shelf Sci. 40, 321–337.

Woo, H., Kim, W., 1997. Floods on the Han River in Korea. Water Int. 22, 230–237.

Woo, S.B., Yoon, B.I., 2011. The classification of estuary and tidal propagation characteristics in thd Gyeong-Gi Bay, South Korea. J. Coast. Res. SI 64, 1624–1628.

Wood, L.J., 2003. Predicting tidal sand reservoir architecture using data from modern and ancient depositional systems. Am. Assoc. Petrol. Geol. Mem. 80, 1–22.

Wright, L.D., 1977. Sediment transport and deposition at river mouths: a synethesis. Geol. Soc. Am. Bull. 88, 857–868.

Wright, L.D., Friedrichs, C.T., 2006. Gravity-driven sediment transport on continental shelves: a status report. Cont. Shelf Res. 26, 2092–2107.

Wright, L.D., Friedrichs, C.T., Kim, S.C., Scully, M.E., 2001. Effects of ambient currents and waves on gravity-driven sediment transport on continental shelves. Mar. Geol. 175, 25–45.

Wright, L.D., Nittrouer, C.A., 1995. Dispersal of river sediments in coastal seas: six contrasting cases. Estuaries 18, 494–508.

Wurpts, R.W., 2005. 15 years experience with fluid mud: definition of the nautical bottom with rheological parameters. Terra Aqua. 99, 22–32.

Yalin, M.S., 1964. Geometrical properties of sand waves. Proc. Am. Soc. Civil Eng. 90, 105–119.

Yalin, M.S., da Silva, A.M.F., 1991. On the formation of alternate bars. In: Soulsby, R., Bettess, R. (Eds.), Sand Transport in Rivers, Estuaries and the Sea. Euromech Colloq. Proc., vol. 262. pp. 171–178.

Yang, B.C., Dalrymple, R.W., Chun, S.S., 2005. Sedimentation on a wave-dominated, open-coast tidal flat, southwestern Korea: summer tidal flat–winter shoreface. Sedimentology 52, 235–252.

Yang, B.C., Dalrymple, R.W., Chun, S.S., 2006a. The significance of hummocky cross-stratification (HCS) wavelengths: evidence from an open-coast tidal flat, South Korea. J. Sediment. Res. 76, 2–8.

Yang, B.C., Dalrymple, R.W., Chun, S.S., Lee, H.J., 2006b. Transgressive sedimentation and stratigraphic evolution of a wave-dominated macrotidal coast, West Korea. Mar. Geol. 235, 35–48.

Yang, C.S., 1989. Active, moribund and buried tidal sand ridges in the East China Sea and the southern Yellow Sea. Mar. Geol. 88, 97–116.

Yang, E.J., Choi, J.K., Hyun, J.H., 2008. Seasonal variation in the community and size structure of nano- and microzooplankton in Gyeonggi Bay, Yellow Sea. Estuarine Coastal Shelf Sci. 77, 320–330.

Yim, W.W.S., Ivanovich, M., Yu, K.F., 1990. Young age bias of radiocarbon dates in pre-Holocene marine deposits of Hong Kong and implications for Pleistocene stratigraphy. Geo Mar. Lett. 10, 165–172.

Yoon, B., Woo, H., 2000. Sediment problems in Korea. J. Hydraul. Eng. 126, 486–491.

Yoon, B.I., Woo, S.-B., 2013. Correlation between freshwater discharge and salinity intrusion in Han River Estuary, South Korea. J. Coast. Res. 65, 1247–1252.

Yoshida, S., Steel, R.J., Dalrymple, R.W., 2007. Changes in depositional processes-an ingredient in a new generation of sequence-stratigraphic models. J. Sediment. Res. 77, 447–460.

Young, I.R., 1999. Seasonal variability of the global ocean wind and wave climate. Int. J. Clim. 19, 931–950.

Yu, O.H., Lee, H.G., Lee, J.H., 2012. Influence of environmental variables on the distribution of macrobenthos in the Han River estuary, Korea. Ocean. Sci. J. 47, 519–528.

Yu, Q., Wang, Y., Gao, J., Gao, S., Flemming, B., 2014. Turbidity maximum formation in a well-mixed macrotidal estuary: the role of tidal pumping. J. Geophys. Res. Oceans 119, 7705–7724. http://dx.doi.org/10.1002/2014JC010228.

Zhang, X., Lin, C.M., Dalrymple, R.W., Li, Y.L., 2014. Facies architecture and depositional model of a macrotidal incised-valley succession (Qiantang River estuary, eastern China), and differences from other macrotidal systems. GSA Bull. 126, 499–522.

GLOSSARY

Chapter Points

This chapter contains a list of important words and phrases that have been used in this book (commonly noted in the text using italic font), commonly without elaboration of their meaning and significance. The definitions and comments included here are intended to give a fuller appreciation of the term or phrase, and to amplify on its meaning as used herein.

Backwater zone. The terminal part of a river whose water-surface gradient periodically flattens out or even dips upriver as a result of water-level fluctuations in the receiving basin (e.g., flood tides, storm surges), which results in a slowing, stopping, or reversal of flow (Lane, 1957; Chatanantavet and Lamb, 2014). In simplistic terms, the length of the backwater zone is defined as the distance inland where sea level intersects the river bed (Paola and Mohrig, 1996). However, when river flow is low, the backwater zone tends to be well developed and can extend hundreds of kilometers upriver, provided the river's gradient is sufficiently low (e.g., Amazon River, Mississippi River). Sediment deposition tends to be promoted in the backwater zone during this time (Nittrouer et al., 2012, Nittrouer, 2013). By contrast, when river flow is high, the backwater zone tends to either be pushed basinward or destroyed. Sediment erosion tends to be promoted in the terminal stretch of the river during this time. The backwater zone is closely related to, but different from, the *tidal-fluvial transition*.

Bar. See *barform*.

Barform. A general term for large-scale morphological features generated by the currents within a sedimentary environment. (The term barform is commonly shortened to "bar," and such features are also called "macroforms" to distinguish them from microforms and mesoforms (i.e., current ripples and *dunes* respectively).) Unlike the case for *bedforms*, there is no universally recognized classification scheme for barforms. Dalrymple and Rhodes (1995) recognize three genetically distinct types: repetitive barforms (e.g., point bars, alternate bars, side bars), elongate bars (e.g., tidal-current ridges, tidal bars), and delta-like bodies. In channelized settings, barforms occur in repetitive series along the length of the channel, with a spacing (λ_B), or meander wavelength, that is proportional to the channel width (w):

$$\lambda_B \sim 6w$$

(Yalin and da Silva, 1992). In more open areas where the barforms are more freestanding without a significant connection to a channel bank, the currents tend to flow obliquely across the barform, and they develop a straight, elongate morphology by means of the so-called Huthnance process (Huthnance, 1982a, b). In somewhat more confined settings, there will be only one elongate bar, but in more open locations such as on shelves, there can be several elongate bars beside each other; in such cases, the spacing is predicted to be ca. 250 times the water depth (Huthnance, 1982b). Dalrymple (2010b) and Dalrymple et al. (2012) have noted that there appears to be a morphological continuum between repetitive bars and elongate tidal bars, with tidally formed elongate bars in the broader, outer part of estuaries occupying locations where repetitive bars would normally occur. In the more confined inner part of estuaries and deltas, elongate tidal bars are attached to tidal point bars, forming a *flood barb* (Robinson, 1960). The genetic relationship between repetitive bars and elongate bars remains to be explored. The third type of barform (delta-like bodies, including *mouth bars*) forms at points of flow expansion, commonly at the end of a channel. Such barforms are solitary.

Bedform. A single *current ripple* or *dune* formed by a current, or a single straight crested or hummocky *wave ripple* (see also *hummocky cross-stratification*) formed by wave action. All such bedforms occur in fields of similar bedforms, termed a bed configuration, which represents a quasi-equilibrium response of the sediment to the local hydrodynamic processes (Southard, 1991). The nature of the bedforms present at any location is dependent on the properties of the sediment (primarily the mean grain size) and the intensity of the process moving the sediment. For currents, the most important variables are the mean current speed, averaged over the water depth in shallow flows, or over the boundary-layer thickness in deeper flows, and the depth of the water (or the boundary-layer thickness). For waves, the controlling variables are the wave orbital diameter and orbital velocity, which are in turn dependent on the wave height and length or period. Water viscosity, which is a function of water temperature, is a second-order parameter for both current- and wave-generated bedforms (Southard, 1991). All bedforms are smaller than any *barform* on which they might occur.

Bedload. That portion of the total *sediment load* carried by a current that moves very close to the fixed bed (typically no higher in the flow than approximately 5 grain diameters). Movement of the grains occurs by means of rolling, sliding, creep as a result of impacts and saltation (small hops). The coarsest sizes in transit travel preferentially within the bedload, although finer grains will also be present in the bedload layer. (See also *suspended load* below.)

Bedload convergence. Zone where the down-transport ends of two *bedload-transport pathways* oriented at a high angle to each other meet; net sediment movement is toward the convergence from both sides. Because the sediment tends to become finer down a transport pathway, either because of abrasion or preferential deposition of the coarser grain sizes (Dalrymple, 2010a), seafloor sediment is generally finer at a bedload convergence than elsewhere in the system. Bedload convergences occur within all river-attached estuaries (bedload is supplied to the estuary from both the shelf and the river; Dalrymple et al., 1992, 2012), and possibly in the mouth bar area of tide-dominated deltas (Dalrymple et al., 2003; Dalrymple and Choi, 2007; Dalrymple, 2010b).

Bedload-transport path. A zone that can be tens to hundreds of kilometers long within a tide-dominated environment in which there is a consistent direction of net (or residual) sediment transport, as a result of differences in the sediment-transport capacity of the ebb and flood currents (Belderson et al., 1982). There are several possible causes of the asymmetry in the transport capacity of the two currents; common ones include deformation of the tidal wave as it propagates into shallow water, which typically generates *flood dominance*, the interaction of the M2 tide (lunar semidiurnal tide) with the M4 tide (first harmonic of the M2 tide), or the superposition of river currents and waves on the tidal flux (Le Bot and Trentesaux, 2004). For a more extended discussion of the generation of bedload-transport pathways in shelf setting, see Reynaud and Dalrymple (2012).

Bubble pulse. The unconcentrated, undesired train of pressure waves generated by an air-gun or sparker energy source in water. Air-gun and sparker sources generate a high-pressure gas bubble in water that oscillates above and below hydrostatic pressure after being generated. This oscillation creates an unconcentrated train of seismic-wave energy that causes stratigraphic surfaces to appear as multiple reflections as opposed to single, discrete reflections. The bubble-pulse effect is especially evident in the top several meters below the seafloor in seismic data from Gyeonggi Bay, where it masks real near-surface reflections.

Clinoform. A sloping depositional surface, commonly S-shaped and consisting of a topset, foreset, and bottomset. Although most commonly associated with sediment bodies that prograde into water bodies, such as deltas or continental margins (Rich, 1951), clinoforms can in reality occur at any scale, in any depositional environment, and can originate in different ways (Helland Hansen and Hampson, 2009). Alluvial fans, current ripples, point bars, and washover fans all have clinoforms and they all grow by the accretion of sediment to their clinoform. The result in each case is a cross-stratified package of strata (i.e., a clinothem sensu Rich, 1951).

Compound dune. A dune with smaller superimposed dunes (Ashley, 1990). (See also the entry for *dune* below.) In tidal settings, compound dunes, which were formerly referred to as *sandwaves* (Dalrymple et al., 1978), are flow transverse, and commonly have low height-to-length ratios. Compound dunes are asymmetric in the direction of the long-term sediment-transport direction; symmetric forms may occur where ebb and flood currents transport equal amounts of sediment. Migration of superimposed dunes down the lees face of the master dune produces compound cross-stratification, in which the cross-beds formed by the smaller dunes rest on erosion surfaces (termed master bedding planes) that dip in the direction of migration of the larger dune (Dalrymple, 1984, 2010b). See Dalrymple and Rhodes (1995) and Dalrymple (2010b) for a more comprehensive description of compound dunes and their structures.

Coriolis effect. A fictitious acceleration needed to explain large-scale fluid motion on the Earth (or any rotating frame of reference) from the perspective of someone at a fixed location on the Earth's surface. The Coriolis effect acts at right angles to the flow, and causes oceanic currents to veer right in the northern hemisphere and left in the southern hemisphere. This effect is a result of the fact that the angular velocity of the Earth increases from zero at the poles to a maximum (467 m/s) at the equator. From the perspective of an observer standing at the north pole, a southward-flowing oceanic current will appear to veer to the right as the Earth rotates out from under it (i.e., as the Earth rotates counter clockwise). Northward-flowing oceanic currents will also appear to veer to the right as the current moves from regions of faster to slower (closer to the north pole) angular speed. The relative motions are reversed in the southern hemisphere. The Coriolis effect is zero at the equator because the equator is normal to the Earth's spin axis and the latitudinal change in angular velocity is small, but the effect becomes significant at $>10°$ latitude north or south.

Delta. A progradational sediment body at the mouth of a river composed predominantly of river-supplied sediment (Bhattacharya, 2010). Deltas are subdivided into three types (river-, wave- and tide-dominated) based on the process that transports and deposits the largest volume of sediment, thereby controlling the basic morphology of the delta (Coleman and Wright, 1975; Galloway, 1975). In river-dominated deltas, the sediment supplied by the river is not redistributed by waves or tidal currents, and takes on a birds-foot or lobate morphology. Wave-dominated deltas have a cuspate shape because the sand deposited initially at the river mouth is redistributed alongshore by wave-generated longshore drift. In tide-dominated deltas such as the Han River delta, tidal currents flowing into and out of the river mouth generate a series of coast-normal elongate tidal *bars*, separated by tidal channels that ultimately connect with the river.

Distributary channel. One of several seaward-bifurcating channels near the mouth of a river. Each distributary channel feeds sediment to a *mouth bar.*

Diurnal. Adjective stemming from *diurnalis*, the late Latin word for "daily." With respect to tides, it refers to any tidal constituent with a period close to 24 h. One of the commonly referenced diurnal periods is 24.82 h, which refers to the time needed for a location on the Earth's surface to make one complete revolution relative to the Moon (i.e., to begin and end directly beneath the Moon). It is slightly longer than 24 h because the Moon is moving in its orbit about the Earth in the same direction as the Earth rotates, but at a much slower speed. Other diurnal tidal periods are related to the rotation of the Earth relative to the Sun. For extended discussions of tidal periods, see Allen (1997) and Kvale (2006, 2012).

Diurnal inequality. With regard to semidiurnal tides, the difference in the amplitude of successive tidal cycles. This is caused by the declination of the Moon relative to the Earth's equator and spin axis, which causes a location to pass alternately closer and farther from the crests of the two tidal bulges, which are located directly beneath the Moon and on the opposite side of the Earth (Kvale, 2006, 2012). The diurnal inequality is zero when the Moon's orbit lies on the Earth's equator.

Dune. An asymmetric, current-generated *bedform* that is larger than 5 cm in height and 60 cm in wavelength, characterized by a higher-angle depositional (lee) side and a lower-angle erosional (stoss) side. Dunes come in a wide range of sizes and shapes; the classification scheme used here is that of Ashley (1990; see below). They are only generated in cohesionless sediment coarser than ~0.15 mm (i.e., upper-fine sand and coarser), at current speeds between about 0.4−2 m/s (Southard and Boguchwal, 1990; Southard, 1991). Dune height tends to scale to flow depth at a ratio of ~1:6 (Yalin, 1964; Flemming, 1988; Dalrymple and Rhodes, 1995), but only ~1/3 of a dune's height tends to be preserved as a cross-set (Leclair and Bridge, 2001). Therefore, by measuring the *average thickness* (not the maximum thickness) of a large number of dune cross-sets in an upward-fining fluvial-channel fill, a first-order approximation of channel depth can be achieved using a cross-set to flow-depth ratio of ~1:20. Although dunes also tend to scale to flow depth in tidal- and tidal–fluvial environments, it is unclear if this same scaling relationship applies, because of the potential existence of systematic differences in dune shape and preservation from fluvial environments (Dalrymple and Rhodes, 1995; Dalrymple and Choi, 2007). Therefore, the relationship between average set thickness and water depth should be used with caution! Dune lee faces in flume studies tend to approach angle of repose (~30°); however, for an unknown reason (possibly the presence of mud in suspension), dunes in some large rivers can display more gently sloping lee faces (<15°) (Best, 2005). In tidal environments, lee-face slopes of large and very large *compound dunes* can be even less (commonly less than 10°, and as low as 1°), a situation first attributed to the bidirectional nature of tidal flows (Allen, 1980), but later ascribed to the disruption of the flow-separation eddy by the presence of the superimposed dunes (Dalrymple and Rhodes, 1995).

The dune classification scheme used here (e.g., Appendix 2) follows that of Ashley (1990). In her comprehensive scheme, dune size (height and wavelength/spacing) is considered to be the first-order ("necessary") parameter for dune classification, whereas dune shape and superposition are second-order ("important") parameters. Other parameters, such as whether the sediment bed is or is not completely covered by sediment, are third-order parameters that are not used here.

Dune size: The size classes listed here are based first on *bedform* length. The corresponding heights were then calculated from the regression line between dune length and height reported by Flemming (1988; see also Dalrymple and Rhodes, 1995). Due to an apparent mathematical error, the heights given in Ashley (1990) do not correspond with those derived from the equation. The corrected values reported in Dalrymple and Rhodes (1995) are given here.

	Small dune	Medium dune	Large dune	Very large dune
Height (m)	0.05–0.25	0.25–0.5	0.5–3	>3
Length (m)	0.6–5	5–10	10–100	>100

Dune shape: *Two-dimensional (2-D) dunes* have relatively straight crests, lack scour pits, and form planar-tabular or trough cross-stratification, whereas *three-dimensional (3-D) dunes* have sinuous to lunate crests and well-developed scour pits, and form trough cross-stratification.

Superposition: *Simple dunes* lack superimposed dunes, whereas *compound dunes* are ornamented by smaller, superimposed dunes.

Ebb-dominated. An adjective describing a situation with a residual sediment transport in the direction of the ebb tide. This term is most commonly used to refer to the transport of *bedload* (i.e., sand), in which case the residual transport occurs because the peak speed of the ebb tidal currents exceeds that of the flood tidal currents: the maximum speed of the flood and ebb currents are more important than their duration because of the power-function relationship between current speed and sediment transport (Dalrymple and Choi, 2003). The hypsometric influence of tidal-flat morphology is also important (Dalrymple et al., 2012). Where the intertidal area is small, average water depth is greater during high tide than low tide, and the crest of the incoming tidal wave moves more quickly than the trough. This tends to generate flood dominated conditions, a situation that is especially common in the outer portions of tide-dominated estuaries where tidal flats are less extensive. By contrast, where the intertidal area is broad, the average water depth at high tide (averaged over tidal flats and channels) is actually less than at low tide, and the crest of the incoming tidal wave moves more slowly than the trough. This tends to generate ebb-dominated conditions, a situation that is especially common in the inner portions of tide-dominated estuaries and deltas where tidal flats are extensive. Ebb dominance is also commonly caused or accentuated by the superposition of a river current on the tidal water motion.

Estuarine circulation. Gravity-driven, landward flow of dense, salty marine water beneath a less dense, seaward-moving freshwater plume at the mouth of a river. (Note that estuarine circulation can occur in both *estuaries* and *deltas* as defined herein.) The density stratification, which is necessary for the generation of estuarine circulation, is best developed in systems where vertical turbulent mixing is minimal (i.e., in systems with relatively limited freshwater discharge and weak tidal currents). Estuarine circulation is commonly not pronounced in macrotidal systems because the strong tidal currents mix the water column, but can occur during river floods when the freshwater discharge is greater, or during neap tides when the tidal mixing is less (e.g., Harris et al., 2004; Uncles et al., 2006a). Mud that settles out from the river plume during these events is transported back toward shore at depth, effectively trapping it in the system.

Estuary. A transgressive coastal environment at the mouth of a river that receives sediment from both fluvial and marine sources, and that contains facies influenced by tide, wave, and fluvial processes. The estuary is considered to extend from the landward limit of tidal facies at its head to the seaward limit of coastal facies at its mouth (Dalrymple et al., 1992, 2012). This definition is considered to be a geologic definition because it is reflected in the spatial distribution of facies in modern environments, and by the vertical (retrogradational) stacking of facies in ancient counterparts. A more general, oceanographic definition of "estuary" that is based on salinity was provided by Pritchard (1967); it states that an estuary is a semi-enclosed coastal body of water in which the salinity is measurably diluted by freshwater derived from land drainage. The geologic definition is used exclusively here.

Flaser. A discontinuous mud drape that is restricted to the trough of a ripple (current or wave) (Reineck and Wunderlich, 1968). Larger-scale examples may also form in the troughs of dunes, especially in the bottom of channels where the suspended sediment concentration is highest.

Flood barb. A short, headward-terminating channel in which the net sediment-transport direction is landward, as a result of dominance by flood tidal currents (Robinson, 1960; Dalrymple, 2010b; Dalrymple et al., 2012). Such channels can occur almost anywhere in a tidal system, from locations in the tidal–fluvial transition where they occur in the seaward end of tide-influenced point bars, to distal locations near the seaward end of estuaries and deltas where they occur between two elongate tidal *barforms* or between an elongate *barform* and the margin of the system.

Flood-dominated. An adjective describing a situation with a residual sediment transport in the direction of the flood tide. This term is most commonly used to refer to the transport of *bedload* (i.e., sand), in which case the residual transport usually occurs because the peak speed of the flood tidal currents exceeds that of the ebb tidal currents: the maximum speed of the flood and ebb currents are more important than their duration because of the power-function relationship between current speed and sediment transport (Dalrymple and Choi, 2003). Many coastal areas are flood dominated because the tidal wave typically becomes asymmetrical as it propagates into shallow water. The crest of the wave experiences less frictional retardation than the trough of the tidal wave because of the greater water depth beneath the wave crest. Consequently, the wave crest moves forward faster than the trough, causing the front of the tidal wave (i.e., the flood tide) to be of shorter duration than the back of the tidal wave (i.e., the ebb tide). This in turn causes peak flood currents to exceed the peak speed of the ebb tide (Friedrichs and Aubrey, 1988).

Fluid mud. A soupy substance composed predominantly of mud and water (little to no sand is present) that has particle concentrations and physical properties somewhere between those of clear water and consolidated mud (Faas, 1991; McAnally et al., 2007); by definition, fluid mud has a suspended sediment concentration of $>10\,g/L$. Fluid mud is commonly recognized at the base of fluvially influenced tidal channels in modern, tide-influenced coastal systems (e.g., the Fly (Wolanski et al., 1995; Dalrymple et al., 2003), Weser (Schrottke et al., 2006; Papenmeier et al., 2013), Amazon (Kineke et al., 1996), Mississippi (Rotondo, 2004), Jiao (Jiang and Mehta, 2000), Severn (Kirby and Parker, 1983), Thames (Inglis and Allen, 1957), and Gironde (Allen et al., 1980)). Alongshore (and upslope) movement of fluid mud and accumulation in the intertidal zone may be an important cause of coastal progradation in the down-drift sector of muddy deltaic systems (e.g., Amazon River delta mudflats, Mississippi River chenier plain (Rotondo, 2004)). Downslope movement of fluid mud under the influence of waves across the subaqueous delta platform is also thought to be an important process in several systems (e.g., Amazon River delta (Kineke et al., 1996), Fly River delta (Crocket, 2003; Dalrymple et al., 2003; Harris et al., 2004), and Eel River delta (Traykovski et al., 2000; Hill et al., 2007)) and may be a significant contributor to deposition on, and progradation of, the prodelta clinoform.

Several factors are thought to promote fluid-mud formation in tide-influenced coastal systems. (1) Flocculation of mud particles commonly occurs either in the river system or at the salinity front, where the salinity is in the range of 1–5‰. This creates aggregates with significantly larger sizes than the constituent particles, which in turn increases the particle *settling velocity* and the rate of sediment delivery to the bed. The influence of this process increases as suspended sediment concentrations increase, because particles are more likely to interact with each other if they are closer together, but becomes less important at suspended sediment concentrations greater than ~10 g/L (Wolanski et al., 1995; Papenmeier et al., 2013) because the particles become so close together that the upward escape of water is impeded and hindered settling starts to have greater importance. (2) "*Estuarine circulation*" (i.e., the landward flow of saltier, denser water beneath the outflowing fresher, buoyant river plume) transports mud that settles from the river plume back toward land. As a result, mud becomes trapped near the salinity front, generating a turbidity maximum. This may be especially important in the Han River system during river floods when abundant light, turbid water is delivered to the basin. (3) The combination of high *settling velocity* and high suspended sediment concentrations combine to cause high fluxes of mud to the bed during slack water at high and low tides. These high downward sediment fluxes appear integral to the formation of fluid muds: at suspended sediment concentrations greater than ~10 g/L downward particle motion is countered by sufficient upward fluid displacement to hinder further settling. The result is a soupy, high-concentration (>10 g/L), fluid mud that requires time to dewater and fully consolidate. (4) Fluid mud tends to form in bathymetric depressions such as channels because a greater amount of suspended sediment can be contained in the thicker water column that occur in deeper water, and because fluid mud tends to flow downslope because it is more dense than the overlying water that has a lower suspended sediment concentration. (5) Once formed, fluid mud may promote deposition of sediment from the fluid mud because it suppresses the near-bed turbulence that is needed to keep mud in suspension (Geyer, 1993; Richardson, 1920). (6) The troughs of the dunes that occur in channel-thalweg locations provide sheltered locations where fluid mud can accumulate preferentially, and can escape resuspension and erosion during the subsequent tide. Fascinating examples of this are provided by Schrottke et al. (2006) and Becker et al. (2013).

High-density muddy suspensions, including fluid mud, do not just occur in situations with no current, but can form and continue to persist beneath currents of significant speed. Recent flume studies (Baas and Best, 2008; Baas et al., 2009, 2011) have shown that the behavior of muddy suspensions is complex, with an evolution of near-bed flow dynamics as the suspended sediment concentration increases: flow behavior changes from fully turbulent flow at concentrations <1 g/L, through transitional turbulent and transitional plug flow, to quasi-laminar plug flow at concentrations >100 g/L. At the lower concentrations, the grains are far enough apart that interparticle forces are overwhelmed by turbulence, but at concentrations of approximately 10 g/L interparticle cohesive forces begin to influence the nature of the flow, damping turbulence and causing the formation of a nonshearing plug near the bed, which can be separated from the stationary bed by a shear layer that can be turbulent (cf. Mackay and Dalrymple, 2011).

The original descriptions of fluid-mud deposits (Bhattacharya and MacEachern, 2009; Ichaso and Dalrymple, 2009) emphasized that they were internally structureless. More recent work (Mackay and Dalrymple, 2011; Reith, 2013) has, however, shown that many tidally formed fluid-mud deposits are not universally structureless, because deposition is inferred to have occurred from transitional turbulent and transitional plug flows that were capable of forming ripple cross-lamination and planar lamination, respectively, in the muddy deposit. The fluid-mud layers that formed in the troughs of dunes, and therefore occur in the bottomset and toeset region of cross-beds, contain complex successions of these lamination types because the fluid mud is interpreted to have persisted over several tidal cycles (Reith, 2013), in the manner documented by Schrottke et al. (2006).

Hummocky cross-stratification. A sedimentary structure that consists of low-angle (<10° dips), undulatory lamination in fine to very fine sand. Strictly speaking, domal draping lamination must be present for the structure to qualify as hummocky cross-stratification (HCS; Cheel and Leckie, 1993); the presence of only concave-up lamina forms is most commonly referred to as swaley cross-stratification (SCS).

These related structures are almost impossible to distinguish in cores, and examples of low-angle, divergent lamination is typically referred to as hummocky cross-stratification. HCS and SCS are generally interpreted to be wave-generated sedimentary structures formed by the aggradation of sediment on an undulatory bed surface consisting of hummocks and swales (Southard et al., 1990; Dumas and Arnott, 2006). The *bedform* responsible for HCS and SCS is believed to be a mound-shaped wave ripple (Yang et al., 2006a; Cummings et al., 2009), that is "orbital" in the sense that its wavelength scales to approximately half of the near-bed wave orbital diameter.

Inclined heterolithic stratification (IHS). Meter- to decameter-thick successions of alternating mud and sand beds that are inclined at a noticeable angle (typically several degrees) relative to the horizontal at the time of deposition (Thomas et al., 1987; Choi et al., 2004a). Although IHS can be produced by several different processes, including delta progradation and tidal–fluvial point bar accretion, it is typically associated with deposits of the latter. In particular, it has been used to describe tidally influenced point bar deposits of the McMurray Formation, the unit that hosts most of Canada's oil sands resources (Jablonski, 2012; Nardin et al., 2013; Musial et al., 2014).

Macrotidal. Adjective used to describe areas of the ocean where the tidal range exceeds 4 m (Davies, 1964). It is an arbitrary subdivision of the spectrum of tidal ranges that has no relation to either physical processes or to the geomorphology and sedimentology of the coastline. Therefore, it should not be confused with the term *tide-dominated*. However, there is a tendency for areas that are macrotidal to be tide-dominated.

Mouth bar. The sandy deposits at the mouth of a distributary channel, composed of the *bedload* sediment supplied predominantly by the river (Bhattacharya, 2010). The morphology of the mouth bar depends on whether the *delta* is river-dominated, wave-dominated, or *tide-dominated*. In the case of tide-dominated deltas such as the Han River delta, tidal currents moving into and out of the river mouth rework the river-supplied sand to create a series of tidal channels and tidal bars (Willis, 2005; Dalrymple, 2010b).

Planar-tabular cross-stratification. A cross-stratified bed bounded by surfaces that are flat (planar), both parallel and perpendicular to the flow direction, and approximately parallel to each other, producing a cross-bed that is tabular. It is generally thought that the cross-strata within the set are straight and meet the basal erosion surface of the set with an angular contact. Although this may occur, it is not universally the case: rounded tangential contacts can occur in planar-tabular cross-stratification, especially in upper-fine sand. Planar-tabular cross-stratification is formed by simple, 2D *dunes* (see *dune classification* above).

Prodelta. The basal, mud-covered portion of the prograding deltaic clinoform (e.g., Coleman, 1981; Roberts and Sydow, 2003).

Ravinement. Erosion produced in shallow and marginal-marine environments by waves and/or tidal currents during transgression.

Ripple. See *current ripple, wave ripple*

Sandwave. See *compound dune.*

Sediment load. Mass (or volume) of sediment transported by a current. Commonly divided into *suspended load* and *bedload.*

Sediment yield. Annual sediment discharge of a river (mass) divided by the area of the catchment upstream of the measurement site (i.e., the amount of sediment eroded per unit area each year; Allen, 1997). Commonly given in $t/km^2/yr$. The sediment yield usually increases as the relief of the catchment increases or as the vegetation cover decreases.

Settling velocity. The steady-state rate of downward fall (i.e., the terminal fall velocity) of a sediment grain. The settling velocity increases as the size of the particle increases and as the density increases. Equant grains settle more rapidly than flat (i.e., blade or disk-shaped) grains, all else being equal.

Storm surge. A vertical rise of coastal water during a storm (Allen, 1997). Storm surges may be generated below the eye of the storm because of the low atmospheric pressure, and/or by favorable winds. A "favorable" wind for the development of a storm surge in the northern hemisphere is a wind that blows parallel to shore, with land to the right. Because the Earth is rotating, such winds will cause a net movement of surface water onshore (to the right of the wind stress) due to the *Coriolis effect*. To satisfy continuity, a return (downwelling) current occurs at depth that may aid the offshore transport of bottom sediment under combined (wave–current) flow. Onshore-moving waves in shallow water also cause a small residual landward movement of water, sometimes referred to as "Stokes drift," which contributes to coastal setup.

Suspended load: The portion of the *sediment load* that travels in either intermittent or continuous suspension. For a sediment grain to be suspended, the upward component of turbulence must exceed the *settling velocity* of the sediment grain. Thus, the finer grain sizes tend to travel by means of suspension whereas larger grains are not suspended and travel as part of the *bedload*. In general terms, as the overall flow velocity increases, the intensity of the turbulence increases, which, in turn, causes an increase in the size of the grains that can be suspended and in the amount of sediment carried in suspension. In most river and tidal channels, the suspended load comprises the bulk of the total *sediment load*.

Tidal flat. Tidal flats are gently sloping sediment surfaces along the shoreline that are alternately exposed and inundated by the tide. Because of their low gradient, energy levels generally decrease toward the land, causing the sediment to be sandy in their lower/outer part and muddy in their upper/inner portion. They can be bordered by salt marshes or mangrove forests at their landward margin. The best known tidal flats occur in back-barrier environments along wave-dominated coasts (e.g., Wadden Sea) and the inner, sheltered parts of tide-dominated estuaries (e.g., Bay of Fundy) and deltas. Such tidal flats are bordered by a tidal channel at their outer, subtidal edge, and are tide-dominated with regard to depositional conditions. Their deposits contain diagnostic tidal sedimentary structures such as flaser, wavy, and lenticular bedding, and can be intensely bioturbated (Dalrymple, 2010b; Flemming, 2012). Tidal rhythmites can be developed along the margins of the tidal channels and gullies that dissect the muddy portions of these tidal flats. Although not nearly as well-known, many modern tidal flats occur along the exposed ocean shoreline in areas with large tidal ranges; see Fan (2012) for an overview. Muddy versions of such open-coast tidal flats are especially important in areas down-drift of large rivers such as the Amazon (Rine and Ginsburg, 1985) and Changjiang (e.g., Fan et al., 2006), but sandy open-coastal tidal flats have also been described (Yang et al., 2005). Because of the lack of protection, open-coast tidal flats experience significant levels of wave action. In fully exposed areas, the tidal-flat deposits can consist entirely of wave-generated storm beds (Yang et al., 2005); such tidal flats can be considered to be "wave-dominated tidal flats," although the term "intertidal shoreface" has also been suggested. See the review in Dalrymple (2010b) for ideas on the transition between such tidal flats and true shorefaces. In areas that are transitional between sheltered, tide-dominated, and exposed wave-dominated tidal flats, there will be a mixture of wave- and tide-generated sedimentary structures. The tidal flats at the head of the large tidal bar in Gyeonggi Bay may be of this intermediate type.

Tidal prism. The volume of water that passes through a given cross-sectional area (e.g., any segment of a tidal channel) during the ebb tide (or flood tide). Tidal prism can be estimated by multiplying the tidal range by the surface area of the region being flooded or drained by the channel in question. The tidal prism is important because it, together with the cross-sectional area of the channel, determines the tidal-current speeds, which in turn determine the effect of the tide on sedimentation. As discussed by Dalrymple (2010b), the area being flooded or drained exerts a stronger control on the tidal prism than the tidal range; therefore, the tidal range is not a fool-proof indication of the degree of tidal influence. Large tidal ranges tend to be associated with large tidal prisms, but areas with large tidal ranges can be wave-dominated if the tidal prism is small because the area is small. This is the case with wave-dominated open-coast tidal flats (Yang et al., 2005). Conversely, areas with small tidal ranges can be tide-dominated if the area term is large, as is the case in many tidal inlets. The infilling of a tidal basin will cause a decrease in the tidal prism and a weakening of the tidal currents, even if tidal range remains unchanged. Such infilling and weakening of the tidal currents will lead to a decrease in the cross-sectional area of the channel; this may be the cause of the narrowing of the large tidal channels observed in Gyeonggi Bay.

Tidal pumping. The process of landward transport of sediment as a result of the *flood dominance* generated by the asymmetry of the tidal wave as it propagates into shallow water (Postma, 1967): the duration of the flood tide becomes shorter than that of the ebb, such that flood tidal currents are faster, on average, than those of the ebb (Friedrichs and Aubrey, 1988). Although this process is usually applied to the transport of *bedload* because of its nonlinear relationship between current speed and discharge, Yu et al. (2014) apply it to the transport of the *suspended load*, noting that the suspended sediment concentration is generally higher during the flood tide, because of the higher current speed, leading to a net landward movement of fine-grained sediment. This tendency for landward movement is accentuated by settling lag and scour lag (van Straaten and Kuenen, 1958; Dalrymple and Choi, 2003). These processes lead to a net landward movement of suspended load on the seaward side of the turbidity maximum, whereas a net seaward movement of suspended sediment occurs on the landward side of the turbidity maximum as a result of river flow (Yu et al., 2014). This convergence of the net transport of suspended sediment is directly comparable to the *bedload convergence* zone.

Tidal ravinement surface. Any erosional surface generated by tidal currents during transgression. Most commonly, such surfaces are created at the base and cutbanks of tidal channels (Dalrymple and Zaitlin, 1994) and are thought (with little detailed justification) to show a channel-like morphology. The seafloor following transgression in tide-dominated settings is commonly ornamented by elongate tidal ridges (Cattaneo and Steel, 2003; Dalrymple, 2010). Tidal ridges either overlie the transgressive ravinement surface (i.e., they were *deposited* during transgression) or underlie the transgressive ravinement surface (i.e., they were *eroded* during the transgression). In some cases, tidal ridges contain an erosional nucleus overlain by a depositional carapace (i.e., the transgressive ravinement surface occurs *within* the tidal ridge). See Chapter 2 for examples of each type of ridge architecture. Contrast this with *wave ravinement surface*.

Tide-dominated. Adjective used to describe a modern depositional environment or an ancient sediment body, where most of the work done (sediment erosion, transport and deposition) is inferred to have been performed by tidal currents based on the geomorphology (Galloway, 1975), stratigraphic architecture (Dalrymple and Choi, 2007; Dalrymple, 2010b; Legler et al., 2013) and/or depositional facies of the deposits (Dalrymple, 2010b; Davis, 2012). In a modern environment, tidal dominance is inferred from the widespread presence of tidal channels and elongate, shore-normal tidal bars. In an ancient succession, tidal dominance would be inferred from the pervasive presence of tidal facies, although this may be misleading. For example, back-barrier deposits are typically tide-dominated even though the larger depositional system consisting of the wave-formed barrier and associated environments is wave-dominated (Flemming, 2012). Because the back-barrier tidal deposits have a significantly higher preservation potential that the wave-formed barrier, there is the potential to misidentify back-barrier deposits as having formed in a tide-dominated setting (Dalrymple, 2010b).

Tropical cyclone. A general term for a volume of air that rotates around an area of low atmospheric pressure that originates over a tropical ocean and is driven principally by heat transfer from the ocean (Emanuel, 2003). They tend to only form over ocean water whose surface temperature exceeds 26° C, and travel in a net poleward direction at speeds between 2 and 10 m/s, quickly losing energy when they move over land or cold water. Tropical cyclones in their formative stage, with maximum winds of 17 m/s or less, are known as tropical depressions; when their wind speeds are in the range of 18–32 m/s, inclusive, they are called tropical storms, whereas tropical cyclones with maximum wind speeds of 33 m/s or greater are called hurricanes in the western North Atlantic and eastern North Pacific regions, typhoons in the western North Pacific, and severe tropical cyclones elsewhere. Tropical cyclones do not form in a nonrotating frame of reference, for example at the equator, where the Coriolis effect is negligible. Instead, they generally form at latitudes of 20–30° north or south of the equator, and migrate westward before turning poleward.

Turbidity maximum. A zone of elevated suspended sediment concentrations in deltas and estuaries, located between less-turbid river and shelf waters (Allen et al., 1980). Suspended sediment concentrations near the bed during slack-water periods can reach values sufficient to form *fluid mud*. The location of the turbidity maximum commonly coincides approximately with the updip limit of salt-water intrusion, but can extend into the freshwater reach. The turbidity maximum is advected back and forth about a central point over several tens of kilometers during ebb and flood tides, and is commonly displaced basinward by several tens of kilometers during periods of high fluvial discharge (Doxaran et al., 2009; Yu et al., 2014). Flocculation associated with advection of fine-grained sediment from freshwater into the zone of brackish water contributes to the formation of the turbidity maximum, but the main processes responsible for the elevated turbidity are *estuarine circulation*, *tidal pumping*, scour and settling lags, and resuspension of bottom sediment by strong tidal currents (Allen et al., 1980; Yu et al., 2014). The turbidity maximum is an area within the estuary or delta where there is a greater abundance of mud in the deposits (Allen et al., 1980; Dalrymple and Choi, 2007; Dalrymple et al., 2012).

Typhoon. See *tropical cyclone*.

Wave ravinement surface. An erosion surface created by landward migration of the shoreface as a result of wave action during transgression. Transgressive ravinement has been well documented in wave-dominated settings. The basic model is that a sandy barrier island forms and migrates landward erosively because of the landward translation of the shoreface (Swift, 1968; Boyd, 2010). Because the shoreface is generally regular in an along-coast direction, wave ravinement surfaces are generally believed to be planar (Walker and Eyles, 1991). Contrast this with *tidal ravinement surface*.

Wave ripple. Any wave-generated bedform, irrespective of size. As with current ripples and dunes, the size and shape of wave ripples is a function of both the sediment itself (primarily mean grain size) and the flow conditions at the bed (oscillatory velocity, orbital diameter) dictated by some combination of wave height, period, and water depth. Flume experiments suggest that larger wave ripples—those with crest-to-crest spacings of decimeters to meters—are "orbital" ripples in that they scale to orbital diameter, the length of the back and forth excursion at the bed (Southard et al., 1990; Cummings et al., 2009). Large wave ripples appear to be invariably straight crested in coarser sand, but can become mound-like ("hummocky") in finer sand (Cummings et al., 2009). The latter are believed to deposit hummocky cross-stratification (Southard et al., 1990). Large wave ripples can be useful when attempting to reconstruct paleoenvironments: they require large orbital diameters, which requires large waves, which requires a large water body (Young et al., 1999; Goda, 2003). Several detailed methods of extracting paleohydraulic information from wave ripples exist (e.g., Komar, 1974; Clifton, 1976; Clifton and Dingler, 1984). As a simple rule of thumb, if the wave ripples in question have a crest-to-crest spacing of 75 cm or more, they are likely to be formed in the open ocean, not an estuary, river, or typical lake (Cummings et al., 2009). Small wave ripples (i.e., height = 1–2 cm, crest-to-crest spacing ~10 cm) are less diagnostic because they can form beneath small waves (orbital ripples) or large waves at low oscillatory velocities (these are anorbital ripples; essentially two-sided current ripples—see Southard et al., 1990).

INDEX